W9-AFO-440

Lecture Notes in Mathematics

Edited by A. Dold and B. Eckmann

1223

Differential Equations in Banach Spaces

Proceedings of a Conference
held in Bologna, July 2–5, 1985

Edited by A. Favini and E. Obrecht

Springer-Verlag
Berlin Heidelberg New York London Paris Tokyo

Editors

Angelo Favini
Enrico Obrecht
Dipartimento di Matematica, Università di Bologna
Piazza di Porta S. Donato 5, 40127 Bologna, Italy

Mathematics Subject Classification (1980): 34G10, 34G20, 45N05, 47D05, 47E05, 47G05

ISBN 3-540-17191-6 Springer-Verlag Berlin Heidelberg New York
ISBN 0-387-17191-6 Springer-Verlag New York Berlin Heidelberg

This work is subject to copyright. All rights are reserved, whether the whole or part of the material is concerned, specifically those of translation, reprinting, re-use of illustrations, broadcasting, reproduction by photocopying machine or similar means, and storage in data banks. Under § 54 of the German Copyright Law where copies are made for other than private use, a fee is payable to "Verwertungsgesellschaft Wort", Munich.

© Springer-Verlag Berlin Heidelberg 1986
Printed in Germany

Printing and binding: Druckhaus Beltz, Hemsbach/Bergstr.
2146/3140-543210

MS-Stk
SCIMON

FOREWORD

This volume contains most contributions to the conference on "DIFFERENTIAL EQUATIONS IN BANACH SPACES", which was held in the Department of Mathematics of the University of Bologna, July 2-5, 1985.

The aim of the meeting was to stimulate the exchange of ideas and informations in this very rapidly developing field.

The contributors, coming from several european countries, Japan and U.S.A., were specialists in different branches of abstract differential equations, so that the conference gave a panorama of different directions of research. Among the topics included are: regular and singular evolution equations both linear and nonlinear of parabolic and hyperbolic type, integro-differential equations, semigroup theory, control theory, wave equations, trasmutation methods, fuchsian differential equations.

During the meeting 23 one hour lectures and 7 short communications took place; 24 of these contributions appear now in this volume, some of them in an enlarged version.

We have pleasure in thanking all participants in the meeting, whose active interest made this conference a success. We are also pleased to thank the institutions which made this meeting possible by their financial support: the Consiglio Nazionale delle Ricerche - Comitato per le Scienze Matematiche, the Ministero della Pubblica Istruzione (fondi 40%) and the University of Bologna.

Finally, we are particularly grateful to those who helped, by their cooperation, and advice, the organization of the meeting and the editing of these proceedings: Giuseppe DA PRATO, Giovanni DORE, Davide GUIDETTI, Eugenio SINESTRARI and, last but not least, Alberto VENNI.

CONTENTS

LIST OF PARTICIPANTS

P. ACQUISTAPACE	Scuola Normale Superiore, PISA
A. AROSIO	University of PISA
M. BARDI	University of PADOVA
B.R. BELLOMO	University of BOLOGNA
M.L. BERNARDI	University of PAVIA
A. BOVE	University of BOLOGNA
R.W. CARROLL	University of ILLINOIS at URBANA-CHAMPAIGN
L. CATTABRIGA	University of BOLOGNA
A. CAVALLUCCI	University of BOLOGNA
J.M. COOPER	University of MARYLAND
G. DA PRATO	Scuola Normale Superiore, PISA
G. DI BLASIO	University of ROMA I
G. DORE	University of BOLOGNA
A. FAVINI	University of BOLOGNA
B. FRANCHI	University of BOLOGNA
F. FRANCHI	University of BOLOGNA
J.A. GOLDSTEIN	TULANE University
R. GRIMMER	University of SOUTHERN ILLINOIS
P. GRISVARD	University of NICE
D. GUIDETTI	University of BOLOGNA
R. LABBAS	University of NICE
E. LANCONELLI	University of BOLOGNA
I. LASIECKA	University of FLORIDA
J.E. LEWIS	University of ILLINOIS at CHICAGO CIRCLE
G. LUMER	University of MONS
A. LUNARDI	University of PISA
G. MANCINI	University of TRIESTE
E. MIRENGHI	University of BARI

R. NAGEL	University of TÜBINGEN
S. NANDA	University of KHARAGPUR
F. NARDINI	University of BOLOGNA
P. NEGRINI	University of BOLOGNA
E. OBRECHT	University of BOLOGNA
S. OHARU	University of HIROSHIMA
P.L. PAPINI	University of BOLOGNA
C. PARENTI	University of BOLOGNA
A. PAZY	HEBREW University of JERUSALEM
P. PLAZZI	University of BOLOGNA
M. POVOAS	University of LISBOA
M.A. POZIO	University of ROMA II
J. PRÜSS	University of PADERBORN
A. PUGLIESE	University of TRENTO
W. SCHAPPACHER	University of GRAZ
V. SCORNAZZANI	University of BOLOGNA
F. SEGALA	University of BOLOGNA
E. SERRA	University of BOLOGNA
E. SINESTRARI	University of ROMA I
H. TAHARA	University of TOKYO
B. TERRENI	University of PISA
R. TRIGGIANI	University of FLORIDA
A. VENNI	University of BOLOGNA
W. von WAHL	University of BAYREUTH
I.I. VRABIE	University of IASI
M. WATANABE	NIIGATA University
A. YAGI	OSAKA University

ON FUNDAMENTAL SOLUTIONS FOR ABSTRACT PARABOLIC EQUATIONS

Paolo Acquistapace

Scuola Normale Superiore
Piazza dei Cavalieri , 7
56100 PISA

Brunello Terreni

Dipartimento di Matematica - Università
Via F. Buonarroti, 2
56100 PISA

0. INTRODUCTION

We consider the linear Cauchy problem

$$(0.1)\quad \begin{cases} u'(t) - A(t)u(t) = f(t), & t\in[0,T] \\ u(0) = x \end{cases}$$

in a Banach space E. Here $x \in E$, $f \in C([0,T],E)$ and for each $t\in[0,T]$ the operator $A(t)$ generates an analytic semigroup $\{e^{sA(t)}\}_{s\geq 0}$ in E; its domain $D_{A(t)}$ may depend on t and be not dense in E.

We make the following assumptions, already introduced in [4]:

HYPOTHESIS I For each $t\in[0,T]$ $A(t):D_{A(t)}\subseteq E \to E$ is a closed linear operator and there exist $\theta_0 \in]\pi/2,\pi[$, $M>0$ such that:

(i) $\rho(A(t)) \supseteq S_{\theta_0} = \{z \in \mathbb{C} : |\arg z| \leq \theta_0\} \cup \{0\}$,

(ii) $\|R(\lambda,A(t))\|_{\mathcal{L}(E)} \leq \frac{M}{1+|\lambda|}$ $\forall \lambda \in S_{\theta_0}$, $\forall t\in[0,T]$.

HYPOTHESIS II There exist $B>0$, $k\in\mathbb{N}$, $\alpha_1,\ldots,\alpha_k,\beta_1,\ldots,\beta_k\in[0,2]$ with $\delta = \min_{1\leq i\leq k} (\alpha_i-\beta_i) > 0$, such that:

$$\|A(t)R(\lambda,A(t))[A(t)^{-1}-A(s)^{-1}]\|_{\mathcal{L}(E)} \leq B\sum_{i=1}^{k} (t-s)^{\alpha_i}|\lambda|^{\beta_i-1}$$

$$\forall \lambda \in S_{\theta_0}-\{0\},\ \forall\ 0\leq s\leq t\leq T.$$

Hypotheses I-II are generally weaker than those known in the literature (see [5, Section 7] for detailed comparisons); in particular, they allow a unified treatment of problem (0.1) in which neither the constancy of the domains $D_{A(t)}$ [9],[8],[2],[3], nor the strong differentiability of the resolvents $t\to R(\lambda,A(t))$ [6],[10],[11],[1] is required. Actually, under these assumptions and provided the data x,f are sufficiently regular, we have proved in [5] existence, uniqueness and sharp regularity results for strict and classical solutions u of (0.1), as well as a representation formula for $A(\cdot)u(\cdot)$ which is obtained without using fundamental solutions. As a consequence, we derived an a-priori estimate for strict solutions of (0.1) of the following kind:

$$(0.2) \qquad \|u'\|_{C([0,T],E)} + \|Au\|_{C([0,T],E)} \le C\{\|x\|_{D_{A(0)}} + \|f\|_{C^\beta([0,T],E)}\}.$$

However this estimate, although interesting of its own, does not seem to be very useful in applications, because (roughly speaking) it involves too strong norms. Thus the aim of this paper is the proof of a better a-priori estimate (i.e. in terms of u, rather than Au) for strict solutions of (0.1). We will also express the solution by the usual variation of parameters formula, finding in particular an explicit representation of the fundamental solution of (0.1).

1. THE A-PRIORI ESTIMATE

A strict (resp. a classical) solution of problem (0.1) is a function u such that u', Au belong to $C([0,T],E)$ (resp $C(]0,T],E)$), $u(0)=x$ and the equation $u'-Au=f$ holds in $[0,T]$ (resp. $]0,T]$).

We prove here the following result:

THEOREM 1.1 *Let $f\in C([0,T],E)$, $x\in D_{A(0)}$ and suppose that $A(0)x+f(0) \in \overline{D_{A(0)}}$. There exists $C>0$ such that if u is a strict solution, then*

$$(1.1) \qquad \|u(t)\|_E \le C\Big(\|x\|_E + \int_0^t \|f(s)\|_E ds\Big) \qquad \forall\ t\in[0,T].$$

Proof First of all we remark that the condition $A(0)x+f(0)\in \overline{D_{A(0)}}$ is necessary for existence of strict solutions and, in fact, if u is a strict solution we must have

$$(1.2) \quad A(t)u(t)+f(t) \in \overline{D_{A(t)}} \quad \forall\ t\in[0,T];$$

this can be proved by the argument used in [5, Proposition 3.7(i)].

Let now $t\in]0,T]$ and define

$$v(s) = e^{(t-s)A(t)}u(s), \quad s\in[0,t].$$

Then for $s\in[0,t[$

$$v'(s) = A(t)e^{(t-s)A(t)}[A(t)^{-1}-A(s)^{-1}]A(s)u(s) + e^{(t-s)A(t)}f(s).$$

Fix $\varepsilon\in]0,1[$, integrate between 0 and $t-\varepsilon t$ and operate with $A(t)$: the result is

$$(1.3) \quad A(t)u(t) - \int_0^t Q(t,s)A(s)u(s)ds = G_\varepsilon(t), \quad t\in]0,T],$$

where we have set

$$(1.4) \quad Q(t,s) = A(t)^2e^{(t-s)A(t)}[A(t)^{-1}-A(s)^{-1}], \quad 0\le s<t\le T,$$

and

$$(1.5) \quad G_\varepsilon(t) = -\int_{t-\varepsilon t}^t Q(t,s)A(s)u(s)ds + A(t)u(t) - A(t)e^{t\varepsilon A(t)}u(t-\varepsilon t)$$
$$+ A(t)e^{tA(t)}x + \int_0^{t-\varepsilon t} A(t)e^{(t-s)A(t)}f(s)ds, \quad t\in]0,T].$$

By [5, Lemma 2.3(i)] it follows that

$$(1.6) \quad \|Q(t,s)\|_{\mathcal{L}(E)} \le K(t-s)^{\delta-1} \quad \forall\ 0\le s<t\le T$$

so that for any fixed $r\in[0,T[$ the Volterra integral operator

$$(1.7) \quad Q_rg(t) = \int_r^t Q(t,s)g(s)ds, \quad t\in[r,T]$$

is well defined in $C([r,T],E)$ or in $L^1(r,T,E)$ (and even in more general spaces, see [5, Proposition 2.4]). Moreover, $(1-Q_r)^{-1}$ exists as a bounded operator in the same spaces [5, Proposition 2.6]. However in this section we are only interested to the case r=0 and we denote the operator Q_0 simply by Q.

Now we turn on the function G_ε and split it into several terms:

$$(1.8)\quad G_\varepsilon(t) = -\int_{t-\varepsilon t}^{t} Q(t,s)A(s)u(s)ds$$
$$- [e^{\varepsilon tA(t)}-1-\varepsilon tA(t)e^{\varepsilon tA(t)}][A(t)u(t)+f(t)] + [e^{\varepsilon tA(t)}-1]f(t)$$
$$- A(t)e^{\varepsilon tA(t)}[u(t-\varepsilon t)-u(t)+\varepsilon tu'(t)] + [A(t)e^{tA(t)}-A(0)e^{tA(0)}]x$$
$$+ A(0)e^{tA(0)}x + \int_0^{t-\varepsilon t} [A(t)e^{(t-s)A(t)}-A(s)e^{(t-s)A(s)}]f(s)ds$$
$$+ \int_0^{t-\varepsilon t} A(s)e^{(t-s)A(s)}f(s)ds = \sum_{i=1}^{8} I_i ;$$

thus by the results of [5] it is easily seen that $G_\varepsilon \in L^\infty(0,T,E) \cap C(]0,T],E)$ for each $\varepsilon \in]0,1[$.

Hence by (1.5) we deduce that

$$A(t)u(t) = [(1-Q)^{-1}G_\varepsilon](t), \quad t\in]0,T].$$

Integrate between εt and t: by using again the equation of (0.1) we get

$$(1.9)\quad u(t) = u(\varepsilon t) + \int_{\varepsilon t}^{t} f(s)ds + \int_{\varepsilon t}^{t} [(1-Q)^{-1}G_\varepsilon](\tau)d\tau.$$

In order to get (1.1), we have to estimate the right member of this equality, by using the splitting (1.8), and then pass to the limit as $\varepsilon\to 0^+$.

First we show that

$$(1.10)\quad \lim_{\varepsilon\to 0^+}\int_{\varepsilon t}^{t} \|[(1-Q)^{-1}(I_1+I_2+I_4)](\tau)\|_E d\tau = 0 \quad \text{uniformly in } t.$$

Indeed, as in [5, Proposition 2.6] it follows that

$$\|(1-Q)^{-1}\|_{\mathcal{L}(L^1(0,t,E))} \le C \quad \forall\, t\in]0,T];$$

on the other hand, by (1.6)

$$\|I_1(\tau)\|_E \le C(\varepsilon T)^\delta \|A(\cdot)u(\cdot)\|_{C([0,T],E)}$$

whereas by (1.2)

$$\begin{cases} \lim_{\varepsilon\to 0^+}\left(\|I_2(\tau)\|_E + \|I_4(\tau)\|_E\right) = 0, \\ \|I_2(\tau)\|_E + \|I_4(\tau)\|_E \le C\|u'\|_{C([0,T],E)} \qquad \forall\ \tau\in]0,T]. \end{cases}$$

Thus (1.10) follows by Lebesgue's Theorem.

Next, it is easily seen that

$$(1.11)\quad \int_{\varepsilon t}^{t} \|[(1-Q)^{-1}I_3](\tau)\|_E d\tau \le C \int_0^t \|f(s)\|_E ds \quad \forall\varepsilon\in]0,1[,\ \forall\ t\in]0,T],$$

and by [5, Lemma 1.10(i)]

$$(1.12)\quad \int_{\varepsilon t}^{t} \|[(1-Q)^{-1}(I_5+I_7)](\tau)\|_E d\tau \le C\ T^\delta\left(\|x\|_E+\int_0^t \|f(s)\|_E ds\right) \ \forall\varepsilon\in]0,1[,\forall t\in]0,T].$$

Finally the terms I_6 and I_8 need a more careful procedure, which rests on the following

LEMMA 1.2 _Let_ $Q = Q_0$ _be defined by_ (1.7). _The_ n-th _iterate_ Q^n $(n\ge 1)$ _is given by_

$$Q^n g(t) = \int_0^t Q_n(t,s)g(s)ds, \quad t\in[0,T]$$

where the kernel $Q_n(t,s)$ _is defined inductively by_

$$Q_1(t,s) = Q(t,s), \quad Q_n(t,s) = \int_s^t Q_{n-1}(t,\sigma)Q(\sigma,s)d\sigma,$$

and satisfies for $0\le s\le\sigma<t\le T$:

$$\text{(i)}\quad \|Q_n(t,s)\|_{\mathcal{L}(E)} \le \frac{K^n\Gamma(\delta)^n}{\Gamma(n\delta)}(t-s)^{n\delta-1} \quad \forall\ n\ge 1,$$

$$\text{(ii)}\quad \|Q(t,\sigma)-Q(t,s)\|_{\mathcal{L}(E)} \le B\,\frac{(\sigma-s)^\delta}{t-s},$$

(iii) $$\int_s^t \frac{\|Q_n(t,\sigma)-Q_n(t,s)\|_{\mathcal{L}(E)}}{\sigma-s}\,d\sigma \le C\,\frac{K^{n-1}\Gamma(\delta)^{n-1}}{\Gamma(n\delta-\delta)}(t-s)^{n\delta-1} \quad \forall\, n \ge 2.$$

Proof The first part is straightforward; the proof of (i) is easily obtained by induction, starting from (1.6). Part (ii) follows readily by [5, Lemma 2.3(iii)].

To prove (iii), we write

$$Q_n(t,\sigma)-Q_n(t,s) = \int_\sigma^t Q_{n-1}(t,r)[Q(r,\sigma)-Q(r,s)]dr - \int_s^\sigma Q_{n-1}(t,r)Q(r,s)dr,$$

and the result follows by (i) and (ii) after straightforward calculations.

Let us estimate the term I_6. We have by Lemma 1.2(iii)-(i)

$$(1.13)\quad \left\|\int_{\varepsilon t}^t [(1-Q)^{-1}I_6](\tau)d\tau\right\|_E \le \left\|\int_{\varepsilon t}^t I_6(\tau)d\tau\right\|_E + \sum_{n=1}^\infty \left\|\int_{\varepsilon t}^t [Q^n I_6](\tau)d\tau\right\|_E$$

$$\le \|(e^{tA(0)}-e^{\varepsilon tA(0)})x\|_E + M\sum_{n=1}^\infty \int_{\varepsilon t}^t \int_0^\tau \frac{\|Q_n(\tau,\sigma)-Q_n(\tau,0)\|_{\mathcal{L}(E)}}{\sigma}d\sigma d\tau \|x\|_E$$

$$+ \sum_{n=1}^\infty \left\|\int_{\varepsilon t}^t Q_n(\tau,0)(e^{\tau A(0)}-1)xd\tau\right\|_E$$

$$\le C(1+T^\delta)\sum_{n=1}^\infty \frac{[K\Gamma(\delta)T^\delta]^n}{n\Gamma(n\delta)}\|x\|_E \quad \forall\, \varepsilon \in]0,1[,\ \forall\, t \in]0,T].$$

Finally the term I_8 is treated in the following way:

$$\left\|\int_{\varepsilon t}^t [(1-Q)^{-1}I_8](\tau)d\tau\right\|_E \le \left\|\int_{\varepsilon t}^t \int_0^{\tau-\varepsilon\tau} A(s)e^{(\tau-s)A(s)}f(s)dsd\tau\right\|_E$$

$$+ \sum_{n=1}^\infty \left\|\int_{\varepsilon t}^t \int_0^\tau \int_0^{\sigma-\varepsilon\sigma} [Q_n(\tau,\sigma)-Q_n(\tau,s)]A(s)e^{(\sigma-s)A(s)}f(s)dsd\sigma d\tau\right\|_E$$

$$+ \sum_{n=1}^\infty \left\|\int_{\varepsilon t}^t \int_0^\tau \int_0^{\sigma-\varepsilon\sigma} Q_n(\tau,s)A(s)e^{(\sigma-s)A(s)}f(s)dsd\sigma d\tau\right\|_E$$

$$= \left\|\int_{\varepsilon t}^t [e^{(t-s)A(s)}-e^{[(\varepsilon t-s)\vee(s\varepsilon/(1-\varepsilon))]A(s)}]f(s)ds\right\|_E$$

$$+ M\sum_{n=1}^\infty \int_{\varepsilon t}^t \int_0^{\tau-\varepsilon\tau}\int_{s/(1-\varepsilon)}^\tau \frac{\|Q_n(\tau,\sigma)-Q_n(\tau,s)\|_{\mathcal{L}(E)}}{\sigma-s}\|f(s)\|_E d\sigma ds d\tau$$

$$+ \sum_{n=1}^\infty \left\|\int_{\varepsilon t}^t \int_0^{\tau-\varepsilon\tau} Q_n(\tau,s)[e^{(\tau-s)A(s)}-e^{(s\varepsilon/(1-\varepsilon))A(s)}]f(s)dsd\tau\right\|_E,$$

where we have used Fubini's Theorem. Consequently we get by Lemma 1.2 (iii)-(i), for each $\varepsilon\in]0,1[$ and $t\in]0,T]$:

$$(1.14)\qquad \left\|\int_{\varepsilon t}^{t}[(1-Q)^{-1}I_8](\tau)d\tau\right\|_E \le C(1+T^\delta)\sum_{n=1}^{\infty}\frac{[K\Gamma(\delta)T^\delta]^n}{n\Gamma(n\delta)}\int_0^t \|f(s)\|_E ds .$$

Recalling (1.8) and collecting (1.10),(1.11),(1.12),(1.13) and (1.14) we can let $\varepsilon\to 0^+$ in (1.9), and (1.1) follows. Theorem 1.1 is completely proved.

REMARK 1.3 A generalized version of Theorem 1.1 can be proved for classical solutions belonging to the class $\bigcup_{0\le\mu<1+\delta} I_\mu(D_A)$; this class was defined in [5, formulas (1.1)-(1.2)]. Correspondingly, the data x,f have to be chosen in $\overline{D_{A(0)}}$ and in $L^1(0,T,E)\cap C(]0,T],E)$ respectively; the proof is essentially the same, but it requires much more technicalities.

2. THE FUNDAMENTAL SOLUTION

The argument used in the proof of Theorem 1.1 can be refined in order to get a deeper result. Namely, we have:

THEOREM 2.1 *Let* $f\in C([0,T],E)$, $x\in D_{A(0)}$ *and suppose that* $A(0)x+f(0) \in \overline{D_{A(0)}}$. *If* u *is a strict solution of problem* (0.1), *then* u *is given by*

$$(2.1)\qquad u(t) = U(t,0)x + \int_0^t U(t,s)f(s)ds,\quad t\in[0,T],$$

where

$$(2.2)\qquad U(t,s) = e^{(t-s)A(s)} + \int_s^t \Big\{\Big[(1-Q_s)^{-1}[A(\cdot)e^{(\cdot-s)A(\cdot)}-A(s)e^{(\cdot-s)A(s)}]\Big](\tau)$$
$$+\int_s^\tau \sum_{n=1}^{\infty}[Q_n(\tau,\sigma)-Q_n(\tau,s)]A(s)e^{(\sigma-s)A(s)}d\sigma$$
$$+\sum_{n=1}^{\infty}Q_n(\tau,s)[e^{(\tau-s)A(s)}-1]\Big\}d\tau ,\qquad 0\le s\le t\le T.$$

Proof As in the proof of Theorem 1.1 we arrive at (1.9); now we try to pass to the limit as $\varepsilon\to 0^+$ directly in this expression, in order to get a representation formula for u(t). Recalling (1.8) and (1.10) we have:

(2.3) $$u(t) = x + \int_0^t f(s)ds + \lim_{\varepsilon\to 0^+} \int_{\varepsilon t}^t [(1-Q)^{-1}(I_3+I_5+I_6+I_7+I_8)](\tau)d\tau \ .$$

Now it is easily seen that

(2.4) $$\lim_{\varepsilon\to 0^+} \int_{\varepsilon t}^t [(1-Q)^{-1} I_5](\tau)d\tau = \int_0^t \Big[(1-Q)^{-1}[A(\cdot)e^{\cdot A(\cdot)} - A(0)e^{\cdot A(0)}]x\Big](\tau)d\tau,$$

whereas

$$\lim_{\varepsilon\to 0^+} \int_{\varepsilon t}^t [(1-Q)^{-1} I_7](\tau)d\tau$$
$$= \int_0^t \Big[(1-Q)^{-1}[\int_0^{\cdot} [A(\cdot)e^{(\cdot-s)A(\cdot)} - A(s)e^{(\cdot-s)A(s)}]f(s)ds]\Big](\tau)d\tau \ ;$$

but if we split the operator $(1-Q)^{-1}$ into its Neumann series, then a simple calculation shows, via Fubini's Theorem, that the last equality can be rewritten as:

(2.5) $$\lim_{\varepsilon\to 0^+} \int_{\varepsilon t}^t [(1-Q)^{-1} I_7](\tau)d\tau$$
$$= \int_0^t \int_s^t \Big[(1-Q_s)^{-1}[A(\cdot)e^{(\cdot-s)A(\cdot)} - A(s)e^{(\cdot-s)A(s)}]\Big](\tau)d\tau f(s)ds \ .$$

Next, concerning I_6 we have:

(2.6) $$\lim_{\varepsilon\to 0^+} \int_{\varepsilon t}^t [(1-Q)^{-1} I_6](\tau)d\tau = \lim_{\varepsilon\to 0^+} \Big\{[e^{tA(0)} - e^{\varepsilon tA(0)}]x$$
$$+ \sum_{n=1}^{\infty} \int_{\varepsilon t}^t \int_0^{\tau} [Q_n(\tau,\sigma) - Q_n(\tau,0)]A(0)e^{\sigma A(0)} x d\sigma d\tau$$
$$+ \sum_{n=1}^{\infty} \int_{\varepsilon t}^t Q_n(\tau,0)[e^{\tau A(0)} - 1]x d\tau\Big\}$$
$$= [e^{tA(0)} - 1]x + \sum_{n=1}^{\infty} \int_0^t \int_0^{\tau} [Q_n(\tau,\sigma) - Q_n(\tau,0)]A(0)e^{\sigma A(0)} x d\sigma d\tau$$
$$+ \sum_{n=1}^{\infty} \int_0^t Q_n(\tau,0)[e^{\tau A(0)} - 1]x d\tau \ .$$

Finally we consider together I_3 and I_8:

$$\lim_{\varepsilon\to 0^+} \int_{\varepsilon t}^t [(1-Q)^{-1}(I_3+I_8)](\tau)d\tau = \lim_{\varepsilon\to 0^+} \Big\{\int_{\varepsilon t}^t \Big[(1-Q)^{-1}[e^{\varepsilon\cdot A(\cdot)} - 1]f(\cdot)\Big](\tau)d\tau$$

$$+\int_{\varepsilon t}^{t}\int_{0}^{\tau-\varepsilon\tau} A(s)e^{(\tau-s)A(s)}f(s)\,ds\,d\tau$$

$$+\sum_{n=1}^{\infty}\int_{\varepsilon t}^{t}\int_{0}^{\tau}\int_{0}^{\sigma-\varepsilon\sigma} [Q_n(\tau,\sigma)-Q_n(\tau,s)]A(s)e^{(\sigma-s)A(s)}f(s)\,ds\,d\sigma\,d\tau$$

$$+\sum_{n=1}^{\infty}\int_{\varepsilon t}^{t}\int_{0}^{\tau}\int_{0}^{\sigma-\varepsilon\sigma} Q_n(\tau,s)A(s)e^{(\sigma-s)A(s)}f(s)\,ds\,d\sigma\,d\tau\Big\}.$$

Using once more Fubini's Theorem, we easily get:

$$\lim_{\varepsilon\to 0^+}\int_{\varepsilon t}^{t}[(1-Q)^{-1}(I_3+I_8)](\tau)\,d\tau = \lim_{\varepsilon\to 0^+}\Big\{\int_{0}^{t}\Big[(1-Q)^{-1}[e^{\varepsilon\cdot A(\cdot)}-1]f(\cdot)\Big](\tau)\,d\tau$$

$$+\int_{0}^{t-\varepsilon t}[e^{(t-s)A(s)}-e^{[(\varepsilon t-s)\vee(s\varepsilon/(1-\varepsilon))]A(s)}]f(s)\,ds$$

$$+\sum_{n=1}^{\infty}\int_{0}^{t}\int_{0}^{\tau}\int_{0}^{\sigma}[Q_n(\tau,\sigma)-Q_n(\tau,s)]A(s)e^{(\sigma-s)A(s)}f(s)\,ds\,d\sigma\,d\tau$$

$$+\sum_{n=1}^{\infty}\int_{0}^{t}\int_{0}^{\tau-\varepsilon\tau}Q_n(\tau,s)[e^{(\tau-s)A(s)}-e^{[(\varepsilon\tau-s)\vee(s\varepsilon/(1-\varepsilon))]A(s)}]f(s)\,ds\,d\tau\Big\}$$

$$= -\int_{0}^{t}[(1-Q)^{-1}f](\tau)\,d\tau + \int_{0}^{t}e^{(t-s)A(s)}f(s)\,ds$$

$$+\sum_{n=1}^{\infty}\int_{0}^{t}\int_{s}^{t}\int_{s}^{\tau}[Q_n(\tau,\sigma)-Q_n(\tau,s)]A(s)e^{(\sigma-s)A(s)}f(s)\,d\sigma\,d\tau\,ds$$

$$+\sum_{n=1}^{\infty}\int_{0}^{t}\int_{s}^{t}Q_n(\tau,s)e^{(\tau-s)A(s)}f(s)\,d\tau\,ds$$

$$+\lim_{\varepsilon\to 0^+}\int_{0}^{t}\Big[(1-Q)^{-1}\Big[[e^{\varepsilon\cdot A(\cdot)}-e^{(\varepsilon\cdot/(1-\varepsilon))A(\cdot)}]f(\cdot)\Big]\Big](\tau)\,d\tau\ .$$

The last limit is 0 since

$$\|e^{\varepsilon\tau A(\tau)}-e^{(\varepsilon\tau/(1-\varepsilon))A(\tau)}\|_{\mathcal{L}(E)} \le M\int_{\varepsilon\tau}^{\varepsilon\tau/(1-\varepsilon)}\frac{d\sigma}{\sigma} = M\log\frac{1}{1-\varepsilon}\ ;$$

hence

$$\lim_{\varepsilon\to 0^+}\int_{\varepsilon t}^{t}[(1-Q)^{-1}(I_3+I_8)](\tau)\,d\tau = \int_{0}^{t}e^{(t-s)A(s)}f(s)\,ds - \int_{0}^{t}f(s)\,ds \tag{2.7}$$

$$+\int_{0}^{t}\sum_{n=1}^{\infty}\int_{s}^{t}Q_n(\tau,s)[e^{(\tau-s)A(s)}-1]\,d\tau f(s)\,ds$$

$$+\int_{0}^{t}\sum_{n=1}^{\infty}\int_{s}^{t}\int_{s}^{\tau}[Q_n(\tau,\sigma)-Q_n(\tau,s)]A(s)e^{(\sigma-s)A(s)}\,d\sigma\,d\tau f(s)\,ds\ .$$

By (2.3),(2.4),(2.5),(2.6) and (2.7) we readily get (2.1) with U(t,s) given by (2.2). The result is proved.

REMARK 2.2 The operator U(t,s) defined by (2.2) enjoys all the usual properties of fundamental solutions. Indeed, it is clear by definition that $U(t,s) \in \mathcal{L}(E)$ for $0 \le s \le t \le T$ and $U(t,t)=1$; however we have

$$\lim_{t \to s^+} U(t,s)x = x \iff \lim_{t \to s^+} e^{(t-s)A(s)}x = x \iff x \in \overline{D_{A(s)}}$$

(see [7, Proposition 1.2]). Moreover by (2.2) and by the representation formula for $A(\cdot)u(\cdot)$ proved in [5] it follows easily that

$$(2.8) \qquad \frac{\partial}{\partial t}U(t,s) = A(t)U(t,s) \qquad \forall\ t \in]s,T].$$

We also have

$$\frac{\partial}{\partial s}U(t,s) = -U(t,s)A(s) \qquad \forall\ s \in [0,t[,$$

in the sense that

$$(2.9) \qquad \lim_{h \to 0} h^{-1}[U(t,s+h)-U(t,s)]A(s)^{-1} = -U(t,s) \qquad \forall\ s \in [0,t[.$$

The proof of (2.9) is not evident; it is necessary to split patiently the ratio $h^{-1}[U(t,s+h)-U(t,s)]$ into 16 terms and to consider each of them separately (but sometimes two terms or more have to be assembled in order to get convergence as $h \to 0$). The result is just (2.9).

REMARK 2.3 Again, a generalized version of Theorem 2.1 holds (with much more tedious proof) for classical solutions belonging to the class $\bigcup_{0 \le \mu \le 1+\delta} I_\mu(D_A)$ with data x,f taken from $\overline{D_{A(0)}}$ and $L^1(0,T,E) \cap C(]0,T],E)$ respectively (compare with Remark 1.3).

REFERENCES

[1] P. ACQUISTAPACE, B. TERRENI, Some existence and regularity results for abstract non-autonomous parabolic equations, J. Math. Anal. Appl. 99 (1984) 9-64.

[2] P. ACQUISTAPACE, B.TERRENI, On the abstract non-autonomous parabolic Cauchy problem in the case of constant domains, Ann. Mat. Pura Appl. (4) 140 (1985) 1-55.

[3] P. ACQUISTAPACE, B.TERRENI, Maximal space regularity for abstract linear non-autonomous parabolic equations, J. Funct. Anal. 60 (1985) 168-210.

[4] P. ACQUISTAPACE, B. TERRENI, Une méthode unifiée pour l'étude des équations linéaires non autonomes paraboliques dans les espaces de Banach, C. R. Acad. Sci. Paris (1) 301 (1985) 107-110.

[5] P. ACQUISTAPACE, B. TERRENI, A unified approach to abstract linear parabolic non-autonomous equations, pre-print Squola Norm. Sup. Pisa (1986).

[6] T. KATO, H. TANABE, On the abstract evolution equations, Osaka Math. J. 14 (1962) 107-133.

[7] E. SINESTRARI, On the abstract Cauchy problem of parabolic type in spaces of continuous functions, J. Math. Anal. Appl. 107 (1985) 16-66.

[8] P. E. SOBOLEVSKII, On equations of parabolic type in Banach space, Trudy Moscow Mat. Obsc. 10 (1961) 297-350 (Russian); English transl.: Amer. Math. Soc. Transl. 49 (1965) 1-62.

[9] H. TANABE, On the equations of evolution in a Banach space, Osaka Math. J. 12 (1960) 363-376.

[10] A. YAGI, On the abstract evolution equations in Banach spaces, J. Math. Soc. Japan 28 (1976) 290-303.

[11] A. YAGI, On the abstract evolution equations of parabolic type, Osaka J. Math. 14 (1977) 557-568.

ON SOME SINGULAR NONLINEAR EVOLUTION EQUATIONS (*)

Marco Luigi BERNARDI

Dipartimento di Matematica - Università di Pavia. I-27100 PAVIA (Italy)

Abstract - We study, in this paper, a class of singular or degenerate nonlinear abstract differential equations of parabolic type. We prove, for such equations, an existence and uniqueness result, in the framework of suitable Banach weighted spaces.

1. INTRODUCTION.

We are concerned, in this paper, with a class of singular or degenerate nonlinear abstract differential equations of parabolic type. Various results are well known about (linear or nonlinear) evolution equations (of parabolic type) of the form

$$(1.1) \qquad (N(u))' + Q(u) = f ,$$

where the operator coefficients N and Q may have singularities or degeneracies of various kinds. For such (and also other) equations and related problems, the main reference (up to 1976) is the book by CARROLL and SHOWALTER [4], where many results and a very extensive bibliography are given.

We concentrate, now, on some subsequent papers, concerning an interesting particular case of (1.1). Let us start by considering the following p.d.e. example (which is a "perturbed" heat equation):

$$(1.2) \qquad \frac{\partial u}{\partial t} + \frac{b(t,x)}{t}u - d(t)\Delta_x u = g \qquad , \quad (x,t) \in \Omega \times]0,T[\ ,$$

where: $0<T<+\infty$; Ω is some open bounded subset of $\mathbb{R}^n$; b is a bounded function; d(t) is a strictly positive function on $]0,T]$, which,however, may vanish or become infinite at t=0. This is a typical example of the general abstract situation considered by BAIOCCHI and BAOUENDI [1]. In

(*) This work was supported in part by the "Istituto di Analisi Numerica del C.N.R." (Pavia, Italy), the "G.N.A.F.A. del C.N.R." (Italy) and the "Ministero della Pubblica Istruzione" (Italy).

fact, they investigated, in a Hilbert space framework, an abstract differential equation of the form:

$$(1.3) \qquad tu'(t) + L(t)u(t) = f(t) \quad , \quad 0<t<T \quad (0<T<+\infty) \ ,$$

where $L(t)$ is a linear unbounded coercive operator on $]0,T]$, which may be singular or degenerate at $t=0$. BAIOCCHI and BAOUENDI [1] used variational methods to obtain, for (1.3), some sharp existence, uniqueness and regularity results, in the framework of suitable Hilbert weighted spaces (the weights involve powers of t and the "behaviour" of $L(t)$, as $t \to 0^+$). We remark that even sharper results were subsequently obtained for the linear equation (1.3), in the particular situation where $L(t)$ is "good" also at $t=0$ (i.e. where $L(t)$ is neither singular nor degenerate at $t=0$). For these results, we refer to: BERNARDI [2] (for the special case where $L(0)$ is a positive selfadjoint operator with compact resolvent); DA PRATO and GRISVARD [6] (who work in a Banach space framework, using their [5] operational methods (semigroups and Dunford's integral techniques)); LEWIS and PARENTI [10] (who study systematically (1.3) in various spaces, using (the Mellin transform and) a suitable representation formula for the solutions of (1.3)). We also mention the recent work of FAVINI [7] (concerning linear equations of the general form (1.1) (and even more general ones), considered by means of the methods of DA PRATO and GRISVARD [5]), whose results also apply well to the just envisaged situation.
Now, we can consider the p.d.e. (1.2), where Δ_x is replaced by some nonlinear operator, e.g. by the generalized Laplace operator

$$(1.4) \qquad \sum_{i=1}^{n} \frac{\partial}{\partial x_i} \left(\left| \frac{\partial u}{\partial x_i} \right|^{p-2} \frac{\partial u}{\partial x_i} \right) \ .$$

This fact suggests to study, more generally, an abstract differential equation of the form

$$(1.5) \qquad tu'(t) + L(t;u(t)) = f(t) \quad , \quad 0 < t < T \ ,$$

where the nonlinear operator L may be singular or degenerate at $t=0$ (and at $t=0$ only). Thus, we study, in this paper, the nonlinear singular differential equation (1.5), where the operator L verifies some suitable properties (see the following section 2 for the precise assumptions). We deal with weak solutions and we work in a suitable Banach weighted space framework (which is a natural generalization of the Baiocchi and Baouendi's one); we obtain an existence and uniqueness result, which agrees well with the Baiocchi and Baouendi's result in the linear case. We have also to remark that KUTTLER [8], [9] obtained some results for general nonlinear degenerate equations like (1.1), where: N is a linear operator, depending on t and possibly degenerate, while the t-dependent Q is nonlinear (with suitable general properties), but without the possibility of being singular or degenerate. Our framework (weighted spaces) and method are different from the Kuttler's ones. However, it could be seen that our results and the Kuttler's ones agree well, in the particular cases where they both apply.

Now, we proceed to give: the functional framework, the basic assumptions on the operator L and some preliminary lemmas (section 2); the existence and uniqueness result (section 3). Some remarks and examples clarify the applicability of our abstract result.

2. SOME FUNCTIONAL SPACES. ASSUMPTIONS. SOME PRELIMINARY RESULTS.

We are studying the abstract differential equation (1.5), taking as main model example the p.d.e. (1.2), where Δ_x is replaced by the differential operator (1.4) : in particular, we can consider the case where u is sought with u=0 on $\partial\Omega\times]0,T[$ (i.e. homogeneous Dirichlet boundary condition). This fact motivates the definitions and assumptions we are going to introduce.

Throughout this paper, we consider, for sake of simplicity only, functions and spaces which are real.

Let V be a Banach space and H be a Hilbert space satisfying:

(2.1) V is reflexive ; $V \subseteq H$ (continuously) ; V is dense in H.

We identify H with its dual space; denoting by V^* the dual space of V, we also get that $H \subseteq V^*$ (continuously and densely). $\| \ \|$, $| \ |$ and $\| \ \|_*$ denote respectively the norms in V, H, V^*. (,) denotes both the scalar product in H and the duality pairing between V^* and V. Moreover, let us take some $p \in \mathbb{R}^1$ verifying

$$(2.2) \qquad 1 < p < +\infty \qquad \text{(and define: } q = p(p-1)^{-1}) .$$

Let now T be given, with $0<T<+\infty$. We consider, moreover, some function a(t) such that:

$$(2.3) \qquad \begin{cases} a(t) :]0,T] \to \mathbb{R}^1_+ \text{ a.e. with, in addition:} \\ a(t),\ (a(t))^{-1} \in L^\infty_{loc}(]0,T]) . \end{cases}$$

Now, we introduce some functional spaces (weighted spaces). If $m \in \mathbb{R}^1$, a(t) verifies (2.3) and p is given as in (2.2), we define (for $0<t_0\leq T$):

$$(2.4) \qquad \begin{cases} Z_m(0,t_0) \equiv \{f | f = f_1+f_2 , \text{ with :} \\ \quad t^{m(p-1)}(a(t))^{-1} f_1(t) \in L^q(0,t_0;V^*) ; \\ \quad t^{mp/2} f_2(t) \in L^2(0,t_0;H)\} ; \end{cases}$$

$$(2.5) \qquad \begin{cases} U_m(0,t_0) \equiv \{u| \ t^m a(t)u(t) \in L^p(0,t_0;V) ; \\ \quad t^{mp/2} u(t) \in L^2(0,t_0;H)\} ; \end{cases}$$

$$(2.6) \qquad W_m(0,t_0) \equiv \{u| \ u(t) \in U_m(0,t_0); \ tu'(t) \in Z_m(0,t_0)\} .$$

$Z_m(0,t_0)$, $U_m(0,t_0)$ and $W_m(0,t_0)$ are, clearly, Banach spaces with respect to their natural norms. Of course, we have that:

$$(2.7) \qquad \begin{cases} Z_m(0,t_0) \subseteq L^q_{loc}(]0,t_0];V^*) + L^2_{loc}(]0,t_0];H) ; \\ U_m(0,t_0) \subseteq L^p_{loc}(]0,t_0];V) \cap L^2_{loc}(]0,t_0];H) \\ W_m(0,t_0) \subseteq \{u| \ u \in L^p_{loc}(]0,t_0];V) \cap L^2_{loc}(]0,t_0];H) ; \\ \qquad u' \in L^q_{loc}(]0,t_0];V^*) + L^2_{loc}(]0,t_0];H)\} . \end{cases}$$

(Note that $]0,t_0]$ is open at 0 and closed at t_0 and remember that a(t) verifies (2.3)).

Now, the following Lemma 1.1 is important for the sequel.

<u>Lemma 1.1</u>. Let (2.1), (2.2) and (2.3) hold and let $m \in \mathbb{R}^1$ be given; let also $t_0 \in]0,T]$. Then, if $u \in W_m(0,t_0)$, it results that $u(t) \in C^0(]0,t_0];H)$ and, moreover, that:

$$(2.8) \qquad t^{\frac{1}{2}(mp+1)}|u(t)| \to 0 \quad , \quad \text{as} \quad t \to 0^+ .$$

<u>Proof</u>. First of all, it is obvious that $u(t) \in C^0(]0,t_0];H)$, thanks to (2.7). It remains only to prove (2.8). Observe that one has (in the sense of $\mathcal{D}'(0,t_0)$):

$$(2.9)\quad \begin{cases} (t^{mp+1}|u(t)|^2)' = (mp+1)t^{mp}|u(t)|^2 + 2t^{mp}(tu'(t),u(t)) = \\ = (mp+1)t^{mp}|u(t)|^2 + 2(t^{m(p-1)+1}(a(t))^{-1}v_1(t),\ t^m a(t)u(t)) + \\ + 2(t^{mp/2+1}v_2(t),\ t^{mp/2}u(t)) \quad , \end{cases}$$

where $u' = v_1+v_2$, according to (2.6) and (2.4). We deduce then, from (2.9), (2.6) and (2.4), that $(t^{mp+1}|u(t)|^2)' \in L^1(0,t_0)$. Hence, (2.8) follows from this conclusion and the fact (see (2.6)) that $t^{mp}|u(t)|^2 \in L^1(0,t_0)$, Q.E.D.

Now, we are going to give the assumptions on the operator L, which appears in the equation (1.5). We prefer to rewrite (1.5) in the following form ($L = A + B$):

$$(2.10)\qquad tu'(t) + A(t;u(t)) + B(t)u(t) = f(t) \quad , \quad 0 < t < T \; ;$$

then, we study (2.10) under the following assumptions on B and A. First of all, we assume that

$$(2.11)\qquad B(t) \in L^\infty(0,T;\mathcal{L}(H,H)) \; .$$

Moreover, we are given a family of (nonlinear) operators $A(t;\cdot): V \to V^*$ (for a.e. $t \in]0,T[$). We assume that:

$$(2.12)\quad \begin{cases} A(t;v(t)) \text{ is } V^*\text{-measurable on }]0,T[, \text{ for every } v(t), \\ \text{which is } V\text{-measurable on }]0,T[\; ; \end{cases}$$

$$(2.13)\qquad A(t;\cdot) \text{ is hemicontinuous a.e. on }]0,T[\; ;$$

$$(2.14)\qquad (A(t;v),v) \geq a^p(t)\|v\|^p \quad , \quad \forall v \in V, \text{ a.e. on }]0,T[\; ;$$

$$(2.15)\quad \begin{cases} \exists\, c_1 > 0 \text{ such that: } \|A(t;v)\|_* \leq c_1 a^p(t)\|v\|^{p-1}, \ \forall v \in V, \\ \text{a.e. on }]0,T[\; ; \end{cases}$$

where $a(t)$ verifies (2.3). Clearly, (2.14) is a coercivity hypothesis, while (2.15) is a boundedness assumption. Note, in particular, that (2.15) and (2.11) imply that $L = A+B$ maps bounded subsets of $U_m(0,t_0)$ onto bounded subsets of $Z_m(0,t_0)$, for every $m \in \mathbb{R}^1$ and $t_0 \in]0,T]$. Moreover, remark that $A(t;0) = 0$ a.e. on $]0,T[$, thanks to (2.15).

Now, we also suppose that:

(2.16) $\left[\begin{array}{l}\text{there exists some } \tilde{\ell} \in \mathbb{R}^1 \text{ such that the operator } A(t;\cdot)+\ell I \\ \text{is } \underline{\text{monotone}} \text{ from } V \text{ to } V^*, \text{ a.e. on }]0,T[.\end{array}\right.$

<u>Remark 2.1</u>. It is important, for the sequel, to observe at once the following fact (which follows from some well known results and procedures; see, e.g., BROWDER [3], LIONS [11]). Let (2.1), (2.2), (2.3), (2.11) ... (2.16) hold; take, moreover, some $\bar{t} \in]0,T[$. Then, it results that:

(2.17) $\left[\begin{array}{l}\text{for every } \tilde{f}(t) \in L^q(\bar{t},T;V^*)+L^2(\bar{t},T;H) \text{ and every } \bar{u} \in H, \\ \text{there exists a unique } \tilde{u}(t) \in L^p(\bar{t},T;V) \cap L^2(\bar{t},T;H) \text{ (with} \\ \text{also } \tilde{u}'(t) \in L^q(\bar{t},T;V^*)+L^2(\bar{t},T;H), \text{ so that } \tilde{u}(t) \in C^0([\bar{t},T]; \\ H)), \text{ which verifies } \tilde{u}(\bar{t}) = \bar{u} \text{ and (2.10) on }]\bar{t},T[.\end{array}\right.$

Moreover, under the assumptions we made till now, we can define the following function (a.e. on $]0,T[$):

(2.18) $\left[\begin{array}{l}\ell_0(t) \equiv \inf \{\ell \mid (L(t;u_1)-L(t;u_2),u_1-u_2) + \ell|u_1-u_2|^2 \geq 0, \\ \qquad \forall u_1,u_2 \in V\},\end{array}\right.$

where $L(t;u) = A(t;u)+B(t)u$. Clearly, under our assumptions, we have that $\ell_0(t) \in L^\infty_{loc}(]0,T])$. We also define:

(2.19) $$\ell_0 = \operatorname{ess}\liminf_{t\to 0^+} \ell_0(t) .$$

Obviously, it results that $\ell_0 < +\infty$, but it may happen, in some cases, that $\ell_0 = -\infty$ (for instance, if A is linear, with p=2 and $\operatorname{ess}\lim_{t\to 0^+} a(t) = +\infty$; see BAIOCCHI and BAOUENDI [1]). (We refer to the final part of the following section 3 for other remarks and examples).

Now, the following Lemma 2.2 is important for the sequel.

<u>Lemma 2.2</u>. Let (2.1), (2.2), (2.3) and (2.11) ... (2.16) hold. Then, for every $\ell > \ell_0$, there exists some $\bar{t} \in]0,T]$ and some $\tilde{c}>0$ such that:

(2.20) $(L(t;v),v)+\ell|v|^2 \geqslant \tilde{c}[a^p(t)\|v\|^p+|v|^2]$, $\forall v\in V$, a.e. on $]0,\bar{t}[$.

Proof. We are considering some $\ell>\ell_0$. Now, we take some $\bar{\ell}\in]\ell_0,\ell[$; thanks to the definitions (2.18) and (2.19), there exists $\bar{t}\in]0,T[$, such that: $\ell_0(t)\leqslant\bar{\ell}$ a.e. on $]0,\bar{t}[$. Hence, we have that:

(2.21) $$(L(t;v),v)+\ell|v|^2 \geqslant (\ell-\bar{\ell})|v|^2, \quad \forall v\in V, \text{ a.e. on }]0,\bar{t}[\ ;$$

from (2.21), it results easily that:

(2.22) $$\begin{cases}(L(t;v),v)+\ell|v|^2 \geqslant \left(1+\dfrac{|\ell|}{\ell-\bar{\ell}}\right)^{-1}(L(t;v),v), \quad \forall v\in V,\\ \text{a.e. on }]0,\bar{t}[.\end{cases}$$

Now, we denote by $c(\ell)$ the positive constant appearing at the right-hand side of (2.22); we also take some real $\bar{b}$ such that $(B(t)v,v)\geqslant\bar{b}|v|^2$ a.e. on $]0,\bar{t}[$ (thanks to (2.11)). Hence, choosing some $\varepsilon\in]0,1[$ and taking into account ((2.11) and) (2.14), we obtain from (2.22) and (2.21) that:

(2.23) $$\begin{cases}(L(t;v),v)+\ell|v|^2 = \varepsilon[(L(t;v),v)+\ell|v|^2]+(1-\varepsilon)[(L(t;v),v)+\\ +\ell|v|^2] \geqslant \varepsilon c(\ell)[a^p(t)\|v\|^p+\bar{b}|v|^2]+(1-\varepsilon)(\ell-\bar{\ell})|v|^2 =\\ = \varepsilon c(\ell)a^p(t)\|v\|^p+[(1-\varepsilon)(\ell-\bar{\ell})+\varepsilon c(\ell)\bar{b}]|v|^2, \text{a.e. on }]0,\bar{t}[.\end{cases}$$

Hence, if we choose ε sufficiently small, (2.23) gives (2.20), Q.E.D.

We finish this section by introducing another notation, which is useful for the sequel. Thus, let us define:

(2.24) $$m_0 \equiv -\frac{1}{p}(2\ell_0+1), \text{ if } \ell_0\in\mathbb{R}^1 \ ; \quad m_0\equiv+\infty, \text{ if } \ell_0=-\infty.$$

(We previously remarked that, under our assumptions on $L = A+B$, we have that $\ell_0<+\infty$; hence, we get that $m_0>-\infty$).

3. THE EXISTENCE AND UNIQUENESS RESULT. SOME EXAMPLES.

We are able, now, to state and to prove our existence and uniqueness result concerning the (nonlinear) singular abstract differential equation (2.10).

Theorem 3.1. Let (2.1), (2.2), (2.3) and (2.11) ... (2.16) hold. Let

m_0 be given by (2.24) (in terms of ℓ_0, defined by (2.19) and (2.18)). Take any $m<m_0$ (strictly). Then, for every $f \in Z_m(0,T)$, there exists a unique $u \in W_m(0,T)$, which solves (2.10).

Remark 3.1. Observe that, in Theorem 3.1, no initial condition (at t=0) on the solution u appears explicitly. However, being $u \in W_m(0,T)$, Lemma 2.1 implies that our solution u(t) has the "initial trend", which is given by (2.8).

We proceed now to prove Theorem 3.1 by several steps.

Proof of Theorem 3.1.

a) (Preliminary remark). First of all, we observe that Theorem 3.1 is proven, if we are able to show that the same result holds in the case where $]0,T[$ is replaced by $]0,\bar{t}[$ (for some $\bar{t} \in]0,T[$). Indeed, suppose that we have proven this result on $]0,\bar{t}[$ and let $w(t) \in W_m(0,\bar{t})$ be the obtained solution. Thanks to Lemma 2.1, we have that $w(t) \in C^0(]0,\bar{t}];H)$. Now, we consider (2.10) on $]\bar{t},T[$, with the initial condition $u(\bar{t})=w(\bar{t})$: thanks to Remark 2.1 (in particular, to (2.17)), this problem has a unique solution $\tilde{u}(t)$, $\bar{t}<t<T$. Hence, if we define u(t) on $]0,T]$ as $u(t) \equiv w(t)$ on $]0,\bar{t}[$ and $u(t) \equiv \tilde{u}(t)$ on $[\bar{t},T]$, this u(t) belongs to $W_m(0,T)$ and is the unique solution of (2.10) (in $]0,T[$), as we claimed. Thus, we have only to prove Theorem 3.1, in the case where $]0,T[$ is replaced by $]0,\bar{t}[$ (for some $\bar{t} \in]0,T[$).

b) (Uniqueness (on $]0,\bar{t}[$)). We are considering some $m<m_0$. Hence, thanks to the definitions (2.24), (2.19) and (2.18), there exists $\bar{t} \in]0,T[$ such that

$$(3.1) \qquad \begin{cases} (L(t;u_1)-L(t;u_2),u_1-u_2) - \frac{1}{2}(mp+1)|u_1-u_2|^2 \geqslant 0 , \\ \forall u_1,u_2 \in V , \text{ a.e. on }]0,\bar{t}[. \end{cases}$$

Now, let $f \in Z_m(0,\bar{t})$ be given and suppose that $u_1, u_2 \in W_m(0,\bar{t})$ both verify (2.10) (on $]0,\bar{t}[$). Then, $v \equiv u_1-u_2$ verifies (on $]0,\bar{t}[$):

$$(3.2) \qquad tv'(t)+L(t;u_1(t))-L(t;u_2(t)) = 0 .$$

Now, we "multiply" both sides of (3.2) by $t^{mp}v(t)$ (in the duality pairing between V* and V); we obtain, with an obvious calculation, that:

$$(3.3)\quad \begin{cases} \frac{1}{2}(t^{mp+1}|v(t)|^2)'+t^{mp}[(L(t;u_1(t))-L(t;u_2(t)),u_1(t)-u_2(t))- \\ -\frac{1}{2}(mp+1)|u_1(t)-u_2(t)|^2] = 0 \quad , \quad \text{a.e. on }]0,\bar{t}[\ . \end{cases}$$

Hence, thanks to (3.1), we deduce from (3.3) that:

$$(3.4)\qquad (t^{mp+1}|v(t)|^2)' \leqslant 0 \qquad \text{a.e. on }]0,\bar{t}[\ ;$$

but, according to Lemma 2.1, $t^{mp+1}|v(t)|^2$ belongs to $C^0([0,\bar{t}])$ and vanishes at $t=0$. Thus, it results that $v\equiv 0$ (i.e. $u_1\equiv u_2$).

c) (Existence (on $]0,\bar{t}[$)). This step is divided into the two substeps c_1) and c_2).

c_1) (Approximation and estimates). We are always considering some $m<m_0$. Hence, thanks to Lemma 2.2, there exists some $\bar{t}\in]0,T]$ and some $\tilde{c}>0$ such that (2.20) holds, where $\ell = -\frac{1}{2}(mp+1)$. From now on, we work on this interval $]0,\bar{t}[$. Let us define $t_k \equiv \frac{\bar{t}}{2k}$ $(k\in\mathbb{N})$. If $f\in Z_m(0,\bar{t})$ is given, let us also define f_k $(k\in\mathbb{N})$ as:

$$(3.5)\qquad f_k(t)\equiv 0 \quad , \quad 0\leqslant t\leqslant t_k \quad , \quad f_k(t) = f(t) \quad \text{a.e. on }]t_k,\bar{t}[\ ;$$

clearly, thanks to (2.7), it results that $f_k = (f_{1,k}+f_{2,k})\in L^q(0,\bar{t};V^*)+ +L^2(0,\bar{t};H)$. It is also obvious that:

$$(3.6)\qquad f_k \to f \quad \text{(strongly)} \quad \text{in } Z_m(0,\bar{t}) \ , \ \text{as } k\to+\infty \ .$$

Now, thanks to Remark 2.1, the problem

$$(3.7)\qquad tu'(t)+L(t;u(t)) = f(t) \ , \ t_k<t<\bar{t} \quad ; \quad u(t_k) = 0 \ ,$$

has a unique solution $\tilde{u}_k(t)$ $(\in L^p(t_k,\bar{t};V)\cap L^2(t_k,\bar{t};H)$, with $\tilde{u}_k'(t)\in L^q(t_k,\bar{t};V^*)+L^2(t_k,\bar{t};H))$. Define, moreover u_k $(k\in\mathbb{N})$ as:

$$(3.8)\qquad u_k(t)\equiv 0 \ , \ 0\leqslant t\leqslant t_k \quad ; \quad u_k(t) = \tilde{u}_k(t) \quad \text{on }]t_k,\bar{t}] \ .$$

Clearly (because $L(t;0) = 0$ a.e., thanks to (2.11) and (2.15)), $u_k(t)$ verifies:

$$(3.9)\qquad tu_k'(t)+L(t;u_k(t)) = f_k(t) \quad , \quad 0 < t < \bar{t} \ .$$

Now, we proceed to estimate $u_k(t)$; we "multiply" both sides of (3.9) by

$t^{mp}u_k(t)$ (in the duality pairing between V^* and V), obtaining:

$$(3.10) \qquad t^{mp+1}(u_k'(t),u_k(t))+t^{mp}(L(t;u_k(t)),u_k(t)) = t^{mp}(f_k(t),u_k(t)).$$

Hence, integrating (3.10) over $[0,t]$ $(0\leq t\leq \bar{t})$ and making some obvious calculations, we get:

$$(3.11) \quad \begin{cases} \frac{1}{2}\, t^{mp+1}|u_k(t)|^2+\int_0^t s^{mp}[(L(s;u_k(s)),u_k(s))-\frac{1}{2}(mp+1)|u_k(s)|^2]ds\leq \\ \leq \int_0^t s^{mp}[\,\|f_{1,k}(s)\|_*\|u_k(s)\|+|f_{2,k}(s)|\,|u_k(s)|\,]ds \quad, 0\leq t\leq \bar{t} \;. \end{cases}$$

Now, we use Lemma 2.2 (with $\ell=-\frac{1}{2}(mp+1)$; remember the beginning of this substep c_1) of this proof); with some other easy calculations, we obtain, from (3.11), that there exists some positive constant $\bar{c}$ (depending on m and p, but <u>not</u> on k) such that (also recall that $f = f_1+f_2$, according to (2.4)):

$$(3.12) \quad \begin{cases} t^{mp+1}|u_k(t)|^2+\int_0^t s^{mp}(a(s))^p\|u_k(s)\|^p ds+\int_0^t s^{mp}|u_k(s)|^2 ds \leq \\ \leq \bar{c}\,[\int_0^{\bar{t}} s^{mp}(a(s))^{-q}\|f_1(s)\|_*^q+\int_0^{\bar{t}} s^{mp}|f_2(s)|^2 ds] \quad, 0\leq t\leq \bar{t} \;. \end{cases}$$

c_2) (<u>Passing to the limit</u>). From the estimate (3.12) and the fact (previously remarked in section 2) that L maps bounded subsets of $U_m(0,\bar{t})$ onto bounded subsets of $Z_m(0,\bar{t})$, we deduce that there exists a subsequence $\{u_j \mid j\in \mathbb{N}\}$ of $\{u_k \mid k\in \mathbb{N}\}$ such that:

$$(3.13) \quad \begin{cases} u_j \to u \quad \text{(weakly) in } U_m(0,\bar{t}) \;; \\ u_j \to u \quad \text{(weakly *) in } \{v \mid t^{\frac{1}{2}(mp+1)}v(t)\in L^\infty(0,\bar{t};H)\} \;; \\ u_j(\bar{t}) \to z \quad \text{(weakly) in } H \;; \\ L(t;u_j) \to h \quad \text{(weakly) in } Z_m(0,\bar{t}) \;. \end{cases}$$

Using some standard procedures (extension to 0 out of $[0,\bar{t}]$ etc.; see, e.g., LIONS [11], chap.2) we can pass to the limit (in (3.9), where k=j), as $j\to+\infty$; we obtain (also recall (3.6)) that u verifies (in the sense of $\mathcal{D}'(]0,\bar{t}[;V^*)$):

$$(3.14) \qquad tu' + h = f \;.$$

Thus, one has, from (3.14), that $tu'\in Z_m(0,\bar{t})$ and hence $u\in W_m(0,\bar{t})$; moreover, it also results that $u(\bar{t}) = z$. It remains only to prove that

(3.15) $$h = L(t;u) .$$

This can be done, by using the monotonicity of $L-\frac{1}{2}(mp+1)I$ from V to V* a.e. on $]0,\bar{t}[$ and the hemicontinuity of L (that is of A; see (2.13)). The procedure here is only a slight modification of the standard one (see, e.g. LIONS [11], chap.2): we give only the main steps, for the reader's convenience. Let us define, for every $v \in U_m(0,\bar{t})$:

(3.16) $$\left[\begin{array}{l} X_j \equiv \int_0^{\bar{t}} t^{mp}[(L(t;u_j(t))-L(t;v(t)),u_j(t)-v(t)) - \\ - \frac{1}{2}(mp+1)|u_j(t)-v(t)|^2]dt \quad , \end{array}\right.$$

and note that $X_j \geq 0$, for every $j \in \mathbb{N}$ and every such v. Now, integrating (3.10) over $[0,\bar{t}]$ and performing there the (usual) integration by parts, use this formula to rewrite suitably X_j. Then, by means of (3.6), (3.13) and $u(\bar{t})=z$, we get:

(3.17) $$\left[\begin{array}{l} \limsup_{j\to+\infty} X_j \leq -\frac{1}{2}\bar{t}^{mp+1}|u(\bar{t})|^2+\int_0^{\bar{t}} t^{mp}[(f(t),u(t))-(h(t),v(t))- \\ - (L(t;v(t)),u(t)-v(t))+ \frac{1}{2}(mp+1)[(u(t),v(t))+(v(t),u(t)-v(t))]]dt \quad . \end{array}\right.$$

On the other hand, we deduce, from (3.14):

(3.18) $$\left[\begin{array}{l} \int_0^{\bar{t}} t^{mp}[(h(t),u(t))- \frac{1}{2}(mp+1)|u(t)|^2]dt = \\ = \int_0^{\bar{t}} t^{mp}(f(t),u(t))dt - \frac{1}{2}\bar{t}^{mp+1}|u(\bar{t})|^2 \quad . \end{array}\right.$$

Hence, from (3.17), (3.18) and $X_j \geq 0$, we obtain that:

(3.19) $$0 \leq \int_0^{\bar{t}} t^{mp}[(h(t)-L(t;v(t)),u(t)-v(t))- \frac{1}{2}(mp+1)|u(t)-v(t)|^2]dt,$$

for every $v(t) \in U_m(0,\bar{t})$. Now, we consider (3.19), taking $v(t)=u(t)-rw(t)$, where $r>0$ and $w(t)$ is any function belonging to $U_m(0,\bar{t})$; dividing by r both sides of the so obtained inequality, we get:

(3.20) $$0 \leq \int_0^{\bar{t}} t^{mp}[(h(t)-L(t;u(t)-rw(t)),w(t))- \frac{r}{2}(mp+1)|w(t)|^2]dt.$$

Hence, let us make $r\to 0^+$ in (3.20); we obtain (thanks to the hemicontinuity of L and the Lebesgue's theorem) that:

(3.21) $$0 \leq \int_0^{\bar{t}} t^{mp}(h(t)-L(t;u(t)),w(t))dt \quad , \quad \forall w(t) \in U_m(0,\bar{t}) \; ;$$

then (3.21) gives, clearly, (3.15).

Hence, Theorem 3.1 is completely proven, Q.E.D.

We finish this paper by seeing how Theorem 3.1 applies to our previous p.d.e. example. However, before doing this, we prefer to present a simple o.d.e. example, where some rather "strange" facts (concerning the result given by Theorem 3.1) can even be viewed.

Example 3.1. Let us take: $V = H = V^* = \mathbb{R}^1$; $B(t) = bI$ (for some $b \in \mathbb{R}^1$); $A(t,u) = (a(t))^p |u|^{p-1} \operatorname{sgn}(u)$ (where $p>1$ and $a(t)$ verifies (2.3')). Clearly, $A(t;\cdot)$ is monotone a.e. on $]0,T[$ (and nonlinear, if $p\neq 2$); (2.11) ...(2.16) hold obviously, in this case. The equation (2.10) is here the following o.d.e.:

$$(3.22) \qquad tu'(t)+(a(t))^p |u(t)|^{p-1} \operatorname{sgn}(u(t))+bu(t) = f(t)\ , \quad 0<t<T\ .$$

Theorem 3.1 applies here (for $m < m_0$). We can obtain here, with some calculations, that $m_0 = \frac{1}{p}(2b-1)$, if $p\neq 2$ (i.e. in the nonlinear case); whereas, if $p=2$ (i.e. in the linear case), m_0 also depends on $a(t)$: precisely, one has here that $m_0 = b-\frac{1}{2}+ \operatorname{ess}\liminf_{t\to 0^+} (a(t))^2$ (hence, one has here that $m_0=+\infty$, if $\operatorname{ess}\lim_{t\to 0^+} a(t) = +\infty$, as we remarked in section 2).

Remark 3.2. Note that the result of Theorem 3.1 is no more true, in general, if we take $m\geq m_0$. This can be seen in the Example 3.1 above, even in the linear case ($p=2$); take, for instance, in (3.22): $p = 2$; $0<T<1$; $a(t) \equiv 1$; $b = -\frac{1}{2}$; $f(t) = t^{-\frac{1}{2}}(\ln t)^{-1}$. Then, one has here that $m_0 = 0$; but, evaluating explicitly the solutions of (3.22) (in this case), one sees that, in the context of Theorem 3.1, the uniqueness fails for $m>0$ and the existence fails for $m = 0$.

Example 3.2. Let us return to the p.d.e. example (see the introduction), which induced us to study the singular abstract differential equation (2.10). Thus, let Ω be a bounded open subset of $\mathbb{R}^n$ and let $p>1$ such that $\frac{1}{p} - \frac{1}{n} \leq \frac{1}{2}$. Take now $V = W_0^{1,p}(\Omega)$ and $H = L^2(\Omega)$ (hence $V^*=W^{-1,q}(\Omega)$). Then, (2.1) clearly holds. Take, moreover:

$$(3.23) \qquad B(t) = b(t,x)I \quad (\text{where } b(t,x) \in L^\infty(\Omega\times]0,T[))\ ;$$

$$(3.24)\quad \begin{cases} A(t;u) = -(\tilde{a}(t))^p \sum_{i=1}^{n} \frac{\partial}{\partial x_i}\left(\left|\frac{\partial u}{\partial x_i}\right|^{p-2} \frac{\partial u}{\partial x_i}\right) \\ \text{(where } \tilde{a}(t) \text{ verifies (2.3)).} \end{cases}$$

Then (2.11) ... (2.16) clearly hold (with $a(t) = c\tilde{a}(t)$; c suitable positive constant). (Note, in particular, that here $A(t;\cdot)$ is monotone from V to V* a.e. on]0,T[). The equation (2.10) is here the following p.d.e.:

$$(3.25)\quad t\frac{\partial u}{\partial t} - (\tilde{a}(t))^p \sum_{i=1}^{n} \frac{\partial}{\partial x_i}\left(\left|\frac{\partial u}{\partial x_i}\right|^{p-2} \frac{\partial u}{\partial x_i}\right) + b(x,t)u = f \ , \ (x,t)\in\Omega\times]0,T[.$$

Theorem 3.1 applies here, for $m<m_0$: it can be seen that m_0 depends only on b(t,x) and p, if $p\neq 2$, but also on $\tilde{a}(t)$ if p=2.

Acknowledgment. This author would like to thank Claudio Baiocchi for some stimulating discussions.

- REFERENCES -

[1] C.BAIOCCHI and M.S.BAOUENDI, Singular evolution equations, J. of Funct.Anal., 25 (1977), 103-120.

[2] M.L.BERNARDI, Su alcune equazioni d'evoluzione singolari, Boll.Un. Mat.Ital., 5, 13-B (1976), 498-517.

[3] F.E.BROWDER, Nonlinear initial value problems, Annals of Math., 82, (1965), 51-87.

[4] R.W.CARROLL and R.E.SHOWALTER, "Singular and Degenerate Cauchy Problems", Math.in Science and Engineering, n.127, Academic Press, New York, 1976.

[5] G.DA PRATO and P.GRISVARD, Sommes d'opérateurs linéaires et équations différentielles opérationnelles, J.Math.Pures et Appl., 54 (1975), 305-387.

[6] G.DA PRATO and P.GRISVARD, On an abstract singular Cauchy problem, Comm. in P.D.E., 3 (11) (1978), 1077-1082.

[7] A.FAVINI, Degenerate and singular evolution equations in Banach space, to appear on Math.Ann.

[8] K.L.KUTTLER JR., A degenerate nonlinear Cauchy problem, Appl.Anal., 13 (1982), 307-322.

[9] K.L.KUTTLER JR., Implicit evolution equations, Appl.Anal., 16(1983), 91-99.

[10] J.E.LEWIS and C.PARENTI, Abstract singular parabolic equations, Comm. in P.D.E., 7 (3) (1982), 279-324.

[11] J.L.LIONS, "Quelques Méthodes de Résolution des Problèmes aux Limites Non Linéaires", Dunod-Gauthier Villars, Paris, 1969.

SOME TRANSMUTATION METHODS FOR CANONICAL SYSTEMS

Robert Carroll
University of Illinois
Urbana, Illinois

1. Introduction. Let H be a separable Hilbert space and L(H) be the space of bounded linear operators $H \to H$. Let $J \in L(H)$ with $J^2 = -I$ and $J^* = -J$ and take $V(\cdot) \in L^1(L(H)) \cap L^\infty(L(H))$ on $[0,\infty)$ with $V^*(x) = V(x)$ and, without loss of generality (cf. [1]) one can take $JV(x) = -V(x)J$. Let P be a projection with $P^* = P$, $PJP = 0$, $JP + PJ = J$, and, for $Q = I-P$, $(P-Q)\exp(\lambda Jx) = \exp(-\lambda Jx)(P-Q)$. There is some redundancy here but prototypically one is dealing with the following situation, extended to H.

EXAMPLE 1.1. Take $H = C^2$ with $J = \begin{pmatrix} 0 & 1 \\ -1 & 0 \end{pmatrix}$, $P = \begin{pmatrix} 1 & 0 \\ 0 & 0 \end{pmatrix}$, $Q = I-P = \begin{pmatrix} 0 & 0 \\ 0 & 1 \end{pmatrix}$, $V = \begin{pmatrix} s & r \\ r & -s \end{pmatrix}$, $r^* = r$ and $s^* = s$, $P_\pm = \frac{1}{2}(I \mp iJ)$, $U_0 = \exp(-\lambda Jx) = P_+\exp(-i\lambda x) + P_-\exp(i\lambda x) = \begin{pmatrix} \mathrm{Cos}\lambda x & -\mathrm{Sin}\lambda x \\ \mathrm{Sin}\lambda x & \mathrm{Cos}\lambda x \end{pmatrix}$, $\Phi_0 = U_0P = \frac{1}{2}\begin{pmatrix} 1 \\ i \end{pmatrix}\exp(-i\lambda x) + \frac{1}{2}\begin{pmatrix} 1 \\ -i \end{pmatrix}\exp(i\lambda x) = \begin{pmatrix} \mathrm{Cos}\lambda x & 0 \\ \mathrm{Sin}\lambda x & 0 \end{pmatrix} \sim \begin{pmatrix} \mathrm{Cos}\lambda x \\ \mathrm{Sin}\lambda x \end{pmatrix}$, etc. Note $P_+ + P_- = I$, $P_+ = \frac{1}{2}\begin{pmatrix} 1 & -i \\ i & 1 \end{pmatrix}$, $P_- = \frac{1}{2}\begin{pmatrix} 1 & i \\ -i & 1 \end{pmatrix}$, $P_+P_- = P_-P_+ = 0$, $H_\pm = P_\pm H$, $JP_\pm = iP_+$, $JP_- = -iP_-$, etc. The operator $K = \frac{1}{2}\begin{pmatrix} 1 & -i \\ -i & -1 \end{pmatrix}$ satisfies $KK^* = P_-$, $K^*K = P_+$, $K^2 = 0$, $P_+K = 0 = KP_-$, $P = \frac{1}{2}(I + K + K^*)$, $KP_+ = K = P_-K$, $KP_- = 0$, etc. To see that $H_- = P_-H = KH$ note that $P_-f = Kh$ when $f = Rh$ for $R = \begin{pmatrix} 1 & 0 \\ 0 & -1 \end{pmatrix}$ and thus $K: H_+ \to H_-$. For a version in C^{2n} see [11] and for philosophy see [1;11;21-23].

Now following [1;2;9;11;12;18;19;21-23] one refers to $[JD_x - V(x)]u = v$ in H or in L(H) as a canonical system. In particular we consider here

$$(1.1) \qquad Qy = [JD_x - V(x)]y = \lambda y$$

which arises in studying e.g. Dirac type systems $-iD_t u = Q(D_x)u$ or transmission line problems $-\hat{I}D_t\varphi = Q(D_x)\varphi$ where $\hat{I} \sim \begin{pmatrix} 1 & 0 \\ 0 & -1 \end{pmatrix}$. We will discuss various aspects of transmutation type connections of Q to $Q_0 = JD_x$. Thus one can produce Parseval type formulas, Goursat problems for transmutation kernels relating the "potential" V to the kernels, Gelfand-Levitan (G-L) equations, scattering formulas, Cauchy problems to characterize transmutations, etc. Our emphasis will be on spectral pairings and we will compare with related results in [6;7;8;11].

2. Basic Information. We begin with a sketch of a procedure outlined in [21] to obtain Parseval formulas for certain solutions of (1.1). The method is modeled in part on the scalar case discussed in [21] and generalized in various ways in [5-7]. Some details of technique and construction are included here to show how transmutation methods transport various types of information around between the Q and the Q_0 theories. Thus the solution $y \in L(H)$ of (1.1) with $y(0) = I$ is denoted by $U(x,\lambda)$ and one can write (cf. [6;7;11;21])

$$(2.1) \qquad U(x,\lambda) = U_0(x,\lambda) + \int_{-x}^{x} K(x,t)U_0(t,\lambda)dt$$

where $U_0(x,\lambda) = \exp(-\lambda Jx)$ satisfies $JD_xU_0 = \lambda U_0$ with $U_0(0,\lambda) = I$. Indeed, setting $\Omega(x,\lambda) = U_0^{-1}(x,\lambda)U(x,\lambda) = I + \int_0^x \exp(2\lambda J\xi)w(x,\xi)d\xi$ one can determine w as a solution of an integral equation via an iterative procedure (cf. [6;11;21]). Writing $w_- = w + JwJ$; $w_+ = w - JwJ$ (so $Jw_+ = w_+J$ and $Jw_- = -w_-J$) one finds that K in (2.1) will be $K = K(x,t) = (1/2)[w_+(x,\frac{1}{2}(x-t)) + w_-(x,\frac{1}{2}(x+t))$ and thus K can be found by standard computations. The integral equation for w can be written in various ways, e.g., noting that $\exp(\lambda J)V = V\exp(-\lambda J)$, (•) $w(t,s) + JV(s) = \int_s^t V(u)Jw(u,u-s)du$. Now using Example 1.1 as a prototypical model one can write from (2.1) (♦) $U(t,\lambda) = U_0(t,\lambda) + \int_0^t K_P(t,s) U_0(s,\lambda)ds$ where $K_P(t,s) = K(t,s) + K(t,-s)(P-Q)$. Set now $\Phi(x,\lambda) = U(x,\lambda)P$ and from (♦)

$$(2.2) \qquad \Phi(t,\lambda) = \Phi_0(t,\lambda) + \int_0^t K_P(t,s)\Phi_0(s,\lambda)ds$$

We note also that $\Phi_0(t,\lambda) = \text{Cos}\lambda xP - \text{Sin}\lambda xJP$ and writing $K_0(t,s) = K(t,s) + K(t,-s)$ with $K_\infty(t,s) = K(t,s) - K(t,-s)$ one has $K_P(t,s) = K_0(t,s)P + K_\infty(t,s)Q$. Next one inverts (2.2) in the spirit of Volterra operators to obtain

$$(2.3) \qquad \Phi_0(x,\lambda) = \Phi(x,\lambda) + \int_0^x L_P(x,t)\Phi(t,\lambda)dt$$

Following calculations in [6;7;11;19;21-23] there results

THEOREM 2.1. Under the hypotheses indicated one has (2.2) - (2.3) with $V(t) = JK_P(t,t) - K_P(t,t)J$, $[JD_t - V(t)]K_P(t,s) = -D_sK_P(t,s)J$, $K_P(t,0) = 0$, $JL_P(x,x) - L_P(x,x)J = -V(x)$, $L_P(x,0)P = L_P(x,0)$, and $D_tL_P(x,t)J + L_P(x,t)V(t) = -JD_xL_P(x,t)$.

Consider next the equation (¶) $JD_xU(x,y) + D_yU(x,y)J = 0$ $(0 \leq y \leq x < \infty)$ with $U(x,0) = f(x)$. One checks easily that this has a unique solution (★) $U(x,y) = \frac{1}{2}[f(x+y) + f(|x-y|)] + \frac{1}{2}J[f(x+y) - f(|x-y|)]J$ ($f(x) \in L(H)$ here). Now define $\tilde{\Phi}(x,\lambda)$ (resp. $\tilde{\Phi}_0(x,\lambda)$) as solutions of $-D_x\tilde{\Phi}J - \tilde{\Phi}V(x) = \lambda\tilde{\Phi}$ (resp. $-D_x\tilde{\Phi}_0J = \lambda\tilde{\Phi}_0$) with initial conditions $\tilde{\Phi}(0,\lambda) = \tilde{\Phi}_0(0,\lambda) = P$. Then one sees easily that $\tilde{\Phi} = \Phi^*$ and in particular $\tilde{\Phi}_0 = P\text{Cos}\lambda x + PJ\text{Sin}\lambda x = PU_0^*(x,\lambda)$. As above one obtains then from (2.3) (cf. [21]) (‡) $\tilde{\Phi}_0(x,\lambda) = \tilde{\Phi}(x,\lambda) + \int_0^x \tilde{\Phi}(t,\lambda)\tilde{L}_P(x,t)dt$ where $L_P^*(x,t) = \tilde{L}_P(x,t)$. Define now

$$(2.4) \qquad f(x,y) = L_P(x,y) + \int_0^y L_P(x,t)\tilde{L}_P(y,t)dt \quad (x \geq y);$$

$$f(x,y) = \tilde{L}_P(y,x) + \int_0^x L_P(x,t)\tilde{L}_P(y,t)dt \quad (x \leq y)$$

Let us note that from Theorem 2.1, $-JD_t\tilde{L}_P(x,t) + V(t)\tilde{L}_P(x,t) = D_x\tilde{L}_P(x,t)J$; $V(x) = \tilde{L}_P(x,x)J - J\tilde{L}_P(x,x)$; $P\tilde{L}_P(x,0) = \tilde{L}_P(x,0)$, etc. Consider then $f(x,y)$ in (2.4) for say $x \geq y$. With some routine calculation one has $JD_xf + D_yfJ = 0$ and hence eventually

PROPOSITION 2.2. The function f defined by (2.4) can be written as $f(x,y) = \frac{1}{2}[L_P(x+y,0) + L_P(|x-y|,0)] + \frac{1}{2}J[L_P(x+y,0) - L_P(|x-y|,0)]J = \frac{1}{2}[\tilde{L}_P(x+y,0) + \tilde{L}_P(|x-y|,0)] + \frac{1}{2}J[\tilde{L}_P(x+y,0) - \tilde{L}_P(|x-y|,0)]J$. It follows from this that $L_P(x,0) + JL_P(x,0)J = \tilde{L}_P(x,0) + J\tilde{L}_P(x,0)J$.

REMARK 2.3. Let us note (cf. [6;20]) that for $1 + L = (1 + K)^{-1}$ and $L^+(x,y) \sim L(y,x)$ $(y \to x)$ one has $1+F = (1 + L)(1 + L^+)$ (definition) so that $F = L + L^+ + LL^+$.

Thus $(1 + K)(1 + F) = 1 + L^+$ or $K + F + KF + L^+$; this is the G-L equation and exhibits F as the standard symmetric kernel.

Next we write $C_0(L(H))$ for continuous functions of compact support with values in L(H) and define (☆) $C(\lambda,f) = \int_0^\infty f(x)\mathrm{Cos}\lambda x dx$ with $S(\lambda,f) = \int_0^\infty f(x)\mathrm{Sin}\lambda x dx$. The standard inversion formulas hold and one has a Parseval relation. Define next for such f

$$(2.5) \qquad \Phi_0(\lambda,f) = \int_0^\infty f(x)\Phi_0(x,\lambda)dx; \; \tilde{\Phi}_0(\lambda,f) = \int_0^\infty \tilde{\Phi}_0(x,\lambda)f(x)dx$$

(recall $\Phi_0 = \mathrm{Cos}\lambda xP - \mathrm{Sin}\lambda xJP = U_0P$ and $\tilde{\Phi}_0 = PU_0^* = P\mathrm{Cos}\lambda x + PJ\mathrm{Sin}\lambda x$). It follows that

$$(2.6) \qquad \int_0^\infty f(x)g(x)dx = (1/\pi)\int_{-\infty}^\infty \Phi_0(\lambda,f)\tilde{\Phi}_0(\lambda,g)d\lambda$$

$$(2.7) \qquad f(x) = (1/\pi)\int_{-\infty}^\infty \Phi_0(\lambda,f)\tilde{\Phi}_0(x,\lambda)d\lambda = (1/\pi)\int_{-\infty}^\infty \Phi_0(x,\lambda)\tilde{\Phi}_0(\lambda,f)d\lambda$$

One uses now the following notation from [21]. Let K^2 denote L^2 functions with compact support and $FK^2 = \hat{K}^2$. One defines Z as entire functions $\hat{f}(\lambda)$ of exponential type with $\hat{f} \in L^1$ for $\lambda \in R$; $\hat{f}_n \to \hat{f}$ in Z means (type $\hat{f}_n$) $\leq$ some fixed σ and $\hat{f}_n \to \hat{f}$ in $L^1(R)$. The dual of Z is Z' and the space of operator valued functions $(\hat{f}(\lambda),R)$: $Z \to L(H)$ is denoted by HZ' (we do not distinguish here between $Z(-\infty,\infty)$ and Z (even) - cf. [21]). If L(x) is a continuous L(H) valued function then $C(\lambda,L) \in HZ'$ via $(\hat{f}(\lambda),C(\lambda,L)) = (1/2)\int_{-\infty}^\infty [L(x) + L(-x)]\int_{-\infty}^\infty \hat{f}(\lambda)\exp(i\lambda x)d\lambda dx$. For $\hat{f} = C(\lambda,f)$ (even) one can write this as $(C(\lambda,f),C(\lambda,L)) = \int_0^\infty [L(x) + L(-x)]\int_0^\infty C(\lambda,f)\mathrm{Cos}\lambda x d\lambda dx$. Similar formulas apply to Sine transforms. Given an orthonormal basis $e_1, e_2, \ldots$ for H let an operator A be written in matrix form as $((A_{ij}))$. For $R \in HZ'$ one has $R_{ij} \in Z'$ and $(\hat{f}(\lambda),R_{ij}) = (\hat{f}(\lambda),R)_{ij}$. If $R \in HZ'$ is positive (i.e. for $\hat{f}(\lambda) \geq 0$, $(\hat{f}(\lambda),R)$ is a nonnegative operator for each $\lambda \in R$) then there is an operator valued measure $d\rho(\lambda)$ such that $(\hat{f}(\lambda)\hat{g}(\lambda),R) = \int_{-\infty}^\infty \hat{f}(\lambda)\hat{g}(\lambda)d\rho(\lambda)$ for $\hat{f},\hat{g} \in \hat{K}^2$. Here $\rho(\lambda) \in L(H)$ and $\rho(\lambda') - \rho(\lambda)$ is a nonnegative operator for $\lambda' > \lambda$ (see [21] - for convenience in what follows we will simply write R in a function notation). Using this background one wants now to show that there is a generalized spectral function $R = ((R_{ij})) \in HZ'$ $(R_{ij} \in Z')$ such that

$$(2.8) \qquad \int_0^\infty f(x)g(x)dx = [\Phi(\lambda,f)PRP\tilde{\Phi}(\lambda,g)]$$

where f,g are suitable functions with values in L(H) and PRP = R with $[F(\lambda)RG(\lambda)] = [\sum_{j,m}(F_{ij}(\lambda)G_{mk}(\lambda),R_{jm})]$ $(\Phi(\lambda,f)$ etc. is defined as in (2.5) - cf. (2.10) below). Further $(L = L_p(x,0), \tilde{L} = \tilde{L}_p(x,0))$ R is given explicitly as

$$(2.9) \qquad R = (1/\pi)[P + \tilde{\Phi}_0(\lambda,L)] = (1/\pi)[P + \Phi_0(\lambda,\tilde{L})]$$

This is then completely analogous to the scalar situation for operators $D^2 - q$ for example (cf. [5-7;21]). We will sketch a proof of (2.8)-(2.9) now following [21] in order to exhibit the ingredients and how they fit together. Thus first one shows

$$(2.10) \qquad \Phi(\lambda,f) = \int_0^\infty f(x)\Phi(x,\lambda)dx = \Phi_0(\lambda,F); \; \tilde{\Phi}(\lambda,g) = \int_0^\infty \tilde{\Phi}(x,\lambda)g(x)dx = \tilde{\Phi}_0(\lambda,G)$$

for suitable f,g where (♠) $f(x) = F(x) + \int_x^\infty F(t)L_p(t,x)dt$ and $g(x) = G(x) +$

$+ \int_x^\infty \tilde{L}_p(t,x)G(t)dt$ (cf. (2.3) and (‡)). To see this we write e.g. $\Phi(\lambda,f) = \int_0^\infty f(x)\Phi(x,\lambda)dx = \int_0^\infty [F(x) + \int_x^\infty F(t)L_p(t,x)dt]\Phi(x,\lambda)dx = \int_0^\infty F(t)[\Phi(t,\lambda) + \int_0^x L_p(t,x)\Phi(x,\lambda)dx]dt = \int_0^\infty F(t)\Phi_0(t,\lambda)dt = \Phi_0(\lambda,F)$. Further one can write (note $L_p(t,x) = 0$ for $x > t$ etc. is implicit here) (■) $\int_0^\infty f(x)g(x)dx = \int_0^\infty F(x)G(x)dx + \int_0^\infty \int_0^\infty F(x)[L_p(x,y) + \tilde{L}_p(y,x) + \int_0^\infty L_p(x,\xi)\tilde{L}_p(y,\xi)d\xi]G(y)dxdy$. Now recall $f(x,y)$ defined by (2.4) and we have (●●) $\int_0^\infty f(x)g(x)dx = \int_0^\infty F(x)G(x)dx + \int_0^\infty \int_0^\infty F(x)f(x,y)G(y)dxdy$. We note here from Remark 2.3 that $\tilde{L}_p^+(y,x) = \tilde{L}_p(x,y)$ and from $(I + K_p)^{-1} = I + L_p$ one has $f(x,y) = (L_p + \tilde{L}_p^+ + L_p\tilde{L}_p^+)(x,y)$ with $K_p + f + K_pf = \tilde{L}_p^+$. Now consider $\varphi_\sigma = \gamma_\sigma L_p(x,0)$ where γ_σ is smooth with $\gamma_\sigma = 1$ for $0 \le x \le 2\sigma$ and $\gamma_\sigma = 0$ for $x \ge 2\sigma+1$; set $\tilde{\varphi}_\sigma = \gamma_\sigma \tilde{L}_p(x,0)$ $(= \varphi_\sigma^*)$ with

$$(2.11) \qquad f_\sigma(x,y) = (1/\pi)\int_{-\infty}^\infty \Phi_0(x,\lambda)\tilde{\Phi}_0(\lambda,\varphi_\sigma)\tilde{\Phi}_0(y,\lambda)d\lambda = (1/\pi)\int_{-\infty}^\infty \Phi_0(x,\lambda)\Phi_0(\lambda,\tilde{\varphi}_\sigma)\tilde{\Phi}_0(y,\lambda)d\lambda$$

Evidently $JD_xf_\sigma + D_yf_\sigma J = 0$ and (assuming both expressions are valid) we can use (2.7) to obtain $f_\sigma(x,0) = (1/\pi)\int_{-\infty}^\infty \Phi_0(x,\lambda)\tilde{\Phi}_0(\lambda,\varphi_\sigma)Pd\lambda = \varphi_\sigma P = \varphi_\sigma = \gamma_\sigma L_p(x,0)$ and $f_\sigma(0,y) = (1/\pi)\int_{-\infty}^\infty P\Phi_0(\lambda,\tilde{\varphi}_\sigma)\tilde{\Phi}_0(y,\lambda)d\lambda = P\tilde{\varphi}_\sigma = \tilde{\varphi}_\sigma = \gamma_\sigma\tilde{L}_p(y,0)$. Consequently from (2.4) and subsequent calculations one has $f_\sigma = f$ for $0 \le x,y \le \sigma$. Note here by Proposition 2.2 $f(x,y)$ involves $L_p(x+y,0)$ etc. so this explains the cutoff of γ_σ at 2σ etc. To see that both forms in (2.11) are equal it is necessary to make some calculations based on Example 1.1 for example and we omit this. One obtains then

<u>THEOREM 2.4</u>. Under the hypotheses indicated, for $0 \le x,y \le \sigma$, f_σ given by (2.11) is the same as f given by (2.4).

Now take $f(x) = g(x) = 0$ for $x > \sigma$ and from (●) $f = F(I + L_p^+)$ implies $F = f(I + K_p^+)$ or $F(x) = f(x) + \int_x^\infty f(t)K_p(t,x)dt$; consequently $F(x) = 0$ for $x > \sigma$ as well, and similarly $G(x) = 0$ for $x > \sigma$. Then from (●●) and (2.11) (■■) $\int_0^\infty f(x)g(x)dx = \int_0^\infty F(x)G(x)dx + (1/\pi)\int_{-\infty}^\infty \Phi_0(\lambda,F)\Phi_0(\lambda,\tilde{\varphi}_\sigma)\tilde{\Phi}_0(\lambda,G)d\lambda$ where $\tilde{\Phi}_0(\lambda,\varphi_\sigma)$ may be used in place of $\Phi_0(\lambda,\tilde{\varphi}_\sigma)$. Now go to (2.10) and use (2.6) to obtain

$$(2.12) \qquad \int_0^\infty f(x)g(x)dx = (1/\pi)\int_{-\infty}^\infty \Phi_0(\lambda,F)[P + \tilde{\Phi}_0(\lambda,\varphi_\sigma)]\tilde{\Phi}_0(\lambda,G)d\lambda = (1/\pi)\int_{-\infty}^\infty \Phi(\lambda,f)[P + \Phi_0(\lambda,\tilde{\varphi}_\sigma)]\tilde{\Phi}(\lambda,g)d\lambda$$

(recall here $\Phi P = \Phi$, $\tilde{\Phi} = P\tilde{\Phi}$, etc.). Next let $\sigma \to \infty$ and the central [ ] term in (2.12) (times $1/\pi$) tends formally to $R = (1/\pi)[P + \Phi_0(\lambda,\tilde{L}_p)]$ or $(1/\pi)[P + \tilde{\Phi}_0(\lambda,L_p)]$ as in (2.9). Clearly $P\Phi_0(\lambda,\tilde{L}_p)P = P\int_0^\infty \tilde{L}_p(x,0)U_0(x,\lambda)PdxP = \Phi_0(\lambda,\tilde{L}_p)$ and $P\tilde{\Phi}_0(\lambda,L_p)P = P\int_0^\infty PU_0^*(x,\lambda)L_p(x,0)dxP = \tilde{\Phi}_0(\lambda,L_p)$ (recall $P\tilde{L}_p = \tilde{L}_p$ and $L_pP = L_p$). Hence formally

<u>THEOREM 2.5</u>. Under the hypotheses indicated the Parseval formula (2.8) holds with (2.9) in the form $\int_0^\infty f(x)g(x)dx = [\Phi(\lambda,f)PRP\tilde{\Phi}(\lambda,g)] = \int_{-\infty}^\infty \Phi(\lambda,f)R\tilde{\Phi}(\lambda,g)d\lambda$ and $f(x) = \int_{-\infty}^\infty \Phi(\lambda,f)R\tilde{\Phi}(x,\lambda)d\lambda = \int_{-\infty}^\infty \Phi(x,\lambda)R\tilde{\Phi}(\lambda,f)d\lambda$.

Let us compare this now to the development of [1;3;4;9;11;12;22;23] as expounded in [11] (cf. also [6;7]). Thus one writes $X = \Phi$ (we continue to use Φ here) and for suitable functions $h(x)$ with values in H

$$(2.13)\qquad h^{\blacktriangle}(\lambda) = \int_0^\infty \Phi^*(x,\lambda)h(x)dx;\ h(x) = (1/\pi)\int_{-\infty}^{\infty} \Phi(x,\lambda)\Delta_\infty(\lambda)h^{\blacktriangle}(\lambda)d\lambda$$

$$(2.14)\qquad \int_0^t h_1^*(x)h_2(x)dx = \int_0^t (h_1,h_2)(x)dx = (1/\pi)\int_{-\infty}^{\infty} h_1^{\blacktriangle *}(\lambda)\Delta_t(\lambda)h_2^{\blacktriangle}(\lambda)d\lambda = (1/\pi)\int_{-\infty}^{\infty} (h_1^{\blacktriangle}(\lambda),\Delta_t(\lambda)h_2^{\blacktriangle}(\lambda))d\lambda$$

Here formally (referring to [6;7;11] for details) $\Delta_t(\lambda) = (E^*E)^{-1}(t,\lambda) = (F^*F)^{-1}(t,\lambda)$ where $E = A - iB$ and $F = A + iB$ for $X = \Phi = \begin{pmatrix} A & 0 \\ B & 0 \end{pmatrix} \sim \begin{pmatrix} A \\ B \end{pmatrix}$ (the framework is set in C^{2n} but one can formally extend to $n \to \infty$ and $H = \tilde{H} \oplus \tilde{H}$ - again we use a function notation for the measure $\Delta_\infty = \lim \Delta_t$ for convenience). The guid͡ing mechanism in this development is the idea of a reproducing kernel in a deBranges space (cf. [6;7;10;11;13-16]) where in the present situation ($a^{\#}(\lambda) = a^*(\lambda^*) = a^*(\bar{\lambda})$)

$$(2.15)\qquad \Lambda_\mu^t(\lambda) = [F^{\#}(t,\lambda)F^{\#*}(t,\mu) - E^{\#}(t,\lambda)E^{\#*}(t,\mu)]/[-2\pi i(\lambda-\mu^*)] =$$

$$= [\Phi^{\#}(t,\lambda)J\Phi^{\#*}(t,\mu)]/[-\pi(\lambda-\mu^*)] = (1/\pi)\int_0^t \Phi^{\#}(s,\lambda)\Phi(s,\mu^*)ds$$

and writing $\langle \check{f},\check{g}\rangle_B = \int_{-\infty}^{\infty} (F^{\#-1}\check{f},F^{\#-1}\check{g})d\lambda$ one has the reproducing property $(h^{\blacktriangle}(\mu),\xi) = \langle h^{\blacktriangle}(\lambda),\Lambda_\mu^t(\lambda)\xi\rangle_B$ (here $h^{\blacktriangle}$ refers to h with h = 0 for x > t). Now given operator valued functions f(x) and g(x) as above write $h_1(x) = f(x)h_1$ and $h_2(x) = g(x)h_2$ for $h_1,h_2 \in H$. Then, using the notation $(X,Y) = X^*Y$ at times, we can write (†) $h_1^{\blacktriangle}(\lambda) = \int_0^\infty \Phi(x,\lambda)^* h_1(x)dx = \int_0^\infty \tilde{\Phi}(x,\lambda)h_1(x)dx = \tilde{\Phi}(\lambda,f)h_1$ and (**) $h_1^{\blacktriangle}(\lambda)^*\check{h}(\lambda) = (h_1^{\blacktriangle}(\lambda),\check{h}(\lambda)) = (\tilde{\Phi}(\lambda,f)h_1, \check{h}(\lambda)) = (h_1,\tilde{\Phi}^*(\lambda,f)\check{h}(\lambda)) = (h_1,\Phi(\lambda,f^*)\check{h}(\lambda))$. It follows that (cf. Theorem 2.5) $\int_0^\infty (fh_1,gh_2)dx = (h_1,\int_0^\infty f^*gh_2dx) = (h_1,\int_{-\infty}^{\infty} \Phi(\lambda,f^*)R\tilde{\Phi}(\lambda,g)h_2d\lambda)$ while from (2.14), (†), (**), etc. one has $\int_0^\infty (fh_1,gh_2)dx = \int_0^\infty h_1^*(x)h_2(x)dx = (1/\pi)\int_{-\infty}^{\infty} h_1^{\blacktriangle *}(\lambda)\Delta_\infty(\lambda)h_2^{\blacktriangle}(\lambda)d\lambda = (h_1,(1/\pi)\int_{-\infty}^{\infty} \Phi(\lambda,f^*)\Delta_\infty(\lambda)\tilde{\Phi}(\lambda,g)h_2d\lambda)$. Thus

<u>THEOREM 2.6</u>. The equations above give a direct identification of $R = (1/\pi)[P + \Phi_0(\lambda,\tilde{L}_P)] = (1/\pi)[P + \tilde{\Phi}_0(\lambda,L_P)]$ with $\Delta_\infty(\lambda)/\pi$. Since in fact $\Delta_\infty(\lambda) = \lim[PU^*(t,\lambda) U(t,\lambda)P]^{-1} = \lim [\Phi^*(t,\lambda)\Phi(t,\lambda)]^{-1}$ as $t \to \infty$ this connects such a limit with $L_P(x,0)$. Thus $\lim [\Phi^*(t,\lambda)\Phi(t,\lambda)]^{-1} = P[I + \int_0^\infty \exp(\lambda Jx)L_P(x,0)dx]$.

<u>REMARK 2.7</u>. One can also give an abstract proof in the present circumstances that the distribution R is in fact represented by a positive operator valued measure (see [21], p. 126 and remarks before (2.8)).

<u>REMARK 2.8</u>. The idea of generalized translation does not play an explicit role in the derivation of Theorem 2.5, in contrast to the scalar formulation of [5-7;21]. The natural object for a "generic" generalized translation appears to be (cf. Theorem 3.3) (▲▲•) $U(x,y) = \int_{-\infty}^{\infty} \Phi(x,\lambda)R\tilde{\Phi}(y,\lambda)d\lambda$ and such objects will be discussed in §3 in more detail. Let us note here that for $\check{F}(y) = \int_{-\infty}^{\infty} \Phi(y,\lambda)R\check{f}(\lambda)d\lambda$, $\int_0^\infty U(x,y)\check{F}(y)dy = \int_{-\infty}^{\infty} \Phi(x,\lambda) R \int_0^\infty\tilde{\Phi}(y,\lambda)\check{F}(y)dyd\lambda = \check{F}(x)$ since $\check{f}(\lambda) = \tilde{\Phi}(\lambda,\check{F})$ (cf. (2.10)). Thus U(x,y) plays the role of $\delta(x-y)$. Similarly from (2.15) and comments thereafter, with $\Delta_t = (F^{\#}F)^{-1}$ one has $(h^{\blacktriangle}(\mu),\xi) = \int_{-\infty}^{\infty} (\Delta_t h^{\blacktriangle}(\lambda),\Lambda_\mu^t(\lambda)\xi)d\lambda = (\int_{-\infty}^{\infty} \Lambda_\mu^t(\lambda)^*\Delta_t h^{\blacktriangle}(\lambda)d\lambda,\xi)$ with $h^{\blacktriangle}(\mu) = \int_{-\infty}^{\infty} \Lambda_\mu^t(\lambda)^*\Delta_t h^{\blacktriangle}(\lambda)d\lambda$ where $\Lambda_\mu^t(\lambda)^* = (1/\pi)\int_0^t \tilde{\Phi}(s,\mu)\Phi(s,\lambda)ds$ (μ real). Thus formally (■•) $\int_{-\infty}^{\infty} \Lambda_\mu^t(\lambda)^*\Delta_t h^{\blacktriangle}(\lambda)d\lambda =$

$= \int_0^t \Phi(s,\mu)(1/\pi)\int_{-\infty}^{\infty} \Phi(s,\lambda)\Delta_t h^{\blacktriangle}(\lambda)d\lambda ds = h^{\blacktriangle}(\mu)$ (note here $h^{\blacktriangle}(\lambda) = \int_0^t \tilde{\Phi}(x,\lambda)h(x)dx$ with Δ_t and $h(s) = (1/\pi)\int_{-\infty}^{\infty} \Phi(s,\lambda)\Delta_t h^{\blacktriangle}(\lambda)d\lambda$). Thus there is a kind of dual role between reproducing kernels and "generic" generalized translations illustrated above via (■•) etc. Of course from the inversion formulas in Theorem 2.5 U is formally $\delta(x-y)$ (from $f(x) = \int_{-\infty}^{\infty} \Phi(x,\lambda)R\int_0^{\infty} \tilde{\Phi}(y,\lambda)f(y)dyd\lambda$) and similarly, writing $\tilde{\Phi}(\mu,f) = \int_0^{\infty} \tilde{\Phi}(x,\mu)\int_{-\infty}^{\infty} \Phi(x,\lambda)$ $R\tilde{\Phi}(\lambda,f)d\lambda dx$, one sees that $(1/\pi)\Lambda_\mu^{\infty}(\lambda)^*\Delta_\infty \sim \int_0^{\infty} \tilde{\Phi}(x,\mu)\Phi(x,\lambda)Rdx \sim \delta(\lambda-\mu)$ as above.

REMARK 2.9. In (2.12) one can think of functions $f,g = 0$ for $x > \sigma-\varepsilon$ (so $F = G = 0$ for $x > \sigma-\varepsilon$) and write $\pi R_\sigma = P + \Phi_0(\lambda,\tilde{\varphi}_\sigma) = P + \tilde{\Phi}_0(\lambda,\varphi_\sigma)$. Then compare this with (2.14) for $t = \sigma$. R_σ seems to depend on γ_σ but one can show that e.g. the term ($t = \sigma$) $\int_{2t}^{2t+1} \exp(\lambda Jx)\gamma_t(x)L_P(x,0)dx = \check{R}_t$ contributes nothing. Thus, using $\exp(\lambda Jx) = P_+$ $\exp(i\lambda x) + P_-\exp(-i\lambda x)$, $\Phi_0(\lambda,F)\check{R}_t\tilde{\Phi}_0(\lambda,G)$ can be written in terms of entire functions A_+ and A_- with $\|A_+\| \leq \check{c}_+\exp(-2\varepsilon \mathrm{Im}\lambda)$ for $\mathrm{Im}\lambda > 0$ and $\|A_-\| \leq \check{c}_-\exp(2\varepsilon \mathrm{Im}\lambda)$ for $\mathrm{Im}\lambda < 0$. Contour integration in the upper (resp. lower) half plane yields $A_+ = A_- = 0$ (cf. [5-7]) and passing $\varepsilon \to 0$ one obtains $\Delta_t = \pi R_t = P[I + \int_0^{2t} \exp(\lambda Jx)L_P(x,0)dx] = [PU^*(t,\lambda)$ $U(t,\lambda)P]^{-1} = [\tilde{\Phi}(t,\lambda)\Phi(t,\lambda)]^{-1}$ (assume here R is a function).

3. Spectral pairings. To begin let us give a spectral formula for $\beta(y,x) = \delta(x-y) + K_P(y,x) = \ker B$ where $B\Phi_0(\cdot,\lambda) = \Phi(t,\lambda)$ as in (2.2). Note that spectral integrals frequently involve distribution ideas (e.g. $(1/2\pi)\int_{-\infty}^{\infty} \exp(i\lambda x)\exp(-i\lambda y)d\lambda = \delta(x-y)$). First we observe from (2.5) and (2.7) that formally

(3.1) $$\pi\delta(\lambda-\mu) = \tilde{\Phi}_0(\lambda,\Phi_0(\cdot,\mu)) = \Phi_0(\lambda,\tilde{\Phi}_0(\cdot,\mu))$$

while formally from Theorem 2.5 (cf. also Remark 2.8)

(3.2) $$\Phi(\lambda,\tilde{\Phi}(\cdot,\mu))R = \delta(\lambda-\mu) = R\tilde{\Phi}(\lambda,\Phi(\cdot,\mu))$$

Thus for $\beta(y,x)$ we consider

(3.3) $$\beta(y,x) = (1/\pi)\int_{-\infty}^{\infty} \Phi(y,\lambda)\tilde{\Phi}_0(x,\lambda)d\lambda$$

Evidently $\langle \beta(y,x),\Phi_0(x,\mu)\rangle = (1/\pi)\int_{-\infty}^{\infty} \Phi(y,\lambda)\tilde{\Phi}_0(\lambda,\Phi_0(\cdot,\mu))d\lambda = \Phi(y,\mu)$ as desired. Next

(3.4) $$\gamma(x,y) = \int_{-\infty}^{\infty} \Phi_0(x,\lambda)R\tilde{\Phi}(y,\lambda)d\lambda$$

provides a kernel for $\mathcal{B} = B^{-1}$ since, using (3.2), $\langle \gamma(x,y),\Phi(y,\mu)\rangle = \int_{-\infty}^{\infty} \Phi_0(x,\lambda)R$ $\tilde{\Phi}(y,\Phi(\cdot,\mu))d\lambda = \Phi_0(x,\mu)$. Thus

THEOREM 3.1. Kernels β and γ as above represent operators B and $\mathcal{B} = B^{-1}$ with $B\Phi_0 = \Phi$ and $\mathcal{B}\Phi = \Phi_0$. Standard arguments will then yield $\beta = \delta + K_P$ and $\gamma = \delta + L_P$. Evidently $[JD_y - V(y)]\beta(y,x) = -D_x\beta(y,x)J$ and $JD_x\gamma(x,y) = -D_y\gamma(x,y)J - \gamma(x,y)V(y)$.

Now go to [11] for some further information; the context in [11] is C^{2n} (for Hilbert space see [1;22;23]) and we spell out some material in the notation of [11] with suitable extensions understood where appropriate. Thus $U \sim \begin{pmatrix} A & C \\ B & D \end{pmatrix}$, $\Phi \sim \begin{pmatrix} A & 0 \\ B & 0 \end{pmatrix} \sim \begin{pmatrix} A \\ B \end{pmatrix}$, $E = A - iB$, $F = A + iB$, and set now (▲▲) $\varepsilon_t = \exp(i\lambda t)E(t,\lambda) = \varepsilon_t(\lambda)$ with $\varphi_t(\lambda) =$

$= \exp(-i\lambda t)F(t,\lambda)$. For example ε_t is invertible in W_+ = Wiener algebra of matrix functions $f(\lambda) = c_f + \hat{k}_f(\lambda)$, $k_f \in L^1$, $k_f = [f - c_f]^\vee = F^{-1}[f - c_f] = (1/2\pi)\int_{-\infty}^{\infty} \exp(-i\lambda x)$ $[f - c_f]d\lambda$ where W_+ involves also $f^\vee(x) = 0$ for $x < 0$. Thus ε_t is analytic and bounded in the upper half plane and in fact $\varepsilon_t(\lambda) \in W_+^I$ $(c_f = I = \varepsilon_t(\infty) - f \sim \varepsilon_t)$. Similarly $\varepsilon_t^\#(\lambda) = \varepsilon_t^*(\lambda^*) \in W_-^I$ and ε_∞ is well defined. Further $\varphi_t(\lambda) \in W_-^I$ and one has $\Delta_t(\lambda) = [\varepsilon_t^\#(\lambda)\varepsilon_t(\lambda)]^{-1} = (E^\# E)^{-1} = (F^\# F)^{-1} = (\varphi_t^\#\varphi_t)^{-1}$ with $\Delta_t \in W^I$. One defines a scattering matrix $(\lambda \in R)$ (♦♦) $S(\lambda) = \varphi_\infty\varepsilon_\infty^{-1} = \varphi_\infty^{*-1}\varepsilon_\infty^*$ (recall $\varepsilon_\infty^*\varepsilon_\infty = \varphi_\infty^*\varphi_\infty$) so that $S^*S = I$ for $\lambda \in R$. Write now $\mathbb{P}_\pm = (I \mp iJ)/2$)

$$(3.5) \qquad Y_+^0 = P_-U_0\binom{I_n}{0} = (1/2)\binom{I_n}{-iI_n}e^{i\lambda t}; \; Y_-^0 = P_+U_0\binom{I_n}{0} = (1/2)\binom{I_n}{iI_n}e^{-i\lambda t}$$

Then $\Phi = [Y_+^0 S(\lambda) + Y_-^0]\varepsilon_\infty + o(1)$ for $t \to \infty$ and $\lambda \in R$. The equations $JD_xY - VY = \lambda Y$ have solutions (we call them Jost solutions) $Y_\pm$ such that (▲) $Y_\pm(t,\lambda) = Y_\pm^0(t,\lambda) + \int_t^\infty M(t,s)$ $Y_\pm^0(s,\lambda)ds$ and one has further (▲♦) $JD_tM(t,s) + D_sM(t,s)J = V(t)M(t,s)$ with $V(t) = M(t,t)J - JM(t,t)$, and (★★) $\Phi(t,\lambda) = Y_+C_+(\lambda) + Y_-C_-(\lambda)$ with $C_+(\lambda) = \varphi_\infty(\lambda)$ and $C_-(\lambda) = \varepsilon_\infty(\lambda)$. Thus $S(\lambda) = C_+C_-^{-1}$ and the situation looks very much like the scalar case.

Now go to $\beta(y,x)$ determined by (3.3). One has $\tilde{\Phi}_0(x,\lambda) = P\exp(\lambda Jx) = P[P_+\exp(i\lambda x) + P_-\exp(-i\lambda x)]$ while one knows $\Phi(y,\lambda)$ is entire of exponential type y (i.e. $\|\Phi\| \leq c\exp(y|\mathrm{Im}\lambda|)$ - recall $V \in L^1$). Thus for $x > y$ $|\exp(i\lambda x)|\exp(y|\mathrm{Im}\lambda|) \leq \exp[-(x-y)\mathrm{Im}\lambda]$ for $\mathrm{Im}\lambda > 0$ and a contour integration in the upper half plane vanishes; similarly for $\mathrm{Im}\lambda < 0$ $|\exp(-i\lambda x)|\exp(y|\mathrm{Im}\lambda|) \leq \exp[-(x-y)|\mathrm{Im}\lambda|]$ and a contour integration in the lower half plane will vanish for $x > y$. Hence formally

THEOREM 3.2. If $\beta(y,x)$ is defined by (3.3) then $\beta(y,x) = 0$ for $x > y$. Similarly $\gamma(x,y)$ defined by (3.4) vanishes for $y > x$.

Proof: For γ set $\pi R = \Delta_\infty = (\varepsilon_\infty^\#\varepsilon_\infty)^{-1} = \varepsilon_\infty^{-1}\varepsilon_\infty^{\#-1} = C_-^{-1}C_-^{\#-1} = (\varphi_\infty^\#\varphi_\infty)^{-1} = \varphi_\infty^{-1}\varphi_\infty^{\#-1} = C_+^{-1}C_+^{\#-1}$. Hence one has $R\tilde{\Phi}(y,\lambda) = R[C_+^\# Y_+^\# + C_-^\# Y_-^\#] = (1/\pi)[C_-^{-1}Y_-^\# + C_+^{-1}Y_+^\#]$. Recall here C_- and C_-^{-1} are analytic for $\mathrm{Im}\lambda > 0$ along with $Y_-^\#$ ($Y_-^\# \sim Y_-^{0\#} \sim \exp(i\lambda y)$ etc.). Thus think e.g. of $\|\Phi_0(x,\lambda)\| \leq c\exp(x|\mathrm{Im}\lambda|)$ without further refinement and for $y > x$, $\|\Phi_0C_-^{-1}Y_-^\#\| \leq \hat{c}\exp[-(y-x)\mathrm{Im}\lambda]$ for $\mathrm{Im}\lambda > 0$ while $\|\Phi_0C_+^{-1}Y_+^\#\| \leq \tilde{c}\exp[-(y-x)|\mathrm{Im}\lambda|]$ for $\mathrm{Im}\lambda < 0$. Contour integration as before yields the result. QED

We can also derive a G-L equation directly from (3.3)-(3.4) as follows. Writing $\Phi(y,\lambda) = \int_0^\infty \beta(y,x)\Phi_0(x,\lambda)dx$, multiply by $R\tilde{\Phi}_0(\xi,\lambda)$ and integrate in λ to obtain $(R = R^*)$

$$(3.6) \qquad \tilde{\beta}(y,\xi) = \gamma^*(\xi,y) = \int_{-\infty}^{\infty} \Phi(y,\lambda)R\tilde{\Phi}_0(\xi,\lambda)d\lambda = \int_0^{\infty} \beta(y,x)\int_{-\infty}^{\infty} \Phi_0(x,\lambda)R\tilde{\Phi}_0(\xi,\lambda)d\lambda dx$$

THEOREM 3.3. The G-L equation for $JD_x - V$ and JD_x can be written as $\tilde{\beta}(y,\xi) = \gamma^*(\xi,y) = \langle \beta(y,x), \mathbb{A}(x,\xi)\rangle$ where $\mathbb{A}(x,\xi) = \int_{-\infty}^{\infty} \Phi_0(x,\lambda)R\tilde{\Phi}_0(\xi,\lambda)d\lambda$.

REMARK 3.4. The G-L equation derived in [11] for the present situation goes as follows (cf. also [6;7] for factorization). Write $\Delta_\infty = I - \hat{h}(\lambda)$, $h \in L^1$, $\hat{h} = Fh$, and set $(Hg)(t) = \int_{-\infty}^{\infty} h(t-s)g(s)ds$. Define an operator N as follows. For $f = \binom{a}{b}$ set $\sqrt{2}(Nf)(s) = a(s) - ib(s)$ for $s > 0$ and $\sqrt{2}(Nf)(s) = a(-s) + ib(-s)$ for $s < 0$. Then $f^\Delta = (1/\sqrt{2})(Nf)^\wedge$ where $f^\Delta = \tilde{\Phi}_0(\lambda,f)$. Next one works from $\Phi = \Phi_0 + \int_0^t K(t,s)\Phi_0 ds =$

$= (I + K)\Phi_0$. Thus $f^{\blacktriangle} = \int \tilde{\Phi} f = \int \tilde{\Phi}_0 (I + K^*) f = [(I + K^*) f]^{\Delta}$ and from the Parseval formula (♣♣) $\int_0^\infty f^* f = (1/\pi) \int_{-\infty}^\infty f^{\blacktriangle *} \Delta_\infty f^{\blacktriangle} d\lambda = (1/2\pi) \int_{-\infty}^\infty [N(I + K^*) f]^\wedge (I - \hat{h}) [N(I + K^*) f]^\wedge d\lambda = \langle (I - H)[N(I + K^*) f], [N(I + K^*) f] \rangle$. Hence $(I + K)[N^*(I - H)N](I + K^*) = I$ and in terms of factorization one writes $((I + L) = (I + K)^{-1})$, $N^*(I - H)N = (I + L)(I + L^*)$. Here for kernel action $(\xi \to y)$ $I+K \sim \beta(y,\xi)$, $K^* \sim \tilde{K}^+$, $L^* \sim \tilde{L}^+$, and $\tilde{\beta} \sim \gamma^{*+}$.

THEOREM 3.5. $N^*(I - H)N = \mathbb{A}$ where $\mathbb{A}$ has kernel $\mathbb{A}(x,\xi)$ $(\xi \to x)$.

Now return to $\mathbb{A}(x,\xi)$ in Theorem 3.3. In the scalar case one represents such an $\mathbb{A}$ in terms of a generalized translation (cf. also Remark 2.8). We must however rethink a little the idea of generalized translation here and note first for $\mathbb{A}$, $JD_x\mathbb{A} = -D_\xi \mathbb{A} J$. Recall now $PRP = R$, $R^* = R$, $\Phi_0 P = \Phi_0$, $P\tilde{\Phi}_0 = \tilde{\Phi}_0$, etc. and write (♣•) $\mathbb{A}(x,0) = \int_{-\infty}^\infty \Phi_0(x,\lambda) R d\lambda = \breve{R}(x)$ with $\pi R = \tilde{\Phi}_0(\lambda,\breve{R}) = \int_0^\infty \tilde{\Phi}_0(x,\lambda)\breve{R}(x)dx$. Note again $R = \Delta_\infty/\pi$ and $\Phi_0 = (P_+\exp(-i\lambda x) + P_-\exp(i\lambda x))P$ so $\breve{R} = (1/\pi)\int_{-\infty}^\infty \Delta_\infty \exp(-i\lambda x) d\lambda P_+ P + (1/\pi)\int_{-\infty}^\infty \Delta_\infty \exp(i\lambda x) d\lambda P_- P$. Recalling $\Delta_\infty = I - \hat{h}$, $\hat{h} = Fh$, we see that $\breve{R}$ and h are intimately related. In fact one can write $R = \begin{pmatrix} \delta(x) \\ 0 \end{pmatrix} - \mathbb{h}(x)$, $\mathbb{h}(x) = (1/\pi)\int_{-\infty}^\infty \Phi_0(x,\lambda)\hat{h}(\lambda)d\lambda$, and $\hat{h} = \tilde{\Phi}_0(\lambda,\mathbb{h}) = \mathbb{h}^{\Delta} = (1/\sqrt{2})(N\mathbb{h})^\wedge$. Consequently from $\hat{h} = Fh$ one has $(1/\sqrt{2})N\mathbb{h} = h$. Now observe that (•♣) $\mathbb{A}(0,\xi) = \int_{-\infty}^\infty R\tilde{\Phi}_0(\xi,\lambda)d\lambda = (\int_{-\infty}^\infty \Phi_0(\xi,\lambda)Rd\lambda)^* = \breve{R}^*(\xi)$; but $R = PRP$ implies that in fact $R = \begin{pmatrix} a & 0 \\ 0 & 0 \end{pmatrix}$, with $a^* = a$, and prototypically then $\breve{R}(x) = \breve{R}^*(x)$ follows with a little calculation. Hence

THEOREM 3.6. Given $\breve{R}$ defined by (♣•) we think of $\mathbb{A}(x,\xi)$ in Theorem 3.3 as a generalized translation and write $\mathbb{A}(x,\xi) = T_x^\xi \breve{R}(x)$. One has $JD_x\mathbb{A} = -D_\xi \mathbb{A} J$ with $\mathbb{A}(x,0) = \breve{R}(x)$ and $\mathbb{A}(0,\xi) = \breve{R}(\xi)$. Evidently $\mathbb{A}^*(x,\xi) = \mathbb{A}(\xi,x)$.

REMARK 3.7. Following the model of $\mathbb{A}$ which represents a kind of generic generalized translation one is led to consider e.g. (•▲) $U(x,\xi) = (1/\pi)\int_{-\infty}^\infty \Phi_0(x,\lambda)\tilde{\Phi}_0(\lambda,f)\tilde{\Phi}_0(\xi,\lambda)d\lambda$ so that $U(x,0) = f(x)P$ while $U(0,\xi) = (1/\pi)\int_{-\infty}^\infty \tilde{\Phi}_0(\lambda,f)\tilde{\Phi}_0(\xi,\lambda)d\lambda$. If $\tilde{\Phi}_0(\lambda,f) = \Phi_0(\lambda,f^*)$ then $U(0,\xi) = f^*(\xi)$ (analogous in a way to $\mathbb{A}$). This involves $f^*\Phi_0 = f^* U_0 P = \Phi_0^* f = PU_0^* f = PU_0^{-1} f$ and is valid when $f = \begin{pmatrix} \alpha & 0 \\ \gamma & 0 \end{pmatrix}$ with $\alpha^* = \alpha$ and $\gamma^* = \gamma$ (note $fP = f$ also). Again $JD_x U = -D_\xi UJ$.

One can now compose $\beta(y,x)$ from (3.3) with $U(x,\xi)$ above in (•▲), using the rule (3.1), to obtain a candidate for determining B via a Cauchy problem ($f = \begin{pmatrix} \alpha & 0 \\ \gamma & 0 \end{pmatrix}$ as above and one refers to [5-7] for the scalar situation). Thus consider

$$\varphi(y,\xi) = \langle \beta(y,x), U(x,\xi) \rangle = (1/\pi)\int_{-\infty}^\infty \Phi(y,\lambda)\tilde{\Phi}_0(\lambda,f)\tilde{\Phi}_0(\xi,\lambda)d\lambda \tag{3.7}$$

Then $\varphi(0,\xi) = f^*(\xi)$ and $\varphi(y,0) = (1/\pi)\int_{-\infty}^\infty \Phi(y,\lambda)\tilde{\Phi}_0(\lambda,f)d\lambda = Bf(y)$. Thus

THEOREM 3.8. If $f = fP = \begin{pmatrix} \alpha & 0 \\ \gamma & 0 \end{pmatrix}$, $\alpha^* = \alpha$, $\gamma^* = \gamma$, as above then $\tilde{\Phi}_0(\lambda,f) = \Phi_0(\lambda,f^*)$ and $U(x,\xi)$ in (•▲) is a "generalized translation" satisfying $JD_x U = -D_\xi UJ$ with $U(x,0) = f(x)$ and $U(0,\xi) = f^*(\xi)$ while $\varphi(y,\xi)$ in (3.7) satisfies $JD_y\varphi - V(y)\varphi = -D_\xi \varphi J$ with $\varphi(0,\xi) = f^*(\xi)$ and $\varphi(y,0) = Bf(y)$.

REMARK 3.9. Let us make a few remarks about reproducing kernels and their relations to various calculations (cf. also [3;4;11;7;13-15]). For example in working with a formula $\varphi(y,\xi) = \langle \beta(y,x),U(x,\xi)\rangle$, without performing the intermediate calculation, if we want to show $[JD_y - V(y)]\varphi = -D_\xi\varphi J$ one would like to write $-D_\xi\varphi J = \langle \beta(y,x), -D_\xi UJ\rangle = \langle \beta(y,x),D_x JU\rangle = \langle -D_x\beta(y,x)J,U(x,\xi)\rangle = [JD_y - V(y)]\varphi$. The step from D_xU to $D_x\beta$ needs some justification for example and what is involved is simply to show (▲●) $\int_0^\infty \tilde{\Phi}_0(x,\lambda)JD_x\Phi_0(x,\mu)dx = -\int_0^\infty D_x\tilde{\Phi}_0(x,\lambda)J\Phi_0(x,\mu)dx$. We recall of course that $\delta(\lambda-\mu) = (1/\pi)\int_0^\infty \tilde{\Phi}_0(x,\lambda)\Phi_0(x,\mu)dx$ from the inversion theory. Now from $JD_x\Phi_0(x,\mu) = \mu\Phi_0(x,\mu)$ with $-D_x\tilde{\Phi}_0(x,\lambda)J = \lambda\tilde{\Phi}(x,\lambda)$ we obtain $D_x[\tilde{\Phi}_0(x,\lambda)J\Phi_0(x,\mu)] = (\mu-\lambda)\tilde{\Phi}_0(x,\lambda)\Phi_0(x,\mu)$. Hence (recall PJP = 0) in particular

$$(3.8)\qquad [\tilde{\Phi}_0(T,\lambda)J\Phi_0(T,\mu)] = (\mu-\lambda)\int_0^T \tilde{\Phi}_0(x,\lambda)\Phi_0(x,\mu)dx = \int_0^T [\tilde{\Phi}_0(x,\lambda)JD_x\Phi_0(x,\mu) + D_x\tilde{\Phi}_0(x,\lambda)J\Phi_0(x,\mu)]dx$$

Since $\pi\delta_T(\lambda,\mu) = \int_0^T \tilde{\Phi}_0(x,\lambda)\Phi_0(x,\mu)dx \to \pi\delta(\lambda-\mu)$ we see that operationally (working on suitable functions $f(\lambda)$) $(\mu-\lambda)\pi\delta_T(\lambda,\mu) \to 0$ and hence (▲●) holds. On the other hand to see exactly the manner in which $\delta_T(\lambda,\mu) \to \delta(\lambda-\mu)$ consider $U_0(T,\lambda) = \exp(-\lambda JT) = P_+ \exp(-i\lambda T) + P_-\exp(i\lambda T)$ where $P_\pm = (1/2)(I \mp iJ)$ and then $\Phi_0 = U_0P \sim (1/2)\binom{1}{i}\exp(-i\mu T) + (1/2)\binom{1}{-i}\exp(i\mu T)$, $J\Phi_0 \sim (1/2)\binom{i}{-1}\exp(-i\mu T) + (1/2)\binom{-i}{-1}\exp(i\mu T)$, and $\Xi^T = \tilde{\Phi}_0(T,\lambda) J\Phi_0(T,\mu) \sim (1/4)[(-2i)\exp(-iT(\lambda-\mu)) + 2i\exp(iT(\lambda-\mu))]$. Hence acting on suitable f, (♠▲) $(1/\pi)\int_{-\infty}^\infty [\Xi^T f(\lambda)/(\mu-\lambda)]d\lambda = (1/2\pi i)\int_{-\infty}^\infty [f(\lambda)\exp(iT(\mu-\lambda))/(\mu-\lambda)]d\lambda - (1/2\pi i)\int_{-\infty}^\infty [f(\lambda) \exp(iT(\lambda-\mu))/(\mu-\lambda)]d\lambda$. Given $f(\lambda)$ entire of bounded exponential type σ, take $T > \sigma$, and μ slightly off the real line as follows. For $\text{Im}\mu > 0$ for example $[\exp(-i\lambda T)f(\lambda)]$ is analytic and suitably bounded in the lower half plane with $(1/(\mu-\lambda))$ analytic so the first integral in the last line of (♠▲) vanishes while the second integral is $f(\mu)$ by an integral in the upper half plane (cf. [5;6] for contour integral techniques). Similar considerations hold for $\text{Im}\mu < 0$. One sees that $\Xi^T/\pi(\mu-\lambda) = \delta_T(\lambda,\mu) \to \delta(\lambda-\mu)$ then, working on entire functions of bounded exponential type.

<u>REMARK 3.10</u>. From (●▲) in Remark 3.7 a generalized translation associated with $Q = JD_x - V$ should be of the form

$$(3.9)\qquad \psi(x,y) = \int_{-\infty}^\infty \Phi(x,\lambda)R\tilde{\Phi}(\lambda,f)\tilde{\Phi}(y,\lambda)d\lambda$$

Then since $[JD_x - V]\Phi = \lambda\Phi$ and $-D_y\tilde{\Phi}J - \tilde{\Phi}V = \lambda\tilde{\Phi}$ we have $(JD_x - V)\psi = -D_y\psi J - \psi V$ with (††) $\psi(x,0) = \int_{-\infty}^\infty \Phi(x,\lambda)R\tilde{\Phi}(\lambda,f)Pd\lambda = f(x)P$ (cf. Theorem 2.5). If one has $R\tilde{\Phi}(\lambda,f) = \tilde{\Phi}(\lambda,g)R$ then (●●●) $\psi(0,y) = \int_{-\infty}^\infty P\Phi(\lambda,g)R\tilde{\Phi}(y,\lambda)d\lambda = Pg(y)$. We anticipate $f = \begin{pmatrix}\alpha & 0\\ \gamma & 0\end{pmatrix}$, $\alpha = \alpha^*$, $\gamma = \gamma^*$, again with $f^* = \begin{pmatrix}\alpha & \gamma\\ 0 & 0\end{pmatrix}$ and $g = f^*$. To have this one would need $\Phi(\lambda,f^*)R = R\tilde{\Phi}(\lambda,f)$ (note here that $[R\tilde{\Phi}(\lambda,f)]^* = \Phi(\lambda,f^*)R$). Now writing $R = \begin{pmatrix}a & 0\\ 0 & 0\end{pmatrix}$ again with $a = a^*$, for f as above $\Phi(\lambda,f^*)R$ involves $\begin{pmatrix}(\alpha A+\gamma B)a & 0\\ 0 & 0\end{pmatrix}$ while $R\tilde{\Phi}(\lambda,f)$ involves $\begin{pmatrix}a(A^*\alpha+B^*\gamma) & 0\\ 0 & 0\end{pmatrix}$. Thus if $(\alpha A + \gamma B)a = a(A^*\alpha + B^*\gamma)$ we have $\Phi(\lambda,f^*)R = R\tilde{\Phi}(\lambda,f)$ (more generally $(\alpha^*A + \gamma^*B)a = a(A^*\alpha + B^*\gamma)$ will do). Summarizing, the function ψ defined by (3.9) satisfies $(JD_x - V(x))\psi = -D_y\psi J - \psi V(y)$ with $\psi(x,0) = f(x)P$ and $\psi(0,y) = Pg(y)$ if $R\tilde{\Phi}(\lambda,f) = \Phi(\lambda,g)R$ (which requires $aA^*f_{11} + aB^*f_{21} = g_{11}Aa + g_{12}Ba$ and $f_{12} = f_{22} = g_{21} = g_{22} = 0$). If $f = \begin{pmatrix}\alpha & 0\\ \gamma & 0\end{pmatrix}$ and $(\alpha^*A + \gamma^*B)a = a(A^*\alpha + B^*\gamma)$ then $\psi(x,0) = f(x)$ and $\psi(0,y) = f^*(y) =$

$= \binom{\alpha^* \ \gamma^*}{0 \ \ 0}$. This is not entirely thrilling and the matter of generalized translation for systems should be examined further (cf. below). Similarly in the spirit of Theorem 3.8 one can also consider (••▲) $\tilde{\varphi}(x,\eta) = \langle \gamma(x,y),\psi(y,\eta)\rangle$ with γ given by (3.4) and by (3.9). Thus $\tilde{\varphi}(x,\eta) = \int_{-\infty}^{\infty} \Phi_0(x,\lambda)R\tilde{\Phi}(\lambda,f)\tilde{\Phi}(\eta,\lambda)d\lambda$ so that $JD_x\tilde{\varphi} = -D_\eta\tilde{\varphi}J - \tilde{\varphi}V(\eta)$ with $\tilde{\varphi}(x,0) = \langle \gamma(x,y),f(y)\rangle P = \mathcal{B}fP$ and $\tilde{\varphi}(0,\eta) = \int_{-\infty}^{\infty} R\tilde{\Phi}(\lambda,f)\tilde{\Phi}(\eta,\lambda)d\lambda$ and thus if $R\tilde{\Phi}(\lambda,f) = \Phi(\lambda,f^*)R$ as above one has $\tilde{\varphi}(0,\eta) = f^*(\eta)$. This situation again is not very satisfactory and suggests further study of the relation of Cauchy problems (and other characterizations such as the minimization procedure of [6-7]) in characterizing transmutations for systems. For preserving the connection of transmutations to Cauchy problems for example one can envision a tensor product context. Thus let $Q_y = [JD_y - V(y)]$ and $Q_x^0 = JD_x$. Then $Q_x^0 \otimes 1$ and $1 \otimes Q_y$ commute over suitable domains in $H \hat{\otimes} H$ ($\wedge$ denotes say a standard Hilbert completion) and one can consider (•■) $\varphi(x,y) = (1/\pi) \int_{-\infty}^{\infty} [\Phi_0(x,\lambda) \otimes \Phi(y,\lambda)]\tilde{\Phi}_0(\lambda,f)d\lambda$. Then $(Q_x^0 \otimes 1)\varphi = (1 \otimes Q_y)\varphi$ while $\varphi(x,0) = (1/\pi) \int_{-\infty}^{\infty} [\Phi_0(x,\lambda)\tilde{\Phi}(\lambda,f) \otimes P]d\lambda = f(x) \otimes P$ and $\varphi(0,y) = (1/\pi)\int_{-\infty}^{\infty} [P \otimes \Phi(y,\lambda)\tilde{\Phi}_0(\lambda,f)]d\lambda = P \otimes \mathcal{B}f(y)$. Of course φ in (3.7) was already essentially satisfactory for $\mathcal{B}$ so consider now a replacement for ψ in (3.9) and $\tilde{\varphi}$ in (••▲). Thus try (cf. [24])

$$(3.10) \qquad \psi(x,y) = \int_{-\infty}^{\infty} [\Phi(x,\lambda) \otimes \Phi(y,\lambda)]R\tilde{\Phi}(\lambda,f)d\lambda$$

Then $(Q_x \otimes 1)\psi = (1 \otimes Q_y)\psi$ while $\psi(x,0) = \int_{-\infty}^{\infty} [\Phi(x,\lambda)R\tilde{\Phi}(\lambda,f) \otimes P]d\lambda = f(x) \otimes P$ with $\psi(0,y) = \int_{-\infty}^{\infty} [P \otimes \Phi(y,\lambda)R\tilde{\Phi}(\lambda,f)]d\lambda = P \otimes f(y)$. Similarly try

$$(3.11) \qquad \tilde{\varphi}(x,\eta) = \int_{-\infty}^{\infty} [\Phi_0(x,\lambda) \otimes \Phi(\eta,\lambda)]R\tilde{\Phi}(\lambda,f)d\lambda$$

Then $(Q_x^0 \otimes 1)\tilde{\varphi} = (1 \otimes Q_y)\tilde{\varphi}$ with $\tilde{\varphi}(x,0) = \int_{-\infty}^{\infty} [\Phi_0(x,\lambda)R\tilde{\Phi}(\lambda,f) \otimes P]d\lambda = \mathcal{B}f(x) \otimes P$ and $\tilde{\varphi}(0,\eta) = \int_{-\infty}^{\infty} [P \otimes \Phi(\eta,\lambda)R\tilde{\Phi}(\lambda,f)]d\lambda = P \otimes f(y)$. Using (3.2) we note also that formally $\tilde{\varphi}(x,\eta) = \langle \gamma(x,y),\psi(y,\eta)\rangle = \int_0^{\infty} [\int_{-\infty}^{\infty} \Phi_0(x,\lambda)R\tilde{\Phi}(y,\lambda)d\lambda][\int_{-\infty}^{\infty} (\Phi(y,\mu) \otimes \Phi(\eta,\mu))R\tilde{\Phi}(\mu,f)d\mu]dy = \int_{-\infty}^{\infty} \int_{-\infty}^{\infty} [\Phi_0(x,\lambda)\delta(\lambda-\mu) \otimes \Phi(\eta,\mu)]R\tilde{\Phi}(\mu,f)d\mu d\lambda$. In the present context we note also a general theorem modeled on the scalar situation (cf. [5-7;19])

<u>THEOREM 3.11</u>. Let the Cauchy problem $(Q_x^0 \otimes 1)\Psi = (1 \otimes Q_y)\Psi$, $\Psi(x,0) = f(x) \otimes P$ have unique solutions in some suitable class of functions or distributions (unspecified here). Writing $P \otimes \mathcal{B}f(y) = \Psi(0,y)$ it follows that $\mathcal{B}Q^0 = Q\mathcal{B}$.

Proof: Set $\Upsilon = (Q_x^0 \otimes 1)\Psi$ so that $(Q_x^0 \otimes 1)\Upsilon = (1 \otimes Q_y)\Upsilon$ (since $(Q_x^0 \otimes 1)$ and $(1 \otimes Q_y)$ commute). Given unique solutions one has then $\Upsilon(x,0) = Q^0f \otimes P$ so that $\Upsilon(0,y) = P \otimes \mathcal{B}Q^0f$ and $\Upsilon(0,y) = (Q_x^0 \otimes 1)\Psi|_{x=0} = (1 \otimes Q_y)\Psi(0,y) = (1 \otimes Q_y)(P \otimes \mathcal{B}f) = P \otimes Q\mathcal{B}f$. QED

<u>REMARK 3.12</u>. Let β be given by (3.3) with $[JD_y - V(y)]\beta = -D_x\beta J$. To see that $\mathcal{B}(JD_xf) = [JD_y - V(y)]\mathcal{B}f$ for suitable f (take $Pf(0) = f(0)$ and e.g. supp f compact) we write $\langle \beta(y,x),JD_xf\rangle = (1/\pi)\int_{-\infty}^{\infty} \Phi(y,\lambda)\int_0^{\infty} \tilde{\Phi}_0(x,\lambda)JD_xfdxd\lambda$. The last integral is $\tilde{\Phi}_0(x,\lambda) Jf|_0^{\infty} - \int_0^{\infty} D_x\tilde{\Phi}_0(x,\lambda)Jf(x)dx$, and at $x = 0$ $PJf(0) = PJPf(0) = 0$. Hence $\langle \beta(y,x),JD_xf\rangle = \mathcal{B}(JD_xf) = -\langle D_x\beta(y,x)J,f\rangle = [JD_y - V(y)]\langle \beta(y,x),f\rangle = [JD_y - V(y)]\mathcal{B}f$.

<u>REMARK 3.13</u>. The connection of the present context to Lax-Phillips scattering (cf. [17]) is spelled out in [11;23].

REMARK 3.14. Let $\widetilde{B}$ be the map with kernel $\widetilde{\beta}(y,\xi)$ ($\xi \to y$) from (3.6). In the scalar situation such a $\widetilde{\beta}$ is a transmutation kernel with important mapping properties for special functions (note $\widetilde{B} \sim B^*$ and cf. [5-7]). Assume now explicitly that the measure represented by R is absolutely continuous (so the function notation is literal); this has also been implicit at times in the preceding. Recall $\Phi = Y_+C_+ + Y_-C_-$, $C_- = \varepsilon_\infty$, $C_+ = \varphi_\infty$, $\pi R = \Delta_\infty = (C_-^*C_-)^{-1} = (C_+^*C_+)^{-1}$, etc. Let us remark explicitly here that terms like ε_∞, φ_∞, $C_\pm$, etc. are to be identified with expressions $\begin{pmatrix} \sim & 0 \\ 0 & 0 \end{pmatrix}$. Then in particular $\Phi(y,\lambda)R = Y_+C_+^{\#-1} + Y_-C_-^{\#-1}$ with $C_+^{\#-1}$ (resp. $C_-^{\#-1}$) analytic and suitably bounded for $\mathrm{Im}\lambda > 0$ (resp. $\mathrm{Im}\lambda < 0$). Recall also $U_0 = P_+\exp(-i\lambda x) + P_-\exp(i\lambda x)$, $\Phi_0 = P_+P\exp(-i\lambda x) + P_-P\exp(i\lambda x)$, $Y_+^0 \sim P_-U_0P = P_-\Phi_0 = P_-P\exp(i\lambda x) = (1/2)\begin{pmatrix} 1 & 0 \\ -i & 0 \end{pmatrix}\exp(i\lambda x)$, and $Y_-^0 \sim P_+U_0P = P_+P\exp(-i\lambda x) = (1/2)\begin{pmatrix} 1 & 0 \\ i & 0 \end{pmatrix}\exp(-i\lambda x)$ (cf. (3.5)). Further from $Y_\pm = (I + M)Y_\pm^0$ one has corresponding $\exp(\pm i\lambda x)$ behavior for $Y_\pm$ and $Y_\pm^0$ as in the proof of Theorem 3.2. Now first, using (3.1) and (3.6), one has formally $\widetilde{B}\Phi_0 = \langle \widetilde{\beta}(y,\xi),\Phi_0(\xi,\mu)\rangle = \int_{-\infty}^{\infty} \Phi(y,\lambda)R \int_0^\infty \widetilde{\Phi}_0(\xi,\lambda)\Phi_0(\xi,\mu)d\xi d\lambda = \pi\Phi(y,\mu)R$ (which is analogous to the scalar situation). On the other hand one can write $\widetilde{\beta}(y,\xi) = \int_{-\infty}^{\infty} [Y_+(y,\lambda)C_+^{\#-1}(\lambda) + Y_-(y,\lambda)C_-^{\#-1}(\lambda)]\, \widetilde{\Phi}_0(\xi,\lambda)d\lambda$. Now write $\Phi_0 = Y_-^0 + Y_+^0$ with $\widetilde{\Phi}_0 = PP_+\exp(i\lambda x) + PP_-\exp(-i\lambda x)$ and use the analyticity and exponential bounds of $Y_+C_+^{\#-1}$ (resp. $Y_-C_-^{\#-1}$) in the upper (resp. lower) half plane to conclude that, for $\xi+y > 0$, $\int_{-\infty}^{\infty} Y_+(y,\lambda)C_+^{\#-1}(\lambda)PP_+\exp(i\lambda\xi)d\lambda = 0 = \int_{-\infty}^{\infty} Y_-(y,\lambda)C_-^{\#-1}(\lambda)PP_-\exp(-i\lambda\xi)d\lambda$. Consequently for $\xi+y > 0$ (▲▲▲) $\widetilde{\beta}(y,\xi) = \int_{-\infty}^{\infty} Y_+(y,\lambda)C_+^{\#-1}(\lambda)PP_-\exp(-i\lambda\xi)d\lambda + \int_{-\infty}^{\infty} Y_-(y,\lambda)C_-^{\#-1}(\lambda)PP_+\exp(i\lambda\xi)d\lambda$. It follows that $\widetilde{\beta}(y,\xi)P_- = \int_{-\infty}^{\infty} Y_+C_+^{\#-1}PP_-\exp(-i\lambda\xi)d\lambda$ and $\widetilde{\beta}(y,\xi)P_+ = \int_{-\infty}^{\infty} Y_-C_-^{\#-1}PP_+\exp(i\lambda\xi)d\lambda$ and consequently (again in analogy to the scalar situation) one has

THEOREM 3.15. Under the hypotheses indicated $\widetilde{B}\Phi_0 = \pi\Phi R$ and $Y_+(y,\lambda)C_+^{\#-1}(\lambda)PP_-P = (1/2\pi)\widetilde{B}[P_-P\exp(i\lambda\xi)] = (1/2\pi)\widetilde{B}[Y_+^0]$ with $Y_-(y,\lambda)C_-^{\#-1}(\lambda)PP_+P = (1/2\pi)\widetilde{B}[P_+P\exp(-i\lambda\xi)] = (1/2\pi)\widetilde{B}[Y_-^0]$.

We note here that $PP_-P = (1/2)\begin{pmatrix} 1 & 0 \\ 0 & 0 \end{pmatrix} = PP_+P$ so that upper block information for $Y_\pm C_\pm^{\#-1}$ seems to be determining (cf. [11] where lower block information is used for determining $C_\pm(\lambda)$). Somewhat more information appears to be present when we do not post multiply (▲▲▲) by P but this is apparently redundant. One would now hope to be able to use similar technique to establish a systems version of the generalized Kontorovič-Lebedev inversion for the scalar case as developed in [6;7]. However there seem to be some new features which we do not yet understand.

REFERENCES

1. V. Adamyan, On the theory of canonical differential operators in Hilbert space, DAN SSSR, 178 (1968), 9-12
2. Z. Agranovič and V. Marčenko, The inverse problem of scattering theory, Gordon-Breach, N.Y., 1963
3. D. Alpay and H. Dym, Hilbert spaces of analytic functions, inverse scattering, and operator models, I, Integ. Eqs. Oper. Theory, 7 (1984), 589-641
4. D. Alpay and H. Dym, Hilbert spaces of analytic functions, inverse scattering, and operator models, II, Integ. Eqs. Oper. Theory, 8 (1985), 145-180
5. R. Carroll, Transmutation, scattering theory, and special functions, North-Holland, Amsterdam, 1982
6. R. Carroll, Transmutation theory and applications, North-Holland, Amsterdam, 1985, to appear
7. R. Carroll, Patterns and structure in systems governed by linear second order differential equations, Acta Applicandae Math., to appear
8. R. Carroll and S. Dolzycki, Transmutation for systems and transmission lines, to appear
9. Yu. Daletskij and M. Krein, Stability of solutions of differential equations in Banach space, AMS Translations, Vol. 43, 1974
10. L. deBranges, Hilbert spaces of entire functions, Prentice-Hall, 1968
11. H. Dym and A. Iacob, Positive definite extensions, canonical equations, and inverse scattering, Topics in Operator Theory, Birkhauser, Basel, 1984, pp. 141-240
12. I. Gokhberg and M. Krein, The theory of Volterra operators in Hilbert space and applications, Moscow, 1967
13. T. Kailath, RKHS approach to detection and estimation problems, I, IEEE Trans. IT-17 (1971), 530-549
14. T. Kailath, R. Geesey, and H. Weinert, Some relations among RKHS norms, Fredholm equations, and innovations representations, IEEE Trans. IT-18 (1972), 341-348
15. T. Kailath and D. Duttweiler, RKHS approach to detection and estimation problems, III, IV, and V, IEEE Trans. IT-18 (1972), 730-745; IT-19 (1973), 19-28 and 29-37
16. T. Kailath and H. Weinert, RKHS approach to detection and estimation problems, II, IEEE Trans. IT-21 (1975), 15-23
17. P. Lax and R. Phillips, Scattering theory and automorphic functions, Bull. Amer. Math. Soc., 2 (1980), 261-295
18. B. Levitan, Inverse Sturm-Liouville problems, Moscow, 1984
19. B. Levitan and I. Sargsyan, Introduction to spectral theory ..., Moscow, 1970
20. J. Loeffel, On an inverse problem in potential scattering theory, Annal. Inst. H. Poincaré, 8 (1968), 339-447
21. V. Marčenko, Sturm-Liouville operators and their applications, Kiev, 1977
22. F. Melik-Adamyan, On canonical differential operators in Hilbert space, Izves. Akad. Nauk Armen. SSR, 12 (1977), 10-31
23. P. Melik-Adamyan, On scattering theory for canonical differential operators, Izves. Akad. Nauk Armen. SSR, 11 (1976), 291-313
24. R. Carroll, Transmutation and operator differential equations, North-Holland, Amsterdam, 1979

Scattering Frequencies for Time - Periodic Scattering Problems

Jeffery Cooper

University of Maryland, College Park, MD

Section 1: Introduction

Scattering frequencies or resonances for the wave equation in presence of a perturbation have been an important subject of research for many years. In this section we indicate some of the results in this area. In section 2 we describe some recent work [3] done with Walter Strauss and Gustavo Perla-Menzala, for the case of a time-periodic potential. Finally in section 3, we consider the case of a moving body and we construct solutions which have increasing energy.

We let $x = (x_1,\dots,x_n)$ denote a point in $\mathbb{R}^n$, with n odd, $n \geq 3$. $|x|$ denotes the Euclidean distance. For the first type of perturbation we consider a real potential $q(x,t)$ which has compact support in x and which has period T in time.

The wave equation with periodic potential is

$$\partial_t^2 u - \Delta u + qu = 0 \quad \text{in } \mathbb{R}^n \times \mathbb{R}_t \tag{1.1}$$

where $u = u(x,t)$.

The second type of perturbation to consider involves a bounded obstacle. Let $\mathcal{O}(t)$ be a smoothly varying family of compact sets in $\mathbb{R}^n$, such that $t \to \mathcal{O}(t)$ has period T. Let $\Omega(t) = \mathbb{R}^n / \mathcal{O}(t)$ denote the exterior region with

$$Q = \bigcup_t \Omega(t) x \{t\} \quad \text{and} \quad \Sigma = \partial Q.$$

We assume that Σ is time-like. That is, $|\nu_t| < |\nu_x|$ at each point of Σ where $\nu = (\nu_x, \nu_t)$ is the space-time unit normal to Σ. The boundary value problem we consider is

$$\begin{cases} \partial_t^2 u - \Delta u = 0 & \text{in } Q \\ u = 0 & \text{on } \Sigma \end{cases} \tag{1.2}$$

Scattering frequencies for either (1) or (2) are complex numbers $\sigma \in \mathbb{C}$ for which there exist special solutions, called scattering eigensolutions of the form

$$u(x,t) = g(x,t)\, e^{i\sigma t} \tag{1.3}$$

where $g(x,t)$ has period T. In addition, we require that u be "outgoing" in the sense that u represents a flow of energy, or radiation, away from the perturbation.

These special solutions can be used for asymptotic expansions near the perturbation of more general solutions:

$$u(x,t) \sim \sum c_j p_j(x,t) e^{i\sigma_j t}.$$

In this context, the scattering eigensolutions are decaying modes if $\mathrm{Im}\sigma_j > 0$, and growing modes if $\mathrm{Im}\sigma_j < 0$.

Finally, the complex numbers $\exp(i\sigma_j T)$ are eigenvalues of the local evolution operator $Z^\rho(T,0)$ of Lax and Phillips. It can also be shown that the scattering frequencies σ_j are poles of the scattering amplitude. For a complete discussion of these matters, the reader is referred to [2].

For a general fixed body or time-independent potential $q(x) \geq 0$, energy conservation and the Rellich uniqueness theorem imply that $\mathrm{Im}\sigma_j > 0$. The scattering frequencies for a fixed sphere can be found: they are the zeros of certain spherical Bessel functions. For the general case, we have only partial results, of which one of the most important is that of Lax and Phillips [6]. In this paper it is shown that for a fixed body, or certain potentials $q = q(x) \geq 0$, there are an infinite number of purely imaginary scattering frequencies σ_j with $\mathrm{Im}\sigma_j \to \infty$. For a survey of results and a discussion of the Singularity Expansion Method used in the engineering literature, the reader is referred to the paper of Dolph and Cho [4].

All the results mentioned so far are obtained by studying integral operators, depending on σ, which are related to the scattering problem. In the last few years, however, a new approach to this problem has been developed using techniques from the propagation of singularities. For fixed bodies which trap rays, scattering frequencies were constructed by Bardos, Guillot and Ralston [1], and Ikawa [5]. Lebeau [7] has used techniques involving analytic wave front sets to determine the location of scattering frequencies for fixed analytic bodies. Melrose [8] has made estimates on the number of scattering frequencies in regions $\{|z| < R\} \subset \mathbb{C}$ for a time - independent potential $q \in C_0^\infty(\mathbb{R}^n)$

Section 2: Time - periodic potentials.

We shall use the method of integral equations. The potentials we consider satisfy the following hypotheses:

(i) $q = 0$ for $|x| \geq \rho > 0$ and for all t;

(ii) $q(x,t+T) = q(x,t)$ for some $T > 0$ and all t;

(iii) $t \to q(\cdot,t)$ is continuous with values in $L^p(\mathbb{R}^n)$ for some $p > n$.

Let P denote the set of such potentials with the norm

$$\sup_{0\leq t\leq T} \|q(t)\|_{L^p(\mathbb{R}^n)}.$$

The following result was proved in [2], but here we give a more elementary proof.

Theorem 1: If $q \in P$, its set of scattering frequencies is a discrete set in $\mathbb{C}$. The set of scattering eigensolutions corresponding to each scattering frequency is finite dimensional.

Theorem 2 is a perturbation result which is the main result of this paper.

Theorem 2: Let $\ell \to q_\ell : \mathbb{R} \to P$ be a continuous family of potentials. Let $\sigma_o(\ell_o)$ be a scattering frequency for q_{ℓ_o}. Then there are scattering frequencies $\sigma(\ell)$ for q_ℓ which form a continuous curve in $\mathbb{C}$ passing through $\sigma_o(\ell_o)$. The scattering frequencies can only disappear when $|\sigma_o(\ell)| \to \infty$.

Using Theorem 2 and known results for time - independent potentials, we can deduce the existence of scattering frequencies for certain $q \in P$.

Corollary: Let $q_o = q_o(x) \in L^\infty(\mathbb{R}^n)$, $q_o = 0$ for $|x|>\rho$, $q_o \geq 0$ and $q_o(x) \geq c > 0$ on some open set. Let $q_1 \in P$ and $q_\varepsilon = q_o + \varepsilon q_1$. Then q_ε has a discrete, non empty, set of scattering frequencies for $\varepsilon > 0$ sufficiently small.

Proof of Corollary: For potentials $q_0 = q_0(x)$ satisfying the conditions of the corollary, it is proved in [3] that there exist an infinite number of purely imaginary scattering frequencies σ_j with $\operatorname{Im}\sigma_j \to \infty$. This is an extension of the result of Lax and Phillips [6]. The corollary then follows immediately from Theorem 2.

We shall need the following notation in the proofs of Theorem 1 and 2. Let H denote the set of pairs $f = [f_1, f_2]$ which are obtained as the completion of $C_0^\infty(\mathbb{R}^n) \times C_0^\infty(\mathbb{R}^n)$ in the energy norm

$$\|f\|_H = \left[\int (|\nabla f_1|^2 + |f_2|^2)\,dx\right]^{1/2}.$$

H is a Hilbert space with the natural inner product. Let $f = [f_1, f_2]$ be a distribution on $\mathbb{R}^n$ such that $\chi f \in H$ for each $\chi \in C_0^\infty(\mathbb{R}^n)$. We define the local energy of f over the ball of radius R to be

$$\|f\|_R = \left[\int (|\nabla f_1|^2 + |f_2|^2)\,dx\right]^{1/2}$$

It is well known that the solutions of

(2.1) $$\partial_t^2 u - \Delta u = 0 \quad \text{in} \quad \mathbb{R}^n \times \mathbb{R}_t$$

generate a unitary group on H, denoted $U_0(t)$, called the free group.

Solutions of the perturbed equation (1.1) generate a two parameter family $U(t,s)$ of bounded evolution operators on H, $U(t,s): H \to H$ $s,t \in \mathbb{R}$.

However, because of the time-dependent nature of the perturbation, energy is not conserved. We can estimate

(2.2) $$\|U(t,s)\| \le \gamma e^{k(t,s)} \quad \text{for } t \ge s$$

where $\|U(t,s)\|$ is the operator norm in $L(H)$, and $\gamma > 0$. The constant k depends on the $\sup_{0\le t\le T} \|q(t)\|_{L^n(\mathbb{R}^n)}$.

The inequality (2.2) is derived by standard methods (for details see [3]).

Next we shall derive an integral equation which must be satisfied by the scattering eigensolutions of (1.1). Suppose that σ is a scattering frequency of (1.1) and that u is a scattering eigensolution. Then

$$\partial_t^2 u - \Delta u = - qu$$

and u is outgoing. Hence, we can write u as a Duhamel integral

$$\underline{u}(t) = -\int_{-\infty}^{t} U_0(t-s)[0,q(s)u(s)]ds \tag{2.3}$$

where $\underline{u}(t) = [u(t),\partial_t u(t)]$. The integral (2.3) converges in the local energy norm for each $R > 0$. This is because qu has support in $|x| \le \rho$, and by virtue of Huyghens' principle,

$$\underline{u}(t)\Big|_{|x|\le R} = -\int_{t-(\rho+R)}^{t} U_0(t,s)[0,q(s)u(s)]ds$$

Equation (2.3) can also be expressed

$$u(t) = -\int_{-\infty}^{t} W(t-s)\, q(s)u(s)ds \tag{2.4}$$

where $W(t-s)h$ denotes the first component of $U_0(t-s)[0,h]$.

Since u is a scattering eigensolution, $g(t) = e^{-i\sigma t}u(t)$ has period T and we deduce that

$$(2.5) \qquad g(t) = -\int_{-\infty}^{t} W(t-s)\ \exp\ (-i\sigma(t-s))q(s)u(s)ds.$$

Now let C be the cylinder $C = \{(x,t): |x|<\rho\}$ and let $C_T = \{(x,t) \in C: 0 < t < T\}$. Define X to be the space of functions $g(x,t)$ on C which have period T with the norm

$$\|g\|_X = \|g\|_{L^r(C_T)} \qquad \text{where} \quad \frac{1}{2} = \frac{1}{p} + \frac{1}{r}\ .$$

Thus we see that if $u(t) = g(t)e^{i\sigma t}$ is a scattering eigensolution, then g satisfies an integral equation in X. To be more precise, for $\sigma \in \mathbb{C}$ and $q \in \mathcal{P}$ we define the operator L_σ on X by

$$(2.6) \quad (L_\sigma g)(t) = -\int_{-\infty}^{t} W(t-\tau)\ \exp\ (-i\sigma(t-\tau))q(\tau)g(\tau)d\tau$$

$$= -\int_{t-(2\rho)}^{t} W(t-\tau)\ \exp\ (-\ i\sigma(t-\tau))q(\tau)d\tau.$$

It is easy to verify that if g has period T, then so does $L_\sigma g$. Now if $g \in X$, the $qg\in L^2(C_T)$ because $\frac{1}{r} + \frac{1}{p} = \frac{1}{2}$. Using the unitarity of $U_0(t)$, we have that $L_\sigma g\in H^1(C_T)\subset X$, and it follows that L_σ is a compact operator in X.

Proof of Theorems 1 and 2

It is clear from (2.6) that $\sigma \to L_\sigma$ is an entire function with values in the space of compact operators on X. If $\ell\to q_\ell: \mathbb{R} \to \mathcal{P}$ is a continuous family of potentials, (2.6) also allows to deduce that $L_{\sigma,\ell}$ is jointly continuous.

From equation (2.5) we see that if σ is a scattering frequency, then $\lambda = 1$ is an eigenvalue of L_σ. Conversely, if $\lambda = 1$ is an eigenvalue of L_σ, then there exists $g \in X$, $g \neq 0$, $L_\sigma g = g$ so that g satisfies (2.5) on C. We can extend $u = \exp\ (i\sigma t)g$ to all of $\mathbb{R}^{n+1}$ by using the values of g on C in the right side of (2.4) and u is a scattering eigensolution.

We are now in a position to use Steinberg's results on analytic families of compact operators. Theorem 3 of [10] when applied to the analytic family L_σ asserts that if for each $\ell \in \mathbb{R}$, there is a $\sigma \in \mathbb{C}$ such that $(L_{\sigma,\ell}-I)^{-1}$ exists, then $\sigma \to (L_{\sigma,\ell}-I)^{-1}$ is meromorphic on $\mathbb{C}$ and the poles of $(L_{\sigma,\ell}-I)^{-1}$ depend continuously on ℓ, appearing or disappearing only when $|\sigma(\ell)| \to \infty$. The inequality (2.2) implies that there are no scattering frequencies of q_ℓ for $\mathrm{Im}\sigma < k_\ell$ (see [1]). Hence by Lemma 2, $L_{\sigma,\ell}-I$ is invertible for $\mathrm{Im}\sigma \le k_\ell$. Therefore Steinberg's result immediately yields our Theorems 1 and 2.

Remark: Theorems 1 and 2 can be easily modified to allow potentials q which have varying periods (see [3]).

Section 3: Growing Modes

For the case of a fixed body or a fixed potential $q = q(x) \ge 0$, we have $\mathrm{Im}\sigma_j > 0$ for each of the scattering frequencies. For a periodically moving body or potential, we have not been able to prove this assertion. Indeed, we suspect it may not be true, and that there may even be scattering frequencies σ_j with $\mathrm{Im}\sigma_j < 0$. This would correspond to a scattering eigensolution $u = g(x,t)e^{i\sigma_j t}$ which grows exponentially as $t \to \infty$. Such a scattering frequency σ_j would yield an eigenvalue $\lambda = \exp(i\sigma T)$ of the localized evolution operation $Z^\rho(T,0)$ with $|\lambda| > 1$.

We have not succeeded in finding an example with a growing scattering eigensolution, but we have constructed an example where the spectral radius

$$r(Z^\rho(T,0)) > 1$$

which of course implies $\|Z^\rho(T,0)\| > 1$. This example can be described as follows. Let A and B be smoothly bounded, compact, convex sets in $\mathbb{R}^3$. Assume $A \subset \{x_1 \le -1\}$ and that $\partial A \cap \{x_1 = -1\}$ is an open set of ∂A. Similarly assume that $B \subset \{x_1 \geq 1\}$ with $\partial B \cap \{x_1 = 1\}$ an open subset of ∂B. Let $A(t) = A + h(t)e_1$, where $e_1 = (1,0,0)$ and $h(t)$ is a real valued, smooth, function of period T with $|h'(t)|<1$. The obstacle is $\mathcal{O}(t) = A(t) \cup B$, and for some $\rho > 0$, $\mathcal{O}(t) \subset \{|x| \leq \rho\}$.

We shall use the geometric optics procedure of Ralston [9] to construct an approximate solution which is concentrated about a (broken) ray which is reflected back and forth between $A(t)$ and B with period T. Specifically we must choose $h(t)$ and construct the broken ray so that reflection with $\partial A(t)$ occurs once during each period. Let $0<v<1$, and choose $T = 2(1+(1+v)^{-1})$. Assume that $h(t) = vt$ in some open interval containing $[0,(1+v)^{-1}]$. Now let the ray be given parametrically by $(-t, 0, 0)$ until the first reflection with $\partial A(t)$. This will occur at $(-(1+v)^{-1},0,0)$ at time $t_o = (1+v)^{-1}$. The period $T = 2(1+(1+v)^{-1})$ is the time needed for the reflected ray to reach ∂B and return to $(-(1+v)^{-1}, 0,0)$ where it reflects again with $\partial A(t)$ at time $T+t_o$. For v sufficently small we can choose h so that $|h(t)|<1$, and $|h'(t)|<1$.

Now for $\xi > 0$ we let $u(x,t,\xi)$ be the solution of the boundary value problem (1.2) with initial conditions

$$(3.1) \qquad \begin{cases} u(x,o;\xi) = a(x)e^{i\xi x_1}, \\ \partial_t u(x,0;\xi) = \left[\dfrac{\partial a}{\partial x_1} + i\xi a\right] e^{i\xi x_1} \end{cases}$$

where $a \in C_o^\infty(\mathbb{R}^3)$ has support in a small ball K_r of radius r, centered at $(0,0,0)$. We chose r so small that the tube of rays parallel to the ray $(-t,0,0)$, $0<t<t_o$, and emanating from K_r all strike $\partial A(t)$ in the planar part $\partial A(t) \cap \{x_1 = vt-1\}$.

Now we construct the geometrical optics approximation to the solution (3.1). Define $W_0(x,t;\xi) = a(x_1+t, x_2, x_3)e^{i\xi(t+x_1)}$ until the first reflection on $\partial A(t)$. The reflected wave will be

$$W_1(x,t;\xi) = -a\left(\left[\frac{1+v}{1-v}\right](x_1-t),x_2,x_3\right) \exp\left(i\left[\frac{1+v}{1-v}\right]\xi\,(x_1-t)\right).$$

In general for j odd, W_j is the reflection from the moving boundary, while for j even, W_j is the reflected wave from the fixed boundary ∂B.

Define the energy in the "x_1 direction" to be

$$E_1(u) = \int_{\Omega(t)} \{|\partial_{x_1}u|^2 + |\partial_t u|^2\}dx.$$

Then it is easy to calculate that $E_1(W_1) = \left[\frac{1+v}{1-v}\right] E_1(W_0)$ while $E_1(W_2) = E_1(W_1)$. In general for j odd,

$$(3.2) \qquad E_1(W_j) = \left[\frac{1+v}{1-v}\right]^{\frac{j+1}{2}} E_1(W_0).$$

We set $W = \sum_{j=0}^{\infty} W_j$. For any $t > 0$ there are at most two nonzero terms in the sum, when reflection is taking place, and the time interval of reflection tends to zero as $t \to \infty$.

From (3.2) we see that the energy of the geometrical optics approxmiation W is growing exponentially. In fact the support of W is always contained in $\{|x| \le \rho\}$, so that $\|W(t;\xi)_\rho\| = \|W(t;\xi)\|$. Furthermore, for t in some open interval about $t = nT$, we have $W = W_{2n}$. Hence

$$(3.3) \qquad \|W(nT;\xi)\|_\rho \ge E_1^{1/2}(W_{2n}(nT;\xi)) \ge \left[\frac{1+v}{1-v}\right]^{n/2} E_1^{1/2}(W_0).$$

$$\ge \xi\|a\| - C_1$$

where $\|a\|$ is the L^2 norm of a and C_1 does not depend on ξ.

We must next compare W with u, the solution of (1.2) and (3.1).

Let $U(t,s)$ be the evolution operator associated with finite energy solutions of (1.2). $U(t,s): H(s) \to H(t)$ where $H(t)$ is the completion of $C_0^\infty(\Omega(t)) \times C_0^\infty(\Omega(t))$ in the energy norm for each t. $H(t)$ is a closed subspace of H. The basic existence theorems [2] allow us only the estimate of the operator norm

$$(3.4) \qquad \|U(t,s)\| \le \gamma e^{k|t-s|} \quad \text{where } \gamma > 0,\ k \ge 0.$$

Now set $V = u-W$. Note that $V = \partial_t V = 0$ at $t = 0$. Thus

$$\underline{V}(x,t;\xi) = -\int_0^t U(t,s)\ [0, \partial_s^2 W - \Delta W]\,ds$$

where $\underline{V} = [V, \partial_t V]$. Using (3.4), we see that the local energy

$$(3.5) \qquad \|V(t;\xi)\|_\rho \le C \int_0^t e^{k(t-s)}\,ds = \frac{C}{k}\,(e^{kt} - 1)$$

where C depends on the L^2 norm of $(\partial_{x_2}^2 + \partial_{x_3}^2)a$ but not on ξ. We combine (3.3) and (3.5) to obtain

where C depends on the L^2 norm of $(\partial_{x_2}^2 + \partial_{x_3}^2)a$ but not on ξ.

We combine (3.3) and (3.5) to obtain

$$(3.6) \qquad \|u(nT;\xi)\|_\rho \ge \left[\frac{1+v}{1-v}\right]^{n/2} (\|a\| - C_1) - \frac{C}{k}(e^{nkT} - 1).$$

Theorem 3: Given $\varepsilon > 0$ and n, there exists a finite energy solution u of (1.2) with C^∞ Cauchy data $[u(0), u_t(0)]$ supported in $\{|x| \le \rho\}$ such that $\|h(0)\| \le 1$ and

$$\|u(nT)\|_\rho \geq C_2 \left[\frac{1+v}{1-v}\right]^{n/2} - \varepsilon$$

where $C_2 > 0$, independent of ε and n.

Proof: Replace u in (3.6) by $\tilde{u} = \frac{1}{\xi}u$. $\|\tilde{u}(0)\|_\rho$ remains bounded as $\xi \to \infty$ and

$$\|u(nT;\xi)\|_\rho \geq \left[\frac{1+v}{1-v}\right]^{n/2} (\|a\| - C\tfrac{1}{\xi}) - \frac{C}{\xi k}(e^{nkT} - 1).$$

Then choose ξ sufficiently large, depending on n and ε.

Corollary: The spectral radius $r(Z^\rho(T,0)) \geq \left[\frac{1+v}{1-v}\right]^{1/2} > 1$.

Proof: $r(Z^\rho(T,0)) = \lim_{n\to\infty} \|Z^\rho(T,0)^n\|^{1/n}$

$$= \lim_{n\to\infty} \|Z^\rho(nT,0)\|^{1/n}.$$

From [2] we recall that $Z^\rho(nT,0) = E^\rho U(nT,0) D^{-\rho}$ where E^ρ and $D^{-\rho}$ are projections in H onto the entering and departing subspaces, respectively. For $f \in H$ with support in $\{|x| \leq \rho\}$, $D^{-\rho}f=f$, and for such an f,

$$\|Z^\rho(nT)f\| \geq \|U(nT,0)f\|_\rho.$$

Use the solutions constructed in Theorem 3 to deduce that

$$\|Z^\rho(nT,0)\| \geq C_2 \left[\frac{1+v}{1-v}\right]^{n/2} - \varepsilon \quad \text{for each } \varepsilon > 0.$$

References

[1] C. Bardos, J. C. Guillot, and J. Ralston, La relation de Poisson pour l'equation des ondes dans un ouvert nonborné, Application à la théorie de la diffusion, Comm. Partial Diff. Eq., 7(1982), 905-958.

[2] J. Cooper and W. Strauss, Abstract scattering for time - periodic systems with applications to electromagnetism, Indiana U. Math. J., 34(1985), 33-83.

[3] J. Cooper, G. Perla-Menzala, and W. Strauss, On the scattering frequencies of time-dependent potentials, to appear in Math. Meth. Appl. Sci.

[4] C. Dolph and S. Cho, On the relationship between the singularity expansion method and the mathematical theory of scattering, IEEE Trans. A. P., 28 (1980), 888-897.

[5] M.Ikawa, On the poles of the scattering matrix for two strictly convex obstacles, J. Math. Kyoto Univ., 23(1983), 127-194.

[6] P. Lax and R. Phillips, Decaying modes for the wave equation in the exterior of an obstacle, Comm. Pure Appl. Math., 22(1969), 737-787.

[7] G. Lebeau, Regularité Gevrey 3 pour la diffraction, Comm. Partial Diff. Eq., 9(1984), 1437-1494.

[8] R. Melrose, Polynomial bound on the number of scattering poles, J. Fun. Anal., 53(1983), 287-303.

[9] J. Ralston, Solutions of the wave equation with localized energy, Comm. Pure Appl. Math., 22(1969), 807-823.

[10] S. Steinberg, Meromorphic families of compact operators, Arch. Rat. Mech. Anal., 13(1968), 372-379.

PERIODIC SOLUTIONS FOR LINEAR INTEGRODIFFERENTIAL EQUATIONS WITH INFINITE DELAY IN BANACH SPACES

Giuseppe Da Prato

Scuola Normale Superiore
Piazza dei Cavalieri 7
56100 Pisa, Italy

Alessandra Lunardi

Dipartimento di Matematica
Università di Pisa
Via Buonarroti 2
56100 Pisa, Italy

0. INTRODUCTION

We are here concerned with the problem

$$(0.1)\quad \begin{cases} u'(t) = Au(t) + \displaystyle\int_{-\infty}^{t} K(t-s)u(s)ds + f(t) \\ \\ u(0) = u(2\pi) \end{cases}$$

where A and $K(t)$ are linear operators (generally unbounded) in a real Banach space X and f is a periodic X-valued function.

Equation (0.1) has been extensively studied in the finite dimensional case, see for instance the survey paper [CL]. In the infinite dimensional case a similar problem has been studied in [NY] in connection with Hopf bifurcation for nonlinear equations.

Using the Fourier series $\sum_{k=-\infty}^{+\infty} u_k e^{ikt}$ of $u(t)$ we get a heuristic formula for the solution of (0.1), that is

$$(0.2)\quad u(t) = \sum_{k=-\infty}^{+\infty} F(ik)f_k e^{ikt} \quad , \quad f_k = \frac{1}{2\pi}\int_0^{2\pi} e^{-iks} f(s)ds$$

where $F(\lambda) = (\lambda - A - \hat{K}(\lambda))^{-1}$ is the resolvent for the initial value problem

$$(0.3) \qquad u'(t) = Au(t) + \int_0^t K(t-s)u(s)ds + f(t) \quad , \quad u(0) = u_0 \quad , \quad t \geq 0$$

which is studied in [DPI] and [L]. In fact it is easy to prove that formula (0.2) gives the solution of (0.1) if X is a Hilbert space, $f \in L^2(0,2\pi;X)$ and $F(\lambda)$ is defined for all $\lambda = ik$, $k \in \mathbb{Z}$. In this case we have also $u', Au \in L^2(0,2\pi;X)$. However in order to apply the linear results to nonlinear problems (as we will do in a subsequent paper), it is more convenient to deal with continuous functions in general Banach space. Therefore our goal is to find a Banach space B of X-valued continuous periodic functions satisfying the so called maximal regularity property, that is, for any $f \in B$ there exists a unique solution of (0.1) $u \in B$ such that $u', Au \in B$.

Due to the difficulties of handling Fourier series in general Banach space, we are only able to show that the space $B = C_{\#}^{1,\theta}(X)$ has the above property ($C_{\#}^{1,\theta}(X)$ is the set of all differentiable 2π-periodic X-valued functions f such that f' is θ-Hölder continuous).

In a subsequent paper we will study Hopf bifurcation for a fully nonlinear integrodifferential equation: to this aim we also consider in section 3 the special case when $F(\lambda)$ has singularities at $\lambda = \pm i$. We are able to show that problem (0.1) has ∞^2 solutions or no solution according to f satisfies or not a suitable compatibility condition.

1. NOTATIONS AND PRELIMINARIES ON THE INITIAL VALUE PROBLEM

Throughout the paper we shall denote by X (with norm $\|\cdot\|$) and D (with norm $\|\cdot\|_D$) two real Banach spaces, D being continuously imbedded in X. We shall collect here some results on the initial value problem

$$(1.1) \quad \begin{cases} u'(t) = Au(t) + \displaystyle\int_0^t K(t-s)u(s)ds + f(t) = Au(t) + (K*u)(t) + f(t), \\ \hfill t \geq 0 \\ u(0) = x \end{cases}$$

where $A : D \to X$, $K(s) : D \to X$ are linear operators. We denote by $\tilde{X} = X + iX$ and $\tilde{D} = D + iD$ the usual complexification of X and D respectively. We set $\tilde{A} : \tilde{D} \to \tilde{X}$, $\tilde{A}(x+iy) = Ax + iAy$ and $\tilde{K}(s) : \tilde{D} \to \tilde{X}$, $\tilde{K}(s)(x+iy) = K(s)x + iK(s)y$ for $x,y \in D$. We assume that

$$(1.2)\quad \begin{cases} \text{The resolvent set } \rho(\tilde{A}) \text{ of } \tilde{A} \text{ contains a sector} \\ S = \{z \in \mathbb{C};\ z \neq \omega,\ |\arg(z-\omega)| < \theta_0\} \text{ with } \omega \in \mathbb{R}, \\ \theta_0 \in]\frac{\pi}{2}, \pi[\text{ and there exists } M > 0 \text{ such that} \\ \qquad \|(z-\omega)(z-A)^{-1}\|_{L(\tilde{X})} \leq M \quad \text{for } z \in S \end{cases}$$

$$(1.3)\quad \begin{cases} \text{For } s \geq 0\ K(s) \in L(D,X), \text{ for each } x \in D \text{ the function } K(\cdot)x \\ \text{is absolutely Laplace transformable. The Laplace transform} \\ \hat{K}(\cdot)x \text{ is analytically extendible to } S \text{ and there exist} \\ \beta, N > 0 \text{ such that the extension satisfies} \\ \qquad |(z-\omega)|^{\beta}\|\hat{K}(z)x\| \leq N\|x\|_D, \quad z \in S, \quad x \in D. \end{cases}$$

Under these assumptions it can be shown the existence of the resolvent operator for problem (1.1) (see [DPI] and [L]). The resolvent operator $R(t)$ is given by

$$R(t) = \frac{1}{2\pi i}\int_{\gamma} e^{zt} F(z)dz \qquad t > 0$$

where γ is a suitable path in $\mathbb{C}$ joining $\infty e^{-i\theta_1}$ with $\infty e^{i\theta_1}$, $\theta_1 \in]\frac{\pi}{2}, \theta_0[$, and

$$(1.4)\quad F(z) = (z - \tilde{A} - \hat{K}(z))^{-1},$$

exists in $L(\tilde{X};\tilde{D})$ for $z \in S$, $|z| \geq R_0$ for some positive R_0.
We have

$$(1.5) \quad \begin{cases} \sup\limits_{z \in S, |z| \geq R_0} \|zF(z)\|_{L(\tilde{X})} < +\infty , \\ \\ \sup\limits_{z \in S, |z| \geq R_0} \|\tilde{A}F(z)\|_{L(\tilde{X})} < +\infty . \end{cases}$$

We denote by ρ_F the set of all $z \in \mathbb{C}$ such that $(z - \tilde{A} - \hat{K}(z))^{-1}$ exists in $L(\tilde{X},\tilde{D})$. We set

$$(1.6) \quad F^+(\lambda) = \frac{1}{2}(F(\lambda) + F(\bar{\lambda})) \quad , \quad F^-(\lambda) = \frac{1}{2i}(F(\lambda) - F(\bar{\lambda})) \quad , \quad \lambda \in \rho_F$$

It is easy to check that $F^+(\lambda)$ and $F^-(\lambda)$ belong to $L(X;D)$.

We give now a regularity result for the solution of (1.1).

PROPOSITION 1.1 Assume (1,2), (1.3). Let $T > 0$, $\theta \in]0,1[$ and $f \in C^{1,\theta}([0,T];X)$, $x \in D$, $Ax + f(0) \in \bar{D}$ (the closure of D in X). Then there exists a unique solution $u \in C^1([0,T],X) \cap C([0,T];D)$ of (1.1). Moreover u', Au, $K*u$ belong to $C^{1,\theta}([\varepsilon,T];X)$ for each $\varepsilon \in]0,T[$.

Proof.
Since $f \in C^{\theta}([0,T];X)$ and $x \in D$, $Ax + f(0) \in \bar{D}$, the function

$$u(t) = R(t)x + \int_0^t R(s)f(t-s)ds \quad , \quad 0 \leq t \leq T$$

is the unique strict solution of (1.1) (see [L], Prop. 2.2).
The functions

$$]0,+\infty[\;\to L(X) \quad , \quad t \to R(t)$$

$$]0,+\infty[\;\to L(X;D) \quad , \quad t \to R(t)$$

are analytic (see [L], Prop. 1.1). We have $u'(t) = v_1(t) + v_2(t)$, $0 \leq t \leq T$, where

$$v_1(t) = R'(t)x + R(t)f(0) + \int_0^t R(s)f'(0)ds \quad , \quad 0 \le t \le T$$

$$v_2(t) = \int_0^t R(t-s)(f'(s) - f'(0))ds \quad , \quad 0 \le t \le T$$

Then v_1 is analytic in $]0,T]$ with values in X and (see [L], Prop. 1.2)

$$Av_1(t) = AR'(t)x + AR(t)x + AR(t)f(0) + R(t)f'(0) - f'(0) -$$

$$- \int_0^t \frac{1}{2\pi i} \int_\gamma e^{\lambda s} \hat{K}(\lambda)F(\lambda)d\lambda f'(0)ds.$$

Therefore Av_1 is analytic in $]0,T]$ with values in X and hence v_1 is analytic in $]0,T]$ with values in D. Finally v_2 belongs to $C^{1,\theta}([0,T];X) \cap C^\theta([0,T];D)$ thanks to Prop. 2.4 of [L]. ■

2. PERIODIC SOLUTIONS (NONRESONANCE CASE)

In this section we shall study the existence of 2π-periodic solutions for the equation

$$(2.1) \qquad u'(t) = Au(t) + \int_{-\infty}^t K(t-s)u(s)ds + f(t)$$

assuming, besides (1.2) and (1.3), the nonresonance condition

$$(2.2) \qquad i\,\mathbb{Z} \in \rho_F \ .$$

Then by (1.5) there exists $M > 0$ such that

$$(2.3) \qquad \|kF(ik)\|_{L(\tilde{X})} + \|\tilde{A}F(ik)\|_{L(\tilde{X})} \le M \quad , \quad k \in \mathbb{Z}$$

We shall find the solution of (2.1) in the form

$$(2.4) \qquad u(t) = \sum_{k=-\infty}^{+\infty} u_k e^{ikt}$$

It is convenient to adopt the following notations: $C_{\#}(X)$ represents the Banach space of all continuous and 2π-periodic functions $\mathbb{R} \to X$; $C^{\theta}_{\#}(X)$ denotes the space of all θ-Hölder continuous and 2π-periodic functions $f : \mathbb{R} \to X$, endowed with the usual norm

$$\|f\|_{C^{\theta}_{\#}(X)} = \sup_{t \in [0,2\pi]} \|f(t)\| + \sup_{0 \le s < t \le 2\pi} \frac{\|f(t) - f(s)\|}{(t-s)^{\theta}}$$

$C^{k}_{\#}(X)$ and $C^{k,\theta}_{\#}(X)$, $k = 1,2,\ldots$ are analogously defined.

We collect now some results on Fourier series in general Banach space.

LEMMA 2.1 Let $u \in C^{\theta}_{\#}(\tilde{X})$ $(0 < \theta < 1)$, and set $u_k = \frac{1}{2\pi}\int_0^{2\pi} e^{-iks}u(s)ds$, $k \in \mathbb{Z}$. Then (2.4) holds and the series is uniformly convergent.

Proof.

Fix $N \in \mathbb{N}$ and set

$$u_N(t) = \sum_{k=-N}^{N} u_k e^{ikt} = \frac{1}{2\pi}\int_0^{2\pi} \frac{u(t-\sigma)}{\sin\frac{\sigma}{2}} \cdot \sin\left(\left(N + \frac{1}{2}\right)\sigma\right)d\sigma$$

Hence

$$u(t) - u_N(t) = \frac{1}{2\pi}\int_0^{2\pi} \frac{\sin((N + \frac{1}{2})\sigma)}{\sigma^{1-\theta+\varepsilon}} \phi(t,\sigma)d\sigma$$

where $\varepsilon \in]0,\theta[$ and

$$\begin{cases} \phi(t,\sigma) = \dfrac{u(t) - u(t-\sigma)}{\sin\frac{\sigma}{2}} \sigma^{1-\theta+\varepsilon}, & t \in \mathbb{R},\ \sigma \in]0,2\pi] \\ \phi(t,0) = 0 & t \in \mathbb{R}. \end{cases}$$

Then ϕ is continuous, and setting

$$\phi_n(t,\sigma) = (2\pi)^{-n} \sum_{h=0}^{n} \binom{n}{h} \sigma^h (2\pi - \sigma)^{n-h} \phi\left(t, \frac{h}{2\pi n}\right)$$

ϕ_n converges uniformly to ϕ as $n \to \infty$. Therefore

$$\|u(t) - u_N(t)\| \leq \frac{1}{2\pi}\int_0^{2\pi} \|\phi(t,\sigma) - \phi_n(t,\sigma)\| \left|\frac{\sin((N+\frac{1}{2})\sigma)}{\sigma^{1-\theta+\varepsilon}}\right| d\sigma +$$

$$+ \frac{1}{2\pi}\left\|\int_0^{2\pi} \phi_n(t,\sigma)\frac{\sin((N+\frac{1}{2})\sigma)}{\sigma^{1-\theta+\varepsilon}}\, d\sigma\right\|$$

so that u_N converges uniformly to u. ■

LEMMA 2.2 For each $\theta \in]0,1[$ there exists $C(\theta) > 0$ such that for all $k \in \mathbb{Z}$

(2.5) $$|k|^{\theta}\|u_k\| \leq C(\theta)\|u\|_{C^{\theta}_{\#}(\tilde{X})} \quad \text{if} \quad u \in C^{\theta}_{\#}(\tilde{X})$$

and

(2.6) $$|k|^{1+\theta}\|u_k\| \leq C(\theta)\|u\|_{C^{1,\theta}_{\#}(\tilde{X})} \quad \text{if} \quad u \in C^{1,\theta}_{\#}(\tilde{X}) .$$

Proof.
For any $u \in C_{\#}(\tilde{X})$ (resp. $u \in C^1_{\#}(\tilde{X})$) we have

$$\|u_k\| \leq \|u\|_{C_{\#}(\tilde{X})} \quad (\text{resp. } \|ku_k\| \leq \|u\|_{C^1_{\#}(\tilde{X})})$$

Then (2.5) follows by interpolation whereas (2.6) follows by (2.5) since $(u')_k = i\,k\,u_k$. ■

We can give now an existence and regularity result for equation (2.1).

THEOREM 2.3 Let (1.2), (1.3) and (2.2) hold. Assume, moreover, that there exists $\theta \in]0,1[$ such that

(2.7) $$\begin{cases} \text{for any } v \in C_{\#}(D) \text{ the function } \phi \text{ defined by} \\ \qquad \phi(t) = \int_{-\infty}^{0} K(t-s)v(s)ds \ , \quad t \geq 0 \\ \text{belongs to } C^{1,\theta}([0,T];X) \text{ for any } T > 0. \end{cases}$$

Then, for each $f \in C^{1,\theta}_{\#}(X)$ there exists a unique 2π-periodic solution u of (2.1) such that $u', Au \in C^{1,\theta}_{\#}(X)$. u is given by

$$(2.8) \quad u(t) = \sum_{k=-\infty}^{+\infty} F(ik)f_k e^{ikt} =$$

$$= F(0)f_0 + 2\sum_{k=1}^{+\infty} [(F^+(ik)f_k^+ - F^-(ik)f_k^-) \cos kt -$$

$$- (F^+(ik)f_k^- + F^-(ik)f_k^+) \sin kt]$$

where $F^{\pm}(ik) \in L(\tilde{X},\tilde{D}) \cap L(X,D)$ are defined by (1.6) and f_k, $f_k^{\pm}$ are given by

$$(2.9) \quad \begin{cases} f_k^+ = \dfrac{1}{2\pi}\displaystyle\int_0^{2\pi} f(t)\cos kt\,dt \ , \ f_k^- = -\dfrac{1}{2\pi}\displaystyle\int_0^{2\pi} f(t)\sin kt\,dt \ , \ k \in \mathbb{Z} \\ f_k = f_k^+ + if_k^- \ , \quad k \in \mathbb{Z} \ . \end{cases}$$

Proof.

By (2.3) and lemma 2.2 we have

$$\|F(ik)f_k\| \le M\,C(\theta)\|f\|_{C^{1,\theta}_{\#}(X)} |k|^{-2-\theta} \quad , \ k \in \mathbb{Z}\setminus\{0\}$$

$$\|AF(ik)f_k\| \le M\,C(\theta)\|f\|_{C^{1,\theta}_{\#}(X)} |k|^{-1-\theta} \quad , \ k \in \mathbb{Z}\setminus\{0\}$$

so that the function u defined by (2.8) is continuously differentiable, $\tilde{A}u = Au$ is continuous and

$$u'(t) = \sum_{k=-\infty}^{+\infty} ikF(ik)f_k\, e^{ikt} \ , \ Au(t) = \sum_{k=-\infty}^{+\infty} \tilde{A}F(ik)f_k\, e^{ikt}$$

$$\int_0^{+\infty} K(s)u(t-s)ds = \sum_{k=-\infty}^{\infty} \int_0^{+\infty} K(s)F(ik)f_k e^{ik(t-s)}ds =$$

$$= \sum_{k=-\infty}^{+\infty} \hat{K}(ik)F(ik)f_k e^{ikt}$$

the last series being absolutely convergent since $\|\hat{K}(ik)\|_{L(\tilde{X},\tilde{D})}$

is bounded. Now, setting

$$u_k = F(ik)f_k \quad , \; k \in \mathbb{Z} \quad ,$$

we have

$$(2.10) \quad iku_k - \tilde{A}u_k - \hat{K}(ik)u_k = f_k \quad , \quad k \in \mathbb{Z}$$

so that u is a solution of (2.1). Setting $u(0) = x$, $g(t) = f(t) + \int_{-\infty}^{0} K(t-s)u(s)ds$, u is the strict solution of the initial value problem (1.1), with f replaced by g. By (2.7), g belongs to $C^{1,\theta}([0,4\pi];X)$; moreover $Ax + g(0) = u'(0) \in \bar{D}$. Therefore, by Proposition 1.1, u' and Au belong to $C^{1,\theta}([2\pi,4\pi];X)$ so that u is a solution of (2.1) with the regularity properties claimed.

Let us prove now uniqueness: if $u \in C^{2,\theta}_{\#}(X) \cap C^{1,\theta}_{\#}(D)$ is a solution of (2.1), then the series

$$u(t) = \sum_{k=-\infty}^{+\infty} u_k e^{ikt} \quad , \quad u'(t) = \sum_{k=-\infty}^{+\infty} iku_k e^{ikt} \quad ,$$

$$Au(t) = \sum_{k=-\infty}^{+\infty} \tilde{A}u_k e^{ikt}$$

are absolutely convergent by lemma 2.2. Plugging u, u', Au in (2.1) we find $u_k = F(ik)f_k$, so that the conclusion follows. ■

3. PERIODIC SOLUTIONS (RESONANCE CASE)

We shall study here periodic solutions of equation (2.1) when ki does not belong to ρ_F for some integer k. More precisely we shall assume the typical hypotheses of Hopf bifurcation:

$$(3.1) \quad \begin{cases} \text{i)} \; ki \in \rho_F \;\; \text{for} \;\; k \in \mathbb{Z} \; , \; k \neq \pm 1 \\ \\ \text{ii)} \; i \;\; \text{is a simple isolated eigenvalue of} \;\; \tilde{A} + \hat{K}(i). \end{cases}$$

As easily seen, (3.1)-ii) implies that $-i$ is a simple isolated eigenvalue of $\tilde{A} + \hat{K}(-i)$.

We denote by $w_0 = x_0 + iy_0$ an eigenvector of $\tilde{A} + \hat{K}(i)$ corresponding to the eigenvalue i; it follows that $\bar{w}_0 = x_0 - iy_0$ is an eigenvector of $\tilde{A} + K(-i)$ with eigenvalue $-i$. Since $\pm i$ are simple eigenvalues, there exist two closed subspaces of $\tilde{X}$, denoted by $\tilde{X}^+$ and $\tilde{X}^-$, such that

$$\tilde{X} = \mathbb{C}w_0 + \tilde{X}^+ = \mathbb{C}\bar{w}_0 + \tilde{X}^- .$$

Thus any $\tilde{x} \in \tilde{X}$ can be uniquely written as

$$\tilde{x} = \langle \tilde{x}, \xi_0 \rangle \, w_0 + x_1$$

where $\xi_0 \in \tilde{X}^*$ (the dual of $\tilde{X}$) and $x_1 \in \tilde{X}^+$. We set $\xi_0 = \phi_0 + i\eta_0$ where ϕ_0, η_0 belong to X^*; we can choose w_0 and ξ_0 such that

$$(3.2) \qquad \langle x_0, \phi_0 \rangle = \langle y_0, \eta_0 \rangle = 1 \ , \quad \langle x_0, \eta_0 \rangle = \langle y_0, \phi_0 \rangle = 0 .$$

Clearly $\pm i - \tilde{A} - \hat{K}(\pm i) : \tilde{D} \cap \tilde{X}^\pm \to \tilde{X}^\pm$ are invertible; we shall denote by $S^\pm \in L(\tilde{X}^\pm, \tilde{D} \cap \tilde{X}^\pm)$ their inverses.

Let now $f \in C^{1,\theta}_{\#}(X)$ and $u \in C^{1,\theta}_{\#}(D) \cap C^{2,\theta}_{\#}(X)$; then u is a solution of (1.1) iff, denoting by $\{u_k\}_{k \in \mathbb{Z}}$ the Fourier coefficients of u, (2.10) holds.

For $k \neq \pm i$ (2.10) has a unique solution u_k given by

$$u_k = F(ik) f_k$$

and, for $k = \pm 1$ (2.10) has a solution iff

$$(3.3) \qquad \langle f_1, \xi_0 \rangle = \langle f_{-1}, \bar{\xi}_0 \rangle = 0$$

where $\bar{\xi}_0 = \phi_0 - i\eta_0$. In this case we get

$$u_1 = c\, w_0 + S^+ f_1 \ , \ u_{-1} = \bar{c}\, \bar{w}_0 + S^- f_{-1} \ , \qquad c \in \mathbb{C} .$$

Then, all the solutions of (2.1) are given by

$$(3.4)\quad u(t) = \sum_{\substack{k \in \mathbb{Z} \\ k \neq \pm 1}} F(ik) f_k e^{ikt} + (c\, w_0 + S^+ f_1) e^{it} + $$

$$+ (\bar{c}\, \bar{w}_0 + S^- f_{-1}) e^{-it} , \qquad c \in \mathbb{C} .$$

Thus, we have proved the following theorem:

<u>THEOREM 3.1</u> <u>Assume</u> (1.2), (1.3), (2.7), (3.1) <u>and let</u> $f \in C^{1,\theta}(X)$. <u>Then equation</u> (2.1) <u>has a</u> 2π-<u>periodic solution</u> $u \in C^{2,\theta}_{\#}(X) \cap C^{1,\theta}_{\#}(D)$ <u>iff</u>

$$(3.5)\quad \begin{cases} \nu(f) = \displaystyle\int_0^{2\pi} \langle f(t), \phi_0 \cos t - \eta_0 \sin t\rangle\, dt = 0 \\ \zeta(f) = \displaystyle\int_0^{2\pi} \langle f(t), \phi_0 \sin t + \eta_0 \cos t\rangle\, dt = 0 \end{cases}$$

■

<u>Remark</u>. Formula (3.4) gives all the solutions of (2.1) with values in $\tilde{D}$: but it is easy to see, by straightforward computations, that, in fact, these solutions have values in D .

<u>EXAMPLE 3.2</u>

Let us consider the special case where $K(t) = be^{-ct}A$, $b \in \mathbb{R}$, $c > 0$. We assume that A satisfies (1.2) and the spectrum of A consists of the sequence $\{-\mu_h\}_{h \in \mathbb{N}} \subset \mathbb{R}$ with $\mu_h \to +\infty$.

We have

$$\hat{K}(\lambda) = \frac{b}{\lambda + c} \tilde{A} , \quad \mathrm{Re}\, \lambda > -c$$

problems. In fact, consider the abstract Cauchy problem

$$\frac{d}{dt} u(t) = Au(t), \quad t \geq 0,$$

$$u(0) = x.$$

If for any $x \in C$ there exists a unique solution $u(t,x)$ that depends continuously on the initial datum x in the sense that if (x_n) is a sequence in C such that $x_n \to x$ then $u(t,x_n)$ converges to $u(t,x)$ uniformly on bounded t-intervals then

$$T(t)x = u(t,x), \quad t \geq 0,$$

defines a semigroup $T(\cdot)$ in X. (Note, that it often occurs that there do not exist solutions for all $x \in X$ but only for x belonging to a certain subset C.) For a detailed discussion of contraction semigroups and semigroups whose Lipschitz-constants are exponentially bounded:

$$\|T(t)\|_{Lip} \leq e^{\omega t} \quad \text{for some } \omega,$$

we refer to ([2]) and ([6]). The asymptotic behavior of linear semigroups is surveyed, for instance, in [7].

2. Stability of Equilibria.

Let $T(\cdot)$ be a nonlinear semigroup in X. The objective of this section is devoted to the investigation of stability properties of equilibrium solutions via the linearization of $T(\cdot)$:

Recall, that $x_0 \in X$ is called an equilibrium solution if $T(t)x_0 = x_0$ for all $t \geq 0$. To begin with, we recall the well-known concept of stability:

Definition. An equilibrium x_0 is called stable (in the sense of Ljapunov) if for any neighborhood U of x_0 there exists a neighborhood V of x_0 such that $T(t)V \subset U$ for all $t \geq 0$. If, in addition, $T(t)x$ converges to x_0 as $t \to \infty$ for any $x \in V$, we call x_0 asymptotically stable. In case that this convergence is exponential we refer to x_0 as an exponentially asymptotically stable equilibrium. If x_0 is not stable, we call it unstable.

The main result is

Theorem 2.1. Let $T(\cdot)$ be a nonlinear semigroup in X and let x_0 be an equilibrium. Suppose that $T(\cdot)$ is Fréchet-differentiable at x_0 with $U(t) = T'(t,[x_0])$ and that the zero solution is exponentially asymptotically stable with respect to this linearized semigroup $U(\cdot)$. Then x_0 is exponentially asymptotically stable with respect to $T(\cdot)$.

This result is an immediate consequence of

Proposition 2.1. Let T be a nonlinear operator in X and let x be a fixed point of T. Let U denote the Fréchet-derivative of T at x. If the spectral radius of U is less than 1 then there exists an $\omega > 0$ such that $e^{\omega n}(T^n y - x) \to 0$ as $n \to \infty$ for all y in a sufficiently small neighborhood of x.

Proof. Without loss of generality, we may assume $x = 0$. By assumption, we have $\lim_{n\to\infty} \|U^n\|^{1/n} < 1$ and hence we can choose a positive integer k so that $\|U^k\| \leq \frac{1}{4}$. Next, choose some $\delta > 0$ such that $\|T^k y - U^k y\| \leq \frac{1}{4}\|y\|$ for $\|y\| \leq \delta$. Then we get for $\|y\| \leq \delta$:

$$\|T^k y\| \leq \|T^k y - U^k y\| + \|U^k y\| \leq \tfrac{1}{2}\|y\| \leq \delta.$$

Taking $\omega < \frac{1}{k} \ln 2$, we deduce that $e^{\omega k} \|T^k y\| < \kappa\|y\|$ with some $\kappa < 1$, so that $\|e^{\omega kn} T^{kn} y\| \to 0$ as $n \to \infty$. Using the fact that $\|T^m y\| \leq M\|y\|$ for sufficiently small y and $m = 1,\dots,k-1$ (since T^m has a Fréchet-derivative at zero), we infer that $\|e^{\omega n} T^n y\|$ converges to 0 as claimed.

Naturally the question arises what happens if the linearized semigroup is not exponentially stable. The basic result is

Proposition 2.2. Let T be a nonlinear operator in X, let x be a fixed point of T and suppose that T is continuously Fréchet-differentiable at x. Let $U = T'[x]$ and assume that X can be decomposed as $X = X_1 \oplus X_2$, where the X_i are U-invariant subspaces of X and there exist a positive integer n and reals $1 < \theta < \eta$ such that for $x_i \in X_i$, $i = 1,2$, we have

$$\|U^n x_1\| \geq \eta\|x_1\|, \quad \|U^n x_2\| \leq \theta\|x_2\|.$$

Then there exists a constant $\varepsilon < 0$ and sequences (n_k) of positive integers, $(x_k) \to x$ such that $\|T^{n_k} x_k - x\| \geq \varepsilon$.

<u>Proof.</u> Again, we may assume without loss of generality that $x = 0$. Let π_i denote the projection of X onto X_i with kernel X_j, $j \neq i$. Moreover, we may also assume that for $y \in X$

$$\|y\| = \|\pi_i y\| + \|\pi_2 y\|$$

as this norm is equivalent to the original one. Next, we choose $\varepsilon > 0$ such that for $\|y\| \leq \varepsilon$ we have

$$\|T^n y - U^n y\| \leq \frac{1}{4}(\eta-\theta)\,\|y\| .$$

Then we obtain for those y that satisfy $\|y\| \leq \varepsilon$ and $\|\pi_2 y\| \leq \|\pi_1 y\|$

$$\|\pi_1 T^n y\| \geq \|\pi_1 U^n y\| - \|\pi_1(T^n y - U^n y)\| \geq \eta\|\pi_1 y\| - \frac{1}{2}(\eta-\theta)\|\pi_1 y\|$$

$$= \frac{1}{2}(\eta+\theta)\|\pi_1 y\| .$$

On the other hand, we also have

$$\|\pi_2 T^n y\| \leq \|\pi_2 U^n y\| + \|\pi_2(T^n y - U^n y)\| \leq \theta\|\pi_2 y\| + \frac{1}{2}(\eta-\theta)\|\pi_1 y\|$$

$$\leq \frac{1}{2}(\eta+\theta)\|\pi_1 y\|$$

so that $T^n y$ again satisfies $\|\pi_1 T^n y\| \geq \|\pi_2 T^n y\|$. Suppose now that $\|T^{nk} y\| \leq \varepsilon$ for all positive integers k. Then we obtain by induction $\|\pi_1 T^{nk} y\| \geq (\eta+\theta)^k\, 2^{-k}\|\pi_1 y\|$ which converges to ∞ as $k \to \infty$ in contrast to the hypothesis. As y can be chosen arbitrarily close to 0 the claim follows.

Applying this result to nonlinear semigroups, we obtain:

<u>Theorem 2.2.</u> Let $T(\cdot)$ be a nonlinear semigroup in X with a fixed point x_0. Let $U(\cdot)$ denote the Fréchet-derivative of $T(\cdot)$ at x_0, and suppose that for some fixed t we have a splitting of X as $X = X_1 \oplus X_2$ where the X_i are invariant with respect to $U(t)$, X_1 is finite-dimensional and with $\omega := \lim\limits_{h\to\infty} \frac{1}{h} \ln \|U(h)\,|_{X_2}\|$ we have

$$\mu := \inf\,\{\|\lambda\| \mid \lambda \in \sigma(U(t)|_{X_1})\} > e^{\omega t}.$$

Then there exist a constant $\varepsilon > 0$ and sequences (x_n) converging to x_0 and (t_n) of positive reals such that $\|T(t_n)x_n - x_0\| \geq \varepsilon$.

Proof. We put $T = T(t)$ in the above proposition and observe that for $e^{\omega t} < \kappa < \nu < \mu$ there exists some positive integer n such that for all $x_i \in X_i$

$$\|U(nt)x_1\| \geq \nu^n \|x_1\|, \quad \|U(nt)x_2\| \leq \kappa^n \|x_2\| .$$

The first inequality follows as X_1 is finite-dimensional whereas the latter one is trivial.

3. Stability of Periodic Orbits.

Again, let $T(\cdot)$ be a nonlinear semigroup in X. We now want to consider stability properties of a nontrivial p-periodic solution x, i.e. $x(t) = T(t)x(0)$, $t \geq 0$ such that $x(p) = x(0)$ and there exists a $t \in (0,p)$ so that $x(t) \neq x(0)$. Such a solution can never be asymptotically stable (in the sense of Ljapunov). Indeed, taking h sufficiently small so that $x(h)$ is close to $x(0)$, we obtain for all positive integers n

$$\|T(np)x(h) - T(np)x(0)\| = \|x(h) - x(0)\|$$

so that

$$\limsup_{t\to\infty} \|T(t)x(h) - T(t)x(0)\| > 0.$$

On the other hand, if $x(\cdot)$ is continuously differentiable and $T(p)$ is Fréchet-differentiable at $x(0)$ with

$$U = T'(p,[x(0)])$$

then we obtain by differentiating $T(p)x(t) = x(t)$ at $t = 0$

$$Ux'(0) = x'(0).$$

From the fact that $T(\cdot)$ is locally Lipschitzian and $x(\cdot)$ being a nontrivial periodic solution, we conclude that $x'(0) \neq 0$. As a consequence, 1 is contained in the point spectrum of U and the assumptions of Theorem 2.1 can never be satisfied. For a detailed discussion of this fact and related topics for the finite-dimensional case we refer to [1].

So, we first have to modify the concept of stability:

Definition. Let $u = \{u(t) \mid 0 \leq t \leq p\}$ be a nontrivial periodic trajectory of a nonlinear semigroup $T(\cdot)$ in X. We call u orbitally **stable** if for any neighborhood U of u there exists a neighborhood V of u so that $T(t)V \subset U$ for all $t \geq 0$. If, in addition $T(t)x$ tends to u as $t \to \infty$ for any $x \in V$, we call u asymptotically orbitally stable. In case u is not orbitally stable we call it orbitally unstable.

For this concept of stability, we get the following criterion:

Theorem 3.1. Let $T(\cdot)$ be a nonlinear semigroup in X and let $x(\cdot)$ be a nontrivial p-periodic orbit. Suppose that $T(p)$ is Fréchet-differentiable at $x(0)$ with $U = T'(p,[x(0)])$. Moreover, assume that X can be decomposed into a pair of U-invariant subspaces $X = X_1 \oplus X_2$, so that X_1 is one-dimensional, $U|_{X_1} = \text{id}$, and the restriction of U to X_2 has spectral radius less than 1. Then $x(\cdot)$ is asymptotically orbitally stable.

Proof. Let Π_i denote the projection of X onto X_i ($i = 1,2,$). We may assume that $\|y\| = \|\Pi_1 y\| + \|\Pi_2 y\|$ for all $y \in X$ so that $\|\Pi_i\| = 1$. Next, choose $x_1 \in X_1$, $x_1^* \in X^*$ such that

$$\Pi_1 y = \langle x_1^*, y\rangle x_1 \quad \text{with} \quad \|x_1\| = \|x_1^*\| = 1.$$

Let M stand for the maximum of the Lipschitz constants of $T(t)$ for $0 \leq t \leq p$. (If $T(\cdot)$ is only locally Lipschitz, we take the Lipschitz constants on a suitable neighborhood of $\{x(t) \mid 0 \leq t \leq p\}$.) To begin with, we prove the infinite-dimensional analogon to the transversality property, i.e. the function

$$\langle x_1^*, x(t) - x(0)\rangle \quad \text{changes sign at} \quad t = 0. \tag{3.1}$$

For $t \geq 0$ set $\delta(t) = \|x(t) - x(0)\|$. Given any real s, choose positive integers k and k^* such that $0 < kp - s < p$ and $0 < s - k^*p < p$, respectively. Then

$$\|x(s+t) - x(s)\| = \|T(s-k^*p)x(t) - T(s-k^*p)x(0)\| \leq M\delta(t)$$

$$\|x(t) - x(0)\| = \|T(kp-s)x(s+t) - T(kp-s)x(s)\|$$

$$\leq M\, \|x(s+t) - x(s)\| ,$$

and consequently,

$$\frac{1}{M}\,\delta(t) \leq \|x(t+s) - x(s)\| \leq M\delta(t). \tag{3.2}$$

As $x(\cdot)$ is a nontrivial orbit, there exist $\varepsilon > 0$ and $0 < \alpha < \alpha+\beta < p$ such that $\|x(t) - x(0)\| > \varepsilon$ for $\alpha \leq t \leq \alpha+\beta$. For $0 < t < \beta$ we select an integer k such that $\alpha < tk \leq \alpha+\beta$. Then

$$\varepsilon \leq \|x(kt) - x(0)\| \leq \sum_{j=0}^{k-1} \|x((j+1)t) - x(jt)\| \leq kM\delta(t) \leq \frac{p}{t}\,M\delta(t)$$

and thus we deduce that

$$\delta(t) \geq \frac{\varepsilon}{Mp}\,t. \tag{3.3}$$

Now, we choose an integer m sufficiently large so that

$$\|U^m\Pi_2\| < \frac{1}{10\ M^2} \tag{3.4}$$

and $\theta \in (0,\beta)$ such that $\delta(s) \leq \delta(\theta)$ for $0 \leq s \leq \theta$ and such that $\|y - x(0)\| \leq M\delta(\theta)$ implies

$$\|T(mp)y - x(0) - U^m(y-x(0))\| \leq \frac{1}{10\ M^2}\,\|y - x(0)\|\,. \tag{3.5}$$

Then we obtain for $|s| < \theta$

$$\|\Pi_2(x(s) - x(0))\| = \|\Pi_2(T(mp)x(s) - x(0))\|$$

$$\leq \|\Pi_2(T(mp)x(s) - x(0) - U^m(x(s) - x(0)))\| + \|\Pi_2U^m(x(s) - x(0))\|$$

$$\leq \frac{1}{10\ M^2}\,\|x(s) - x(0)\| + \frac{1}{10\ M^2}\,\|x(s) - x(0)\| = \frac{1}{5\ M^2}\,\|x(s) - x(0)\|\,.$$

As

$$\|\Pi_1(x(s) - x(0))\| \geq \|x(s) - x(0)\| - \|\Pi_2(x(s) - x(0))\|$$

we get

$$\|\Pi_1(x(s) - x(0))\| \geq \frac{4}{5}\,\|x(s) - x(0)\| \geq 4M^2\,\|\Pi_2(x(s) - x(0))\|. \tag{3.6}$$

In particular, we infer from (3.2) and (3.3) that for $s \neq 0$

$$|\langle x_1^*, x(s) - x(0)\rangle| = \|\Pi_1(x(s) - x(0))\| \geq \frac{4}{5M}\,\delta(|s|) > 0, \tag{3.7}$$

Thus the function $\langle x_1^*, x(s) - x(0)\rangle$ has its only zero for $s \in [-\theta,\theta]$ at $s = 0$. For $-\theta \leq t \leq 0$, we have

$$\frac{\delta(\theta)}{M} \leq \|x(t+\theta) - x(t)\| = \|T(mp)x(t+\theta) - T(mp)x(t)\|$$

$$= \|T(mp)x(t+\theta) - x(0) - U^m(x(t+\theta) - x(0)) - T(mp)x(t) + x(0)$$

$$+ U^m(x(t) - x(0)) + \Pi_1 U^m(x(t+\theta) - x(t)) + \Pi_2 U^m(x(t+\theta) - x(t))\|$$

$$\leq \frac{1}{10\ M^2}\|x(t+\theta) - x(0)\| + \frac{1}{10\ M^2}\|x(t) - x(0)\|$$

$$+ |\langle x_1^*, x(t+\theta) - x(t)\rangle| + \frac{1}{10\ M^2}\|x(t+\theta) - x(t)\|$$

$$\leq \frac{3\delta(\theta)}{10\ M} + |\langle x_1^*, x(t+\theta) - x(t)\rangle|,$$

and thus

$$|\langle x_1^*, x(t+\theta) - x(t)\rangle| \geq \frac{7\delta(\theta)}{10\ M} > 0.$$

So the function $\langle x_1^*,x(t+\theta) - x(t)\rangle$ has constant sign for $-\theta \leq t \leq 0$, and in particular

$$\text{sign} \langle x_1^*,x(0) - x(-\theta)\rangle = \text{sign} \langle x_1^*,x(\theta) - x(0)\rangle,$$

and consequently, $\langle x_1^*,x(s) - x(0)\rangle$ changes sign in $s = 0$. Thus the transversality property is proven. It should be pointed out, that (3.1) and (3.7) imply that for each $\rho \in [-\frac{4}{5M}\delta(\theta), \frac{4}{5M}\delta(\theta)]$ there exists some $s \in [-\theta,\theta]$ with $\langle x_1^*, x(s) - x(0)\rangle = \rho$. As a consequence, we have: For any $z \in X$ with $\|z\| \leq \frac{4}{5M}\delta(\theta)$ there exists some $s \in [-\theta,\theta]$ with

$$\Pi_1(x(s) - x(0)) = \Pi_1 z. \tag{3.8}$$

We are now in the position to prove the theorem. We set $\varepsilon = \frac{4}{5M^2}\delta(\theta)$ and verify the following statement:

If $y \in X$ is such that for some $t \in [0,p]$ we have $\|y - x(t)\| \leq \varepsilon$ then

$$\inf_{s\in[0,p]} \|T((m+1)p)y - x(s)\| \leq \frac{1}{2}\|y - x(t)\|.$$

It is clear that this implies that x is asymptotically orbitally stable. We first put $z = T(p-t)y - x(0)$. As

$$\|z\| = \|T(p-t)y - T(p-t)x(t)\| \leq M\,\|y - x(t)\| \leq \frac{4}{5M}\,\delta(\theta),$$

we may choose $s \in [-\theta,\theta]$ according to (3.8) such that

$$\Pi_1(x(s) - x(0)) = \Pi_1 z.$$

Now we obtain

$$\|T((m+1)p)y - x(s+t)\| \leq M\,\|(T(m+1)p - t)y - x(s)\|$$

$$\leq M(\,\|(T(m+1)p - t)y - x(0) - U^m z\| + \|\Pi_1(x(0) - x(s) + U^m z)\|$$

$$+ \|\Pi_2(x(0) - x(s))\| + \|\Pi_2 U^m z\|\,)$$

By (3.5)(which can be applied with y replaced by $T(p-t)y$ as $\|T(p-t)y - x(0)\| \leq \frac{4}{5M}\,\delta(\theta) \leq M\delta(\theta)$), we deduce that

$$M\|T((m+1)p - t)y - x(0) - U^m z\| \leq \frac{M}{10\,M^2}\,\|z\| \leq \frac{1}{10}\,\|y - x(t)\|\,.$$

From the choice of s we have

$$\Pi_1(x(0) - x(s) + U^m z) = 0.$$

By (3.6)

$$M\|\Pi_2(x(0) - x(s))\| \leq \frac{1}{4M}\|\Pi_1(x(0) - x(s))\| = \frac{1}{4M}\,\|\Pi_1 z\|$$

$$\leq \frac{1}{4}\,\|y - x(t)\|\,.$$

By (3.4)

$$M\|\Pi_2 U^m z\| \leq \frac{M}{10\,M^2}\,\|z\| \leq \frac{1}{10}\,\|y - x(t)\|\,.$$

Summing up these estimates, we obtain finally

$$\|T((m+1)py - x(t+s)\| \leq \left(\frac{2}{10} + \frac{1}{4}\right)\|y - x(t)\| \leq \frac{1}{2}\,\|y - x(t)\|$$

as claimed.

Remark. Suppose that $T(\cdot)$ is a nonlinear semigroup in X and $x(\cdot)$ is a p-periodic orbit such that for each nonnegative s and t $T(t)$

is Fréchet-differentiable at $x(s)$ with Fréchet-derivative with $T'(t,[x(s)]) = U(t,s)$. If X can be decomposed into $X = X_1 \oplus X_2$ as in the previous theorem, then there exists such a decomposition $X = X_1(t) \oplus X_2(t)$ into $U(p+t,t)$-invariant subspaces for all $t \geq 0$.

Proof. First note, that from the chain-rule we get

$$U(t,s)U(s,r) = U(t,r).$$

Now, choose x_1 and x_1^* as in the proof of the previous theorem. For $0 \leq t \leq p$ we set

$$x_1(t) = U(t,0)x_1 \quad \text{and} \quad x_1^*(t) = U^*(p,t)x_1^*.$$

(For t greater than p we perform a periodic continuation.) Put

$$X_1(t) = \text{span } x_1(t) = U(t,0)X_1 \quad \text{and}$$

$$X_2(t) = \ker x_1^*(t) = \ker U^*(p,t)x_1^*(0).$$

Note that the invariance of X_1 and X_2 impy that

$$U^*(p,0)x_1^*(0) = x_1^*(0)$$

$x \in X_1(t) \cap X_2(t)$ implies that

$$x = \lambda U(t,0)x_1(0) \quad \text{and} \quad \langle x_1^*, U(p,t)x\rangle = 0$$

and hence $0 = \langle x_1^*, \lambda U(p,t)U(t,0)x_1(0)\rangle = \lambda$, and thus $x = 0$, i.e. $X_1(t) \cap X_2(t) = \{0\}$. As $U^*(p,t)x_1^*|_{X_1(t)} \neq 0$ (since $\langle U^*(p,t)x_1^*, U(t,0)x_1\rangle = 1$), we conclude that $X_1(t) + X_2(t) = X$. So, it remains to verify that both, $X_1(t)$ and $X_2(t)$ are invariant with respect to $U(t+p,t)$. This fact follows easily as

$$U(p+t,t)x_1(t) = x_1(t) \quad \text{and} \quad \langle x_1^*, U(p,t)x\rangle = 0$$

implies that

$$\langle x_1^*, U(p,t)U(p+t,t)x\rangle = \langle x_1^*, U(2p,t+p)U(t+p,t)x\rangle$$

$$= \langle x_1^*, U(2p,p)U(p,t)x\rangle = \langle x_1^*, U(p,t)x\rangle = 0.$$

In passing, we note that even the stronger property

$$U(s,t)X_i(t) \subset X_i(s)$$

holds, since

$$U(s,t)x_1(t) = U(s,t)U(t,0)x_1(0) = U(s,0)x_1(0) = x_1(s)$$

and

$$\langle x_1^*, U(p,t)x\rangle = 0 \quad \text{implies that} \quad \langle x_1^*, U(p,s)U(s,t)x\rangle = 0.$$

Theorem 3.2. Let $T(\cdot)$ be a nonlinear semigroup on X and suppose that there exists an $x_0 \in X$ such that $T(p)x_0 = x_0$ for some $p > 0$. Let $T(p)$ be Fréchet-differentiable at x_0 with Fréchet derivative denoted by U and assume that there exists a decomposition of X as $X = X_1 \oplus X_2$ with projections $\pi_i : X \to X_i$, $i = 1,2$, such that

$$\|U^k\pi_1 x\| > \theta_1\|\pi_1 x\| \quad \text{and} \quad \|U^k\pi_2 x\| < \theta_2\|\pi_2 x\|$$

for some $k \geq 1$ with $1 \leq \theta_2 < \theta_1$ and $U^k X_i \subset X_i$. Then the orbit $\gamma = \{T(t)x_0;\ 0 \leq t \leq p\}$ is not orbitally stable, in particular, there exists some $\varepsilon > 0$ and a sequence (y_0^n) converging to x_0 such that for some sequence m_n of integers we have

$$\operatorname{dist}(\gamma, T(m_n p)y_0^n) \geq \varepsilon.$$

Proof. Let M be the Lipschitz constant of $T(p)$. We choose ε sufficiently small so that $\|y - x_0\| < 8M\varepsilon$ implies

$$\|T(p)y - x_0 - U(y-x_0)\| < \rho\|y - x_0\| \quad \text{with} \quad \rho = \tfrac{1}{8}(\theta_1-\theta_2).$$

In particular, we obtain in case $\|T(t)x_0 - x_0\| < 8M\varepsilon$

$$\|T(t)x_0 - x_0 - U(T(t)x_0 - x_0)\| < \rho\|T(t)x_0 - x_0\|$$

and consequently,

$$\begin{aligned}\rho\,\|T(t)x_0 - x_0\| &\geq -\|\pi_1(T(t)x_0 - x_0)\| + \|U\pi_1(T(t)x_0 - x_0)\| \\ &\geq (\theta_1-\theta_2)\,\|\pi_1(T(t)x_0 - x_0)\|,\end{aligned}$$

i.e. $\|\pi_1(T(t)x_0 - x_0)\| \le \frac{1}{8}\|T(t)x_0 - x_0\|$. Take now y_0 such that $\|x_0 - y_0\| < \varepsilon$ and $\|\pi_1(y_0 - x_0)\| \ge \|\pi_2(y_0 - x_0)\|$ and suppose that for all positive integers m we have

$$\operatorname{dist}(T(mp)y_0,\gamma) < \varepsilon.$$

In the sequel we set $y_m = T(mp)y_0$ and put $\kappa = \frac{1}{2}(\theta_1+\theta_2)$. By induction, we shall show that

$$\|y_m - x_0\| < 4M\varepsilon,\ \|\pi_1(y_m - x_0)\| \ge \|\pi_2(y_m - x_0)\|,$$

$$\operatorname{dist}(y_m,\gamma) \ge \tfrac{1}{2}\|\pi_1(y_m - x_0)\|,\ \|\pi_1(y_m - x_0)\| \ge \kappa^m\|\pi_1(y_0 - x_0\|.$$

(Which in an obvious contradiction to $\operatorname{dist}(y_m,\gamma) < \varepsilon$ for large m.) So, suppose that $\|y_m - x_0\| < 4M\varepsilon$, $\|\pi_1(y_m - x_0)\| \ge \|\pi_2(y_m - x_0)\|$ and $\|\pi_1(y_m - x_0)\| \ge \kappa^m\|\pi_1(y_0 - x_0)\|$. (Note, that these inequalities are valid for $m = 0$.) Then, let $t > 0$ be chosen so that $\|T(t)x_0 - y_m\| = \operatorname{dist}(y_m,\gamma)$. We obtain

$$\begin{aligned}\|T(t)x_0 - y_m\| &\ge \|\pi_1(T(t)x_0 - y_m)\| \\ &\ge \|\pi_1(y_m - x_0)\| - \|\pi_1(T(t)x_0 - x_0)\| \\ &\ge \|\pi_1(y_m - x_0)\| - \tfrac{1}{8}\|T(t)x_0 - x_0\| \\ &\ge \tfrac{1}{2}\|\pi_1(y_m - x_0)\|,\end{aligned}$$

(where we used the estimate

$$\|T(t)x_0 - x_0\| \le \|y_m - T(t)x_0\| + \|y_m - x_0\| \le 2\|y_m - x_0\|.)$$

Therefore, $\operatorname{dist}(y_m,\gamma) \ge \frac{1}{2}\|\pi_1(y_m - x_0)\|$. In particular, $\|y_m - x_0\| \le 2\|\pi_1(y_m - x_0)\| \le 4\operatorname{dist}(y_m,\gamma) \le 4\varepsilon$, so that $\|y_{m+1} - x_0\| \le 4M\varepsilon$. We next compute

$$\begin{aligned}\|\pi_1(y_{m+1} - x_0)\| &\ge -\|\pi_1(y_{m+1} - x_0 - U(y_m-x_0)\| + \|\pi_1U(y_m-x_0)\| \\ &\ge \theta_1\|\pi_1(y_m - x_0)\| - \tfrac{1}{8}(\theta_1 - \theta_2)\|y_m - x_0\| \\ &\ge (\theta_1 - \tfrac{1}{4}(\theta_1-\theta_2))\|\pi_1(y_m-x_0)\|\end{aligned}$$

$$> \kappa \|\pi_1(y_m - x_0)\| > \kappa^{m+1} \|\pi_1(y_0 - x_0)\| .$$

On the other hand, we get

$$\|\pi_2(y_{m+1} - x_0)\| \le \|\pi_2 U(y_m - x_0)\| + \|\pi_2(y_{m+1} - x_0 - U(y_m - x_0)\|$$
$$\le \theta_2 \|\pi_2(y_m - x_0)\| + \frac{1}{8}(\theta_1 - \theta_2) \|y_m - x_0\|$$
$$\le \kappa \|\pi_1(y_m - x_0)\| \le \|\pi_1(y_{m+1} - x_0)\|$$

and hence the induction is valid. Choosing now any sequence (y^n) satisfying $y^n \to x_0$ and $\|\pi_1(y^n - x_0)\| > \|\pi_2(y^n - x_0)\|$ we obtain the claim.

References.

[1] Amann, H.: Gewöhnliche Differentialgleichungen, De Gruyter, Berlin-New York 1983.

[2] Brezis, H.: Operateurs maximaux monotones et semigroups de contractions dans les espaces de Hilbert, North Holland 1973.

[3] Crandall, M., P. Rabinowitz: Mathematical theory of bifurcation in "Bifurcation Phenomena in Mathematical Physics", C. Bardos, D. Bessis, eds., NSI, 1980, 3-46.

[4] Hale, J., L. Magalhaes, W. Oliva: An introduction to infinite dimensional dynamical systems - Geometric theory, Applied Math. Sciences, 47, Springer 1984.

[5] Henry, D.: Geometric theory of semilinear parabolic equations, Springer Lecture Notes 840, 1981.

[6] Pazy, A.: Semigroups of linear operators and applications to partial differential equations, Applied Math. Sciences 44, Springer 1983.

[7] Schappacher, W.: Asymptotic behavior of linear semigroups, Quaderni Bari, 1983.

[8] Webb, G.F.: Theory of age-dependent population dynamics, to appear 1985.

ON A CLASS OF SEMILINEAR PARABOLIC EQUATIONS IN L^1

G. Di Blasio
Dipartimento di Matematica
Università di Roma 'La Sapienza'
P.le A. Moro 2, 00185 Roma, Italy

INTRODUCTION

Let $\Omega \subset R^n$ be an open bounded set with smooth boundary $\partial\Omega$ and let L be a linear second order elliptic operator on Ω. In this paper we study the problem

$$(P)\quad \begin{cases} u_t(t,x) + Lu(t,x) + \gamma(x,u(t,x)) = v(t,x) \\ u(t,x) = 0 \quad , \quad t > 0 \quad , \quad x \in \partial\Omega \\ u(0,x) = u_o(x) \quad , \quad x \in \Omega \end{cases}$$

where $\gamma(x,.)$ is an increasing function in R for a.e. $x \in \Omega$, $v \in L^1(\,]0,T[\, \times \Omega)$ and $u_o \in L^1(\Omega)$. Under the assumption that $|\gamma(x,u)| \leq a(x)\phi(u)$ with $a \in L^1(\Omega)$ and ϕ bounded for bounded u we prove that there exists a unique generalized solution of (P). Moreover we give conditions upon v and u_o guaranteeing the existence of regular solutions of (P). These results generalize previous ones obtained by the author in [5].

Problems of similar kind where studied by several authors if γ does not depend on x. If γ depends also on x we recall the paper of Brézis and Strauss [2] where it is proved the existence of generalized solutions of (P) if $|\gamma(x,u)| \leq a(x)\phi(u)$ with $a \in L^2(\Omega)$ and ϕ bounded for bounded u.

1. STATEMENT OF THE PROBLEM

Let $\Omega \subset R^n$ be an open bounded set with smooth boundary $\partial\Omega$. On Ω we consider a second order uniformly elliptic differential operator of the form

$$Lu = \sum_{h,k}^{1,n} \frac{\partial}{\partial x_h}\left(a_{h,k}\frac{\partial}{\partial x_k} u\right) + \sum_{h=1}^{n} \frac{\partial}{\partial x_h} a_h u + a_0 u$$

where $a_{h,k}$, $a_h \in C^1(\bar{\Omega})$ and $a_0 \in C(\bar{\Omega})$. Next denote by E the operator defined as

$$D(E) = \{u \in W_0^{1,1}(\Omega) : Lu \in L^1(\Omega)\} \quad ; \; Eu = Lu$$

where Lu is understood in the sense of distributions. It is known (see [1] and [8]) that $-E$ generates an analytic semigroup in $L^1(\Omega)$.

Further let $\gamma : \Omega \times R \to R$, $(x,u) \to \gamma(x,u)$ be a function satisfying the properties

(i) $\gamma(.,u)$ is measurable for each $u \in R$

(ii) $\gamma(x,.)$ is continuous and increasing and $\gamma(x,0) = 0$, for a.e. $x \in \Omega$

(iii) $|\gamma(x,u)| \leq a(x)\phi(u)$, with $a \geq 0$, $a \in L^1(\Omega)$ and ϕ bounded for bounded u

We shall denote by g the operator defined by

$$D(g) = \{u \in L^1(\Omega) : \gamma(.,u(.)) \in L^1(\Omega)\}, \quad g(u)(x) = \gamma(x,u(x))$$

Given $f \in L^1(0,T;L^1(\Omega))$ and $u_0 \in L^1(\Omega)$ we shall prove the existence of generalized solutions for the problem

$$(1) \quad \begin{cases} u' + Eu + g(u) = f(t) \\ u(0) = u_0 \end{cases}$$

Moreover we shall give conditions on f and u_o guaranteeing the existence of regular solutions of (1).

2. THE APPROXIMATING PROBLEM

We begin with some facts and definitions which will be used in the sequel.

For $1 \leq p \leq \infty$ we set $X_p = L^p(\Omega)$ and denote by $|.|_p$ the norm in X_p. Moreover for $1 \leq p < \infty$ we denote by X_p^* the dual space of X_p and by $<u,u^*>_p$ the value of $x^* \in X_p^*$ at $u \in X_p$. Further for $1 \leq p < \infty$ we denote by $\partial|.|_p$ the subdifferential of the norm in X_p, i.e. the function (which is multivalued if $p = 1$)

$$\partial|u|_p = \{u^* \in X_p^* : |u + v|_p \geq |u|_p + <v,u^*>_p\}$$

Next let $F : D(F) \subset X_p \to X_p$. We say that F is accretive if for each $\lambda > 0$ and $u,v \in D(F)$ we have

$$|u - v|_p \leq |u - v + \lambda(F(u) - F(v))|_p$$

We recall that if F is accretive then for each $u,v \in D(F)$ there exists $z_p \in \partial|u - v|_p$ verifying

(2) $\quad <F(u) - F(v), z_p>_p \geq 0$

moreover if F is accretive and continuous then (2) is verified for each $z_p \in \partial|u - v|_p$.

Now let us set for fixed $\varepsilon > 0$

$$\gamma_\varepsilon(x,u) = \frac{\gamma(x,u)}{1 + \varepsilon\,|\gamma(x,u)|/a(x)}$$

It is easy to see that

(i) $\quad |\gamma_\varepsilon(x,u)| \leq a(x)\varepsilon^{-1}$

(ii) $\quad \gamma_\varepsilon(x,.)$ is continuous and increasing

Therefore the operator $g_\varepsilon : X_1 \to X_1$ defined by

$$g_\varepsilon(u)(x) = \gamma_\varepsilon(x,u(x))$$

is continuous and accretive. Moreover it can be seen that

$$\lim_{\varepsilon \to 0} g_\varepsilon(u) = g(u)$$

uniformly for u on bounded subsets of X_∞. Next denote by $\gamma_\varepsilon^\lambda$ the function

$$\begin{aligned}(3)\qquad \gamma_\varepsilon^\lambda(x,u) &= (1/\lambda)(I - (I + \lambda\gamma_\varepsilon(x,.))^{-1})(u) = \\ &\gamma_\varepsilon(x,.)(I + \lambda\gamma_\varepsilon(x,.))^{-1}(u)\end{aligned}$$

where I is the identity. It is easy to see that $\gamma_\varepsilon^\lambda(x,.)$ is increasing and Lipschitz continuous with Lipschitz constant $2/\lambda$. Therefore the operator $g_\varepsilon^\lambda : X_1 \to X_1$ defined by

$$g_\varepsilon^\lambda(u)(x) = \gamma_\varepsilon^\lambda(x,u(x))$$

is Lipschitz continuous and accretive. Moreover we have

$$\lim_{\lambda \to 0} g_\varepsilon^\lambda(u) = g_\varepsilon(u)$$

for each $u \in X_1$.

We shall consider the following approximating problem

$$(4)\qquad \begin{cases} u' + Eu + g_\varepsilon^\lambda(u) = f(t) \\ u(0) = u_0 \end{cases}$$

together with the integrated version

$$(5)\qquad u(t) = T(t)u_0 + \int_0^t T(t - s)(-g_\varepsilon^\lambda(u(s)) + f(s))ds$$

where $T(t)$ is the semigroup generated by $-E$. To study (4) and (5) we need the intermediate spaces $D(\theta,1)$ between $D(E)$ and X_1 and the intermediate space $W^{\theta,1}(0,T;X_1)$ between $W^{1,1}(0,T;X_1)$ and $L^1(0,T;X_1)$ defined as follows ($\theta \in]0,1[$)

$$D(\theta,1) = \{x \in X_1 : |x|_{\theta,1} = \int_0^{+\infty} t^{-\theta} |ET(t)x|_1 dt < +\infty\}$$

endowed with the norm

$$|x|_{D(\theta,1)} = |x|_1 + |x|_{\theta,1}$$

and

$$W^{\theta,1}(0,T;X_1) = \{u \in L^1(0,T;X_1) : N_{\theta,1}(u) = \int_0^T \int_0^T \frac{|u(t) - u(s)|_1}{|t - s|^{1+\theta}} dsdt < +\infty\}$$

endowed with the norm

$$|u|_{W^{\theta,1}} = |u|_{L^1(X_1)} + N_{\theta,1}(u)$$

We have:

THEOREM 1. Let $f \in W^{\theta,1}(0,T;X_1)$ and $u_o \in D(\theta,1)$, for some $\theta \in]0,1[$. Then there exists a unique u satisfying the following properties

(i) $u \in L^1(0,T;D(E)) \cap W^{1,1}(0,T;X_1)$

(ii) $Eu, u' \in W^{\theta,1}(0,T;X_1)$; $u' \in L^1(0,T;D(\theta,1))$

(iii) u satisfies (4)

If in addition $f \in L^1(0,T;X_p)$ and $u_o \in X_p$ for some $p > 1$, then $u(t) \in X_p$ and denoting by ω the constant

$$\omega = \max \{\sup a_o , \sup (a_o + \sum_{h=1}^{n} \frac{\partial}{\partial x_h} a_h)\}$$

we have

(iv) $|u(t)|_p \leq \exp(\omega t)\ |u_o|_p + \int_o^t \exp(\omega(t-s))|f(s)|_p ds$

PROOF. Since g_ε^λ is Lipschitz continuous it can be proved by a fixed point argument that there exists a unique $u \in C(0,T;X_1)$ verifying (5). Since $T(t)$ is analytic we have that (see Theorem 1 (i) of the Appendix) $u \in W^{\alpha,1}(0,T;X_1)$ for each $\alpha \in]0,1[$. Therefore by the Lipschitz continuity of g_ε^λ we have

$$\int_o^T \int_o^T |g_\varepsilon^\lambda(u(t)) - g_\varepsilon^\lambda(u(s))|_1 |t-s|^{-1-\alpha} dsdt \leq 2/\lambda\ N_{\alpha,1}(u)$$

so that $g_\varepsilon^\lambda(u) \in W^{\alpha,1}(0,T;X_1)$, for each $\alpha \in]0,1[$. Therefore assertions (i), (ii) and (iii) follow from Theorem 1 (ii)-(iv) of the Appendix. To prove (iv) let us note that the restriction $-E_p$ of $-E$ to X_p generate an analytic semigroup for $p \in]1,\infty[$. Moreover the restriction of g_ε^λ to X_p is Lipschitz continuous and accretive. Therefore, by an argument similar to the one used above, it can be proved that if $f \in C^1(0,T;X_p)$ and $u_o \in C_o^2(\Omega)$ the solution of (4) belongs to $L^1(0,T;D(E_p)) \cap W^{1,p}(0,T;X_p)$, for $p \in]1,\infty[$. Moreover since $E_p + \omega I$ is accretive in X_p (see [1]) we have, for $z_p(t)\ \partial|u(t)|_p$

$$<E_p u(t) + \omega u(t), z_p(t)>_p \geq 0$$

so that

$$<E_p u(t), z_p(t)>_p \geq -\omega|u(t)|_p$$

Therefore from (4) we get

$$D_-|u(t)|_p \leq <u'(t), z_p(t)>_p \leq \omega|u(t)|_p + <f(t), z_p(t)>_p \leq$$

$$\omega|u(t)|_p + |f(t)|_p$$

which implies (iv) for $p < \infty$, and hence for $p = \infty$, for f and u_o regular. To complete the proof it suffices to approximate f and u_o with regular data and then to pass to the limit.

The following theorem concerns the dependence of the solutions of (4) upon the data.

THEOREM 2. Let $f,h \in W^{\theta,1}(0,T;X_1)$, let $u_o, v_o \in D(\theta,1)$ and let u,v be the corresponding solutions of (4). Then we have

(i) $$|u(t) - v(t)|_1 \leq \exp(\omega t)|u_o - v_o|_1 + \int_o^t \exp(\omega(t - s))|f(s) - h(s)|_1 ds$$

(ii) $$\int_o^t |g_\varepsilon^\lambda(u(s)) - g_\varepsilon^\lambda(v(s))|_1 ds \leq \exp(\omega t)|u_o - v_o|_1 + \int_o^t \exp(\omega(t - s))|f(s) - h(s))|_1 ds$$

PROOF. Since the function $\gamma_\varepsilon^\lambda(x,.)$ defined by (3) is increasing in R we have

$$(\gamma_\varepsilon^\lambda(x,u) - \gamma_\varepsilon^\lambda(x,v))\text{sign}(u - v) = |\gamma_\varepsilon^\lambda(x,u) - \gamma_\varepsilon^\lambda(x,v)|$$

Therefore there exists $z(t) \in \partial|u(t) - v(t)|_1$ such that

$$D_-|u(t) - v(t)|_1 \leq \langle u'(t) - v'(t), z(t)\rangle_1 =$$

$$\langle -Eu(t) + Ev(t) - g_\varepsilon^\lambda(u(t)) + g_\varepsilon^\lambda(v(t)) + f(t) - h(t), z(t)\rangle_1 \leq$$

$$\omega|u(t) - v(t)|_1 - |g_\varepsilon^\lambda(u(t)) - g_\varepsilon^\lambda(v(t))|_1 + |f(t) - h(t)|_1$$

so that

$$D_- \exp(-\omega t)|u(t) - v(t)|_1 \leq \exp(-\omega t)(-|g_\varepsilon^\lambda(u(t)) - g_\varepsilon^\lambda(v(t))|_1 + |f(t) - h(t)|_1)$$

and hence

$$\exp(-\omega t)|u(t) - v(t)|_1 \leq |u_o - v_o|_1 + \int_o^t \exp(-\omega s)(-|g_\varepsilon^\lambda(u(s)) - g_\varepsilon^\lambda(v(s))|_1 + |f(s) - h(s)|_1)ds$$

which implies (i) and (ii).

3. CONVERGENCE OF THE APPROXIMATE SOLUTIONS

We now investigate the behavior of the solutions of (4) when $\lambda \to 0$. We have:

THEOREM 3. Let $f \in W^{\theta,1}(0,T;X_1)$ and $u_0 \in D(\theta,1)$. Moreover denote by u_λ the solution of (4) given by theorem 1. Then there exists u satisfying the following properties

(i) $u_\lambda \to u$ in $C(0,T;X_1)$

(ii) u satisfies

(6) $$u(t) = T(t)u_0 + \int_0^t T(t - s)(-g_\varepsilon(u(s)) + f(s))ds$$

Moreover if $f,h \in W^{\theta,1}(0,T;X_1)$, if $u_0,v_0 \in D(\theta,1)$ and u,v are the corresponding solutions of (6) then

(iii) $$|u(t) - v(t)|_1 \leq \exp(\omega t)|u_0 - v_0|_1 + \int_0^t \exp(\omega(t - s))|f(s) - h(s)|_1 ds$$

(iv) $$\int_0^t |g_\varepsilon(u(s)) - g_\varepsilon(v(s))|_1 ds \leq \exp(\omega t)|u_0 - v_0|_1 + \int_0^t \exp(\omega(t - s))|f(s) - h(s)|_1 ds$$

If in addition $f \in L^1(0,T;X_p)$ and $u_0 \in X_p$ for some $p > 1$ then

(v) $$|u(t)|_p \leq \exp(\omega t)|u_0|_p + \int_0^t \exp(\omega(t - s))|f(s)|_p ds$$

PROOF. Let u_μ be the solution of (4) with λ replaced by μ. We have

$$(u_\lambda - u_\mu)' = -E(u_\lambda - \mu_\mu) - g_\varepsilon^\lambda(u_\lambda) + g_\varepsilon^\lambda(u_\mu) - (g_\varepsilon^\lambda - g_\varepsilon^\mu)(u_\mu)$$

so that there exists $z(t) \in \partial|u_\lambda(t) - u_\mu(t)|_1$ verifying

$$D_-|u_\lambda(t) - u_\mu(t)|_1 \le \langle u'_\lambda(t) - u'_\mu(t), z(t)\rangle_1 =$$

$$-\langle E(u_\lambda(t) - u_\mu(t)) + g_\varepsilon^\lambda(u_\lambda(t)) - g_\varepsilon^\lambda(u_\mu(t)), z(t)\rangle_1 +$$

$$\langle (g_\varepsilon^\lambda - g_\varepsilon^\mu)(u_\mu(t)), z(t)\rangle_1 \le$$

$$\omega|u_\lambda(t) - u_\mu(t)|_1 + |(g_\varepsilon^\lambda - g_\varepsilon^\mu)(u_\mu(t))|_1 \le$$

$$\omega|u_\lambda(t) - u_\mu(t)|_1 + |\mu - \lambda| |g_\varepsilon(I + \lambda g_\varepsilon)^{-1}(I + \mu g_\varepsilon)^{-1} g_\varepsilon(u_\mu(t))|_1$$

where we used (3). Hence from (i) of sect. 2 we get

$$|u_\lambda(t) - u_\mu(t)|_1 \le \text{const}|\lambda - \mu|$$

so that $\{u_\lambda\}$ is Cauchy in $C(0,T;X_1)$. Consequently there exists $u = \lim u_\lambda$ in $C(0,T;X_1)$ and we have

$$|g_\varepsilon^\lambda(u_\lambda(t)) - g_\varepsilon(u(t))|_1 \le$$

$$|g_\varepsilon^\lambda(u_\lambda(t)) - g_\varepsilon(u_\lambda(t))|_1 + |g_\varepsilon(u_\lambda(t)) - g_\varepsilon(u(t))|_1 \le$$

$$\lambda|g_\varepsilon(I + \lambda g_\varepsilon)^{-1}(u_\lambda(t))|_1 + |g_\varepsilon(u_\lambda(t)) - g_\varepsilon(u(t))|_1$$

from which it follows that $g_\varepsilon^\lambda(u_\lambda(t)) \to g_\varepsilon(u(t))$ in X_1. Since u satisfies (5) we have that u satisfies (6). Finally assertions (iii)-(v) are consequence of Theorem 1 (iv) and Theorem 2 (i) and (ii).

Next we investigate the behavior of the solutions of (6) when $\varepsilon \to 0$. We have:

THEOREM 4. Let $u_o \in D(\theta,1) \cap X_\infty$ and $f \in W^{\theta,1}(0,T;X_1) \cap L^1(0,T;X_\infty)$. Moreover denote by u_ε the solution of (6) given by Theorem 3. Then there exists u verifying

(i) $u_\varepsilon \to u$ in $C(0,T;X_1)$

(ii) $|u(t)|_\infty \le \exp(\omega t)|u_o|_\infty + \int_0^t \exp(\omega(t - s))|f(s)|_\infty ds$

(iii) $u(t) = T(t)u_o + \int_0^t T(t - s)(-g(u(s)) + f(s))ds$

Moreover if $f,h \in W^{\theta,1}(0,T;X_1) \cap L^1(0,T;X_\infty)$, if $u_o, v_o \in D(\theta,1) \cap X_\infty$ and u,v are the corresponding solutions of (iii), then

(iv) $$|u(t) - v(t)|_1 \le \exp(\omega t)|u_o - v_o|_1 + \\ + \int_o^t \exp(\omega(t - s))|f(s) - h(s)|_1 ds$$

(v) $$\int_o^t |g(u(s)) - g(v(s))|_1 \, ds \le \exp(-\omega t)|u_o - v_o|_1 + \\ \int_o^t \exp(\omega(t - s))|f(s) - h(s)|_1 ds$$

PROOF. Let u_ε be the solution of (6) and let u_η be the solution of (6) with ε replaced by η. We have

$$u_\eta(t) = T(t)u_o + \int_o^t T(t - s)(-g_\varepsilon(u_\eta(s)) + (g_\varepsilon - g_\eta)(u_\eta(s)) + f(s))ds$$

Therefore using Theorem 3 (iii) we get

$$|u_\varepsilon(t) - u_\eta(t)|_1 \le \int_o^t \exp(\omega(t - s))|(g_\varepsilon - g_\eta)(u_\eta(s))|_1 ds$$

Now property (v) of Theorem 3 implies that

$$|u_\eta(t)|_\infty \le \exp(\omega t)|u_o|_\infty + \int_o^t \exp(\omega(t - s))|f(s)|_\infty ds$$

so that from the definition of g_ε we have

$$|(g_\eta - g_\varepsilon)(u_\eta(s))|_1 \le \text{const}|\varepsilon - \eta|$$

Therefore $\{u_\varepsilon\}$ is Cauchy in $C(0,T;X_1)$. Consequently there exists $u = \lim u_\varepsilon$ in $C(0,T;X_1)$. Moreover we have

$$|u(t)|_\infty \le \exp(\omega t)|u_o|_\infty + \int_o^t \exp(\omega(t - s))|f(s)|_\infty ds$$

so that $u(t) \in D(g)$ and moreover

$$|g_\varepsilon(u_\varepsilon(t)) - g(u(t))|_1 \le |(g_\varepsilon - g)(u_\varepsilon(t))|_1 + |g(u_\varepsilon(t)) - g(u(t))|_1 \to 0$$

as $\varepsilon \to 0$, since g is continuous on bounded subsets of X_∞. Therefore from (6) we get that u satisfies (iii). Finally assertions (ii), (iv) and (v) are consequence of Theorem 3 (iii)-(iv).

4. EXISTENCE AND REGULARITY RESULTS

Using the results of the preceding section we are able to study the problem

$$(7)\quad \begin{cases} u'(t) + Eu(t) + g(u(t)) = f(t) \\ u(0) = u_o \end{cases}$$

together with the integrated version

$$(8)\quad u(t) = T(t)u_o + \int_o^t T(t - s)(-g(u(s)) + f(s))ds$$

We begin with the following result.

THEOREM 5. Let $u_o \in X_1$ and $f \in L^1(0,T;X_1)$. Then there exists a unique $u \in C(0,T;X_1)$ such that $u(t) \in D(g)$ for a.e. $t \in]0,T[$, satisfying (8). Moreover we have

(i) $u \in L^1(0,T;D(\theta,1)) \cap W^{\theta,1}(0,T;X_1)$, for each $\theta \in]0,1[$

(ii) $|u(t)|_1 \leq \exp(\omega t)|u_o|_1 + \int_o^t \exp(\omega(t - s))|f(s)|_1 ds$

Furthermore if $u_o, v_o \in X_1$, if $f,h \in L^1(0,T;X_1)$ and u,v are the corresponding solutions of (8) then

(iii) $|u(t) - v(t)|_1 \leq \exp(\omega t)|u_o - v_o|_1 + \int_o^t \exp(\omega(t - s))|f(s) - h(s)|_1 ds$

and

(iv) $\int_o^t |g(u(s)) - g(v(s))|_1 ds \le \exp(\omega t)|u_o - v_o|_1 +$

$$\int_o^t \exp(\omega(t - s))|f(s) - h(s)|_1 ds$$

PROOF. Let $\{u_{o,n}\} \subset D(\theta,1) \cap X_\infty$ and $\{f_n\} \subset W^{\theta,1}(0,T;X_1) \cap L^1(0,T;X_\infty)$ be such that $u_{o,n} \to u_o$ in X_1 and that $f_n \to f$ in $L^1(0,T;X_1)$. Moreover let u_n be the function given by Theorem 4 (iii), solution of the equation

$$(9) \quad u_n(t) = T(t)u_{o,n} + \int_o^t T(t - s)(-g(u_n(s)) + f_n(s))ds$$

By Theorem 4 (iv) and (v) we have

$$|u_n(t) - u_m(t)|_1 \le \exp(\omega t)|u_{o,n} - u_{o,m}|_1 + \int_o^t \exp(\omega(t - s))|f_n(s) - f_m(s)|_1 ds$$

and

$$\int_o^t |g(u_n(s)) - g(u_m(s))|_1 ds \le \exp(\omega t)|u_{o,n} - u_{o,m}|_1 + \int_o^t \exp(\omega(t - s))|f_n(s) - f_m(s)|_1 ds$$

so that $\{u_n\}$ is Cauchy in $C(0,T;X_1)$ and $\{g(u_n)\}$ is Cauchy in $L^1(0,T;X_1)$. Consequently there exist $u = \lim u_n$ and $w = \lim g(u_n)$. These properties and the continuity of $\gamma(x,.)$ imply that there exists a subsequence n_k such that $\gamma(x,u_{n_k}(t)(x)) \to \gamma(x,u(t)(x))$ for a.e. $(t,x) \in]0,T[\times \Omega$. This in turn implies that $w = g(u)$ and hence $u(t) \in D(g)$ for a.e. $t \in]0,T[$ and $g(u_n) \to g(u)$ in $L^1(0,T;X_1)$. Finally letting $n \to \infty$ we obtain from (9) that u satisfies (8) and assertion (i) (see Theorem 1 (i) of the appendix). To prove uniqueness let $u_o, v_o \in X_1$, let $f,h \in L^1(0,T;X_1)$ and let u,v be solutions of (8). Set

$$\hat{f} = -g(u) + f \quad , \quad \hat{h} = -g(v) + h$$

moreover let $\{u_{o,n}\}$, $\{v_{o,n}\} \subset D(E)$ and $\{f_n\}, \{h_n\} \subset C^1(0,T;X_1)$ be such that

$$u_{o,n} \to u_o \text{ and } v_{o,n} \to v_o \text{ , in } X_1$$

and

$$f_n \to \hat{f} \text{ and } h_n \to \hat{h} \text{ , in } L^1(0,T;X_1).$$

Since the data $(u_{o,n}, f_n)$ and $(v_{o,n}, h_n)$ are regular there exist $u_n, v_n \in C^1(0,T;X_1) \cap C(0,T;D(E))$ verifying

$$(10) \quad \begin{cases} u' + Eu_n = f_n \\ u_n(0) = u_{o,n} \end{cases} \qquad \begin{cases} v'_n + Ev_n = h_n \\ v_n(0) = v_{o,n} \end{cases}$$

Since $E + \omega I$ is accretive in X_1 there exists $z_n(t) \in \partial|u_n(t) - v_n(t)|_1$ verifying

$$\langle Eu_n(t) - Ev_n(t), z_n(t)\rangle_1 \geq -\omega|u_n(t) - v_n(t)|_1$$

Therefore from (10) we have

$$\begin{aligned} D_-|u_n(t) - v_n(t)|_1 &\leq \langle u'_n(t) - v'_n(t), z_n(t)\rangle_1 \leq \\ &\omega|u_n(t) - v_n(t)|_1 + \langle f_n(t) - h_n(t), z_n(t)\rangle_1 \leq \\ &\omega|u_n(t) - v_n(t)|_1 + \langle \hat{f}(t) - \hat{h}(t), z_n(t)\rangle_1 + \\ &|f_n(t) - \hat{f}(t)|_1 + |h_n(t) - \hat{h}(t)|_1 \end{aligned}$$

from which it follows that

(11) $|u_n(t) - v_n(t)|_1 \leq \exp(\omega t)|u_{o,n} - v_{o,n}|_1 +$

$$\int_o^t \exp(\omega(t - s))(<\hat{f}(s) - \hat{h}(s), z_n(s)>_1 + |f_n(s) - \hat{f}(s)|_1 +$$

$$|h_n(s) + \hat{h}(s)|_1)ds$$

Now property $u_n - v_n \to u-v$ in $C(0,T;X_1)$ implies (see e.g. [3]) that there exist $\{n_k\}$ and $z(t) \in \partial|u(t) - v(t)|_1$ such that

$$\int_o^t \exp(\omega(t - s))<\hat{f}(s) - \hat{h}(s), z_{n_k}(s)>_1 ds \to$$

$$\int_o^t \exp(\omega(t - s))<\hat{f}(s) - \hat{h}(s), z(s)>_1 ds \leq$$

$$\int_o^t \exp(\omega(t - s))(|f(s) - h(s)|_1 - |g(u(s)) - g(v(s))|_1\, ds)$$

Therefore using (11) for n replaced by n_k and passing to the limit we get that u and v satisfy (iii) and (iv). Since the uniqueness and (ii) are consequence of (iii) the proof is complete.

Finally concerning problem (7) we have the following result.

THEOREM 6. Let $u_o \in D(\theta,1) \cap X_\infty$ and $f \in W^{\theta,1}(0,T;X_1) \cap L^1(0,T;X_\infty)$, for some $\theta \in]0,1[$. Then there exists a unique u satisfying the properties

(i) $u \in L^1(0,T;D(E)) \cap W^{1,1}(0,T;X_1) \cap L^1(0,T;D(g))$

(ii) $Eu, u' \in W^{\theta,1}(0,T;X_1)$, $u' \in L^1(0,T;D(\theta,1))$

(iii) u satisfies (7).

PROOF. Let u be the solution of (8) given by Theorem 5. We shall prove that the function $g(u(.))$ belongs to $W^{\theta,1}(0,T;X_1)$.

Using Theorem 5 (iv) with $v_o = u(\sigma)$ and $h(s) = f(s + \sigma)$, for fixed $\sigma \in]0,T-t[$, we have

$$(12)\quad \int_o^t |g(u(s+\sigma)) - g(u(s))|_1 ds \le \exp(\omega T)(|u_o - u(\sigma)|_1 +$$

$$\int_o^t |f(s+\sigma) - f(s)|_1 ds)$$

Now

$$N_{\theta,1}(g(u)) \le 2\int_o^T dt \int_o^t |g(u(t)) - g(u(s))|_1 (t-s)^{-1-\theta} ds =$$

$$2\int_o^T dt \int_o^t |g(u(t)) - g(u(t-s))|_1 s^{-1-\theta} ds =$$

$$2\int_o^T s^{-1-\theta} ds \int_o^{T-s} |g(u(t+s)) - g(u(t))|_1 dt$$

so that from (12)

$$N_{\theta,1}(g(u)) \le 2\int_o^T s^{-1-\theta} ds \exp(\omega T)(|u(s) - u_o|_1 +$$

$$\int_o^{T-s} |f(t+s) - f(t)|_1 dt) \le$$

$$\text{const}(\int_o^T s^{-1-\theta}|u(s) - u_o|_1 ds + N_{\theta,1}(f))$$

Moreover from (8)

$$\int_o^T s^{-1-\theta}|u(s) - u_o|_1 ds \le \int_o^T s^{-1-\theta}|T(s)u_o - u_o|_1 ds +$$

$$\int_o^T s^{-1-\theta}\int_o^s |T(t-\sigma)(-g(u(\sigma)) + f(\sigma))|_1 d\sigma = I_1 + I_2$$

Now we have

$$I_1 = \int_o^T |\int_o^s ET(\sigma)u_o d\sigma|_1 s^{-1-\theta} ds \le \int_o^T ds \int_o^s |ET(\sigma)u_o|_1 {}^{-1-\theta} d\sigma \le$$

$$T|u_o|_{\theta,1}$$

moreover

$$I_2 \leq \exp(\omega T)(\int_0^T s^{-1-\theta} ds \int_0^s (|g(u(\sigma))|_1 + |f(\sigma)|_1) d\sigma = J_1 + J_2$$

Furthermore from property (iii) of sect. 1 and Theorem 4 (ii) we have

$$J_1 \leq \text{const} \int_0^T s^{-1-\theta} ds \int_0^s d\sigma$$

Finally

$$J_2 \leq \text{const}(\int_0^T s^{-1-\theta} ds \int_0^s (|f(\sigma) - f(s)|_1 + |f(s)|_1) d\sigma \leq$$

$$\text{const}(N_{\theta,1}(f) + \int_0^T |f(s)|_1 s^{-\theta} ds) \leq \text{const} |f|_{W^{\theta,1}(X_1)}$$

where we used the fact (see e.g. [6, Lemma 7 of the Appendix])

$$\int_0^{+\infty} |f(s)|_1 s^{-\theta} ds \leq \text{const} |f|_{W^{\theta,1}(X_1)}$$

Hence we have proved that the solution of (8) satisfies the property $g(u(.)) \in W^{\theta,1}(0,T;X_1)$. Therefore assertions (i)-(iii) follow from the regularity results of parabolic equations (see Theorem 1 of the appendix).

APPENDIX

In this section we recall the results concerning linear parabolic equations which are used in the preceding sections. For the proofs we refer to [4] and [6].

Let X be a Banach space with norm $\|.\|$ and let $A : D(A) \subset X \to X$ be the infinitesimal generator of an analytic semigroup $\exp(tA)$ on X. For $\theta \in]0,1[$ set

$$|x|_{\theta,1} = \int_0^{+\infty} \|A \exp(tA)x\| t^{-\theta} dt$$

and denote by $D_A(\theta,1)$ the space

$$D_A(\theta,1) = \{x \in X : |x|_{\theta,1} < +\infty\}$$

endowed with the norm

$$|x|_{D_A(\theta,1)} = \|x\| + |x|_{\theta,1}$$

It is easy to see that $D(A) \hookrightarrow D_A(\theta,1) \hookrightarrow X$ and that $D_A(\theta,1) \hookrightarrow D_A(\hat{\theta},1)$ for $\hat{\theta} < \theta$.

Furthermore for $u \in L^1(0,T;X)$ and $\theta \in]0,1[$ set

$$N_{\theta,1}(u) = \int_0^T \int_0^T \|u(t) - u(s)\| |t - s|^{-1-\theta} dsdt$$

and denote by $W^{\theta,1}(0,T;X)$ the Sobolev space

$$W^{\theta,1}(0,T;X) = \{u \in L^1(0,T;X) : N_{\theta,1}(u) < +\infty\}$$

endowed with the norm

$$|u|_{W^{\theta,1}(X)} = |u|_{L^1(X)} + N_{\theta,1}(u)$$

It can be seen that $W^{1,1}(0,T;X) \hookrightarrow W^{\theta,1}(0,T;X) \hookrightarrow L^1(0,t;X)$.

The spaces $D_A(\theta,1)$ and $W^{\theta,1}(0,T;X)$ play an importan role in studying the abstract parabolic equation

$$(1) \quad \begin{cases} u' = Au + f(t) \\ u(0) = u_o \end{cases}$$

with $f \in L^1(0,T;X)$ and $u_o \in X$. We have

THEOREM 1. Let $f \in L^1(0,T;X)$, let $u_o \in X$ and set

$$u(t) = \exp(tA)u_o + \int_0^t \exp((t - s)A)f(s)ds$$

then for each $0 < \varepsilon \leq \theta < 1$ we have

(i) $u \in L^1(0,T;D_A(\theta,1)) \cap W^{\theta,1}(0,T;X) \cap W^{\theta-\varepsilon,1}(0,T;D_A(\varepsilon,1))$

If $f \in W^{\theta,1}(0,T;X)$ and $u_o \in D_A(\theta,1)$ for some $\theta \in]0,1[$, then we have

(ii) $u \in W^{1,1}(0,T;X) \cap L^1(0,T;D_A)$

(iii) $u', Au \in W^{\theta,1}(0,T;X)$, $u' \in L^1(0,T;D_A(\theta,1))$

(iv) u satisfies (1).

REFERENCES

[1] H. Amann, Dual semigroups and second order linear elliptic boundary value problems, Israel J. Math. 45, 225-254 (1983)

[2] H. Brézis and W. Strauss, Semilinear second order elliptic equations in L_1, J. Math. Soc. Japan 25, 565-590 (1973)

[3] M.G. Crandall, The semigroup approach to first order quasilinear equations in several space variable, Israel J. Math. 12, 108-132 (1972)

[4] G. Da Prato and P. Grisvard, Somme d'opérateurs linéaires et équations différentielles opérationelles, J. Math. Pures Appl. 54, 305-387 (1975)

[5] G. Di Blasio, Perturbations of second order elliptic operators and semi-linear evolution equations, Nonlinear Anal. TMA 3, 293-304 (1977)

[6] G. Di Blasio, Linear parabolic evolution equations in L^p-spaces, Ann. Mat. Pura Appl. IV, 55-104 (1984)

[7] F.J. Massey, Semilinear parabolic equations with L^1 initial data, Indiana Univ. Math. J. 26, 399-412 (1977)

[8] A. Pazy, Semigroups of linear operators and application to partial differential equations, Applied Math. Sc. 44, Springer-Verlag, New York 1983

ON A SINGULAR NON-AUTONOMOUS EQUATION IN BANACH SPACES

Giovanni Dore and Davide Guidetti
Dipartimento di Matematica
Università di Bologna
Bologna, Italia

INTRODUCTION

In this paper we consider the following singular evolution equation:

$$t\,u'(t) - \Lambda(t)\,u(t) = f(t) \qquad t \in [0,T]$$

in the parabolic case; here u and f are supposed to be continuous functions with values in a complex Banach space E and the operators $\Lambda(t)$ infinitesimal generators of analytic semigroups on E fulfilling assumptions of Tanabe-Sobolevskiĭ type.

This equation has been studied by many authors. We can mention the following: in the case of bounded $\Lambda(t)$ Gluško-Kreĭn [11] and Gluško [10], in the hyperbolic case Da Prato-Grisvard [7] and Coppoletta [4], in the parabolic case Baiocchi-Baouendi [1] and Bernardi [3] (Hilbert space framework with variational techniques), Friedman-Schuss [9] and Schuss [13] (regularity of weak solutions in Hilbert spaces), Lewis-Parenti [12] (L^p theory in Hilbert spaces), Sobolevskiĭ [14] (autonomous equations in Banach spaces), Dore-Venni [8] (autonomous equations on $\mathbb{R}^+$ in Banach spaces).

The methods we employ are inspired by Da Prato-Grisvard [6].

1 - NOTATION AND PRELIMINARY RESULTS

Let E, F be two complex Banach spaces, such that F is continuously, densely imbedded in E, $\{\Lambda(t) : t \in [0,T]\}$ a family of operators belonging to $\mathscr{L}(F,E)$. We say that the operators $\Lambda(t)$ satisfy the hypothesis H_k ($k \in \mathbb{N} \cup \{0\}$) if:

a) $t \to \Lambda(t) \in C^{k,\alpha}([0,T], \mathscr{L}(F,E))$ for some $\alpha \in]0,1[$.

b) $\exists\, \omega \in]\frac{\pi}{2},\pi[$ $\exists\, \varepsilon \in]0,1[$ $\forall\, t \in [0,T]$ $k-\varepsilon+\Sigma \subset \rho(\Lambda(t))$ with $\Sigma = \{z \in \mathbb{C} : |\mathrm{Arg}\, z| \leq \omega\}$ and $\rho(\Lambda(t))$ the resolvent set of $\Lambda(t)$ as operator in E. Moreover $\exists\, C \in \mathbb{R}^+$ $\forall\, t \in [0,T]$ $\forall\, z \in k-\varepsilon+\Sigma$ $\|(\Lambda(t)-z)^{-1}\| \leq \frac{C}{|z|+1}$.

We pose $X_k = C^k([0,T],E)$ with norm

$\|u\|_k = \sum_{j=0}^{k-1} \|u^{(j)}(0)\|_E + \max\{\|u^{(k)}(t)\|_E : t \in [0,T]\}$ for $k \geq 1$ and $\|u\|_0 = \max\{\|u(t)\|_E : t \in [0,T]\}$ and $X_{k,0} = \{u \in X_k : u(0) = \ldots = u^{(k-1)}(0) = 0\}$ for $k \geq 1$, $X_{0,0} = X_0$ with the norm induced by X_k.

We define the following operators in X_k:

$D(A) = \{u \in X_k : t \to tu'(t) \in X_k\}$ (the derivative is in the sense of distributions)

$Au(t) = tu'(t)$

$D(B) = C^k([0,T],F)$

$Bu(t) = \Lambda(t)u(t)$

We denote by A_0, B_0 respectively the part of A and B in $X_{k,0}$.

We want to study the strong and weak solutions of the equation

$$(A-B)u = f \quad \text{and} \quad (A_0-B_0)u = f,$$

and compare the two kinds of solution.

2 - STRONG SOLUTIONS OF THE EQUATION $(A_0 - B_0)u = f$.

Let $u, f \in X_{k,0}$. We say that u is a <u>strong solution</u> of the equation $(A_0-B_0)u = f$ if there exists a sequence (u_n) in $D(A_0) \cap D(B_0)$ converging to u such that $(A_0-B_0)u_n$ converges to f. This is the same as saying that, if we call $\overline{A_0-B_0}$ the (possibly multivalued) operator the graph of which is the closure of the graph of A_0-B_0, then $u \in D(\overline{A_0-B_0})$ and $f \in (\overline{A_0-B_0})u$.

It can be easily proved that:

PROPOSITION 2.1 - $\rho(A_0) = \{z \in \mathbb{C} : \mathrm{Re}\, z < k\}$ <u>and</u>

$$(A_0-z)^{-1}v(t) = t^z \int_0^t s^{-z-1} v(s)\, ds \quad \forall\, t \in [0,T].$$

<u>Moreover</u> $\|(A_0-z)^{-1}\| = \frac{1}{k-\mathrm{Re}\, z}$.

$\rho(B_0) \supset k-\varepsilon+\Sigma$ <u>and</u> $(B_0-z)^{-1}v(t) = (\Lambda(t)-z)^{-1}v(t)$ <u>for</u> $t \in [0,T]$.

<u>Moreover</u> $\|(B_0-z)^{-1}\| \leq \frac{\mathrm{const}}{|z|+1}$.

<u>Finally</u> $(A_0-z)^{-1}(D(B_0)) \subset D(B_0)$ <u>for</u> $\mathrm{Re}\, z < k$.

Now we are going to define an "approximate" solution of the equation.

Let γ be the clockwise oriented boundary of $k-\varepsilon+\Sigma$ and define

$$S = \frac{-1}{2\pi i}\int_\gamma (A_0 - z)^{-1}(B_0 - z)^{-1}\,dz$$

Owing to proposition 2.1 S is well-defined in $\mathscr{L}(X_{k,0})$.

PROPOSITION 2.2 - _If_ $u \in D(B_0)$ _then_ $Su \in D(A_0)\cap D(B_0)$ _and_ $(A_0 - B_0)Su = u + Ru$, with $R \in \mathscr{L}(X_{k,0})$ _defined as_

$$R = \frac{1}{2\pi i}\int_\gamma [B_0,(A_0 - z)^{-1}](B_0 - z)^{-1}\,dz$$

Moreover $\sigma(R) = \{0\}$.

PROOF We start by proving that, by renorming suitably $X_{k,0}$, R is a bounded operator with arbitrarily small norm. This proves that $\sigma(R) = \{0\}$.

For $u \in X_{k,0}$, $\delta \in \mathbb{R}^+$ define: $\|u\|_\delta = \max\{e^{-\delta t}\|u^{(k)}(t)\|_E : t \in [0,T]\}$.

For $z \in \gamma$, $u \in X_{k,0}$, $t \in [0,T]$, one has:

$$[B_0,(A_0 - z)^{-1}](B_0 - z)^{-1}u(t) = t^z \int_0^t s^{-z-1}(\Lambda(t) - \Lambda(s))(\Lambda(s) - z)^{-1}u(s)\,ds$$

so that

$$e^{-\delta t}\|([B_0,(A_0 - z)^{-1}](B_0 - z)^{-1}u)^{(k)}(t)\| \le$$

$$\le e^{-\delta t}\sum_{j=0}^k \binom{k}{j}\int_0^1 \|s^{-z-1}(\Lambda^{(k-j)}(t) - s^{k-j}\Lambda^{(k-j)}(ts))s^j((B_0 - z)^{-1}u)^{(j)}(ts)\|\,ds$$

$$\le \text{const}\int_0^1 s^{-\mathrm{Re}\,z-1}t^\alpha(1-s)^\alpha e^{-\delta t(1-s)}s^k\,ds\,\|u\|_\delta .$$

This is majorized uniformly in δ by $t^\alpha B(k-\mathrm{Re}\,z,\alpha+1)\|u\|_\delta$ that is summable on γ so that Ru is well-defined and if t is little enough then $\|e^{-\delta t}(Ru)^{(k)}(t)\| \le \varepsilon$.

On the other hand if $t \in [\eta,T]$, $\eta>0$, it is

$$\int_\gamma\int_0^1 s^{k-\mathrm{Re}\,z-1}t^\alpha(1-s)^\alpha e^{-\delta t(1-s)}\,ds\,|dz| \le$$

$$\le T^\alpha \int_\gamma\int_0^1 s^{k-\mathrm{Re}\,z-1}(1-s)^\alpha e^{-\delta\eta(1-s)}\,ds\,|dz| \xrightarrow[\delta\to\infty]{} 0.$$

So that if δ is sufficiently large $\|Ru\|_\delta \le 2\varepsilon\|u\|_\delta$ and so $\sigma(R) = \{0\}$.

Now let $u \in D(B_0)$. By prop. 2.1 $(A_0 - z)^{-1}(B_0 - z)^{-1}u \in D(B_0)$ and

$$B_0(A_0 - z)^{-1}(B_0 - z)^{-1}u = [B_0,(A_0 - z)^{-1}](B_0 - z)^{-1}u + (A_0 - z)^{-1}(B_0 - z)^{-1}B_0u$$

so that $Su \in D(B_0)$ and $B_0Su = SB_0u - Ru$.

Moreover, as $u \in D(B_0)$, $Su = -\frac{1}{2\pi i}\int_\gamma \frac{1}{z+1}(A_0 - z)^{-1}(B_0 - z)^{-1}(B_0 + 1)u\,dz + {}$ $+ \frac{1}{2\pi i}\int_\gamma \frac{1}{z+1}(A_0 - z)^{-1}u\,dz = -\frac{1}{2\pi i}\int_\gamma \frac{1}{z+1}(A_0 - z)^{-1}(B_0 - z)^{-1}(B_0 + 1)u\,dz + {}$ $+ (A_0 + 1)^{-1}u$, so that $Su \in D(A_0)$ and $(A_0 + 1)Su = u +$ $- \frac{1}{2\pi i}\int_\gamma \frac{1}{z+1}(B_0 - z)^{-1}(B_0 + 1)u\,dz + S(B_0 + 1)u = u + S(B_0 + 1)u$.

This implies that $A_0Su = u + SB_0u$ and so $Su \in D(A_0)\cap D(B_0)$ and $(A_0 - B_0)Su = u + Ru$.

PROPOSITION 2.3 - _Assume that the hypothesis_ H_k _are satisfied and moreover_ $F = E$ _(that is the operators_ $\Lambda(t)$ _are continuous on_ E). _Then_ $A_0 - B_0$

is invertible and $(A_0-B_0)^{-1} = S(1+R)^{-1}$, *with* S *and* R *defined above*.

PROOF By prop. 2.2, as $D(B_0) = X_{k,0}$ the equation $(A_0-B_0)u = v$ has the solution $u = S(1+R)^{-1}v \quad \forall\, v \in X_{k,0}$.
Now we want to prove that the equation $(A_0-B_0)u = 0$ has only the trivial solution.
In fact, if $(A_0-B_0)u = 0$, then $\forall\, t>0 \quad tu'(t) -\Lambda(0)u(t) = (\Lambda(t) -\Lambda(0))u(t)$
Assume $k = 0$.
Since the autonomous equation $tu'(t) -\Lambda(0)u(t) = f(t)$ $(f \in X_0)$ has in X_0 the only solution $u(t) = \int_0^t (\frac{t}{s})^{\Lambda(0)} s^{-1} f(s)\, ds$, it is

$$\|u(t)\|_E \le M\int_0^t (\tfrac{t}{s})^{-\rho} s^{-1}\|f(s)\|_E\, ds, \text{ with } \rho = -\max\{\mathrm{Re}\, z:\ z\in\sigma(\Lambda(0))\},$$

$\rho \ge 0$, and $M>0$.
Thus, if $f(t) = (\Lambda(t) - \Lambda(0))u(t)$,

$$\|u(t)\|_E \le \text{const.}\int_0^t (\tfrac{t}{s})^{-\rho} s^{\alpha-1}\|u(s)\|_E\, ds \le \text{const. } t^{\alpha}\max\{\|u(s)\|_E : s\in[0,t]\},$$

which implies $u(t) = 0$ in a neighbourhood of 0. Finally Peano-Picard theory implies that $u(t) = 0$ on the whole $[0,T]$.
If $k \ge 1$, from $(A_0- B_0)u = 0$, derivating k times, it follows

$$tu^{(k+1)}(t) - (\Lambda(0) - k)u^{(k)}(t) = (\Lambda(t) - \Lambda(0))u^{(k)}(t) + \sum_{j=0}^{k-1}\binom{k}{j}\Lambda^{(k-j)}(t)u^{(j)}(t)$$

and as the operators $\Lambda(t) - k$ satisfy the condition H_0, it follows

$$\|u^{(k)}(t)\| \le \text{const.}\, t^{\alpha} \max\{\|(\Lambda(s)- \Lambda(0))u^{(k)}(s) + \sum_{j=0}^{k-1}\binom{k}{j}\Lambda^{(k-j)}(s)u^{(j)}(s)\|_E : s\in[0,T]\}.$$

As $u(0) = \ldots = u^{(k-1)}(0) = 0$, $\|(\Lambda(s) - \Lambda(0))u^{(k)}(s) + \sum_{j=0}^{k-1}\binom{k}{j}\Lambda^{(k-j)}(s)u^{(j)}(s)\|_E \le \text{const } s^{\alpha}\max\{\|u^{(k)}(\sigma)\|_E : \sigma\in[0,s]\}$ so that $\|u^{(k)}(t)\|_E \le \text{const } t^{2\alpha}\max\{\|u^{(k)}(s)\|_E : s\in[0,T]\}$, which, as above, implies $u^{(k)}=0$ and, as $u\in X_{k,0}$, $u=0$.

PROPOSITION 2.4 - *The inclusion* $\overline{(A_0 - B_0)} \ni 0$ *has only the trivial solution*.

PROOF We define $\Lambda_j(t) = j\Lambda(t)(j - \Lambda(t))^{-1}$ and $B_ju(t) = \Lambda_j(t)u(t)$. One can verify that for j sufficiently large the operators $\Lambda_j(t)$ satisfy the hypothesis H_k, with estimates that are uniform in j, so that in view of prop. 2.3, the operators $A_0 - B_j$ have equicontinuous inverses.
Now, let $\overline{A_0 - B_0}u \ni 0$. It exists a sequence (u_n) in $D(A_0)\cap D(B_0)$ such that $u_n \to u$ in $X_{k,0}$, $v_n = (A_0- B_0)u_n \to 0$ in $X_{k,0}$.
We have $(A_0- B_j)u_n = (B_0- B_j)u_n + v_n$.
By the equicontinuity of the operators $(A_0- B_j)^{-1}$, one has

$$\|u_n\| \le \text{const}(\|(B_0- B_j)u_n\| + \|v_n\|) \quad \text{for every } j,\ n.$$

But $(B_0- B_j)u_n \to 0$, and so, $\|u_n\| \le \text{const}\|v_n\|$, which implies $u = 0$.

PROPOSITION 2.5 - $A_0- B_0$ *is closable in* $X_{k,0}$.

PROOF Owing to prop.2.2 and prop.2.4, for every $\lambda > 0$, the (a priori multivalued) operator $\overline{A_0 - B_0} + \lambda$ has a bounded (single valued) inverse $J_\lambda = (\overline{A_0 - B_0} + \lambda)^{-1}$.

Now we prove that $\|(\overline{A_0 - B_0} + \lambda)^{-1}\|_{\mathscr{L}(X_{k,0})} \leq \frac{\text{const}}{\lambda}$

To this end, we estimate

$$S_\lambda u = -\frac{1}{2\pi i}\int_\gamma (A_0 - z)^{-1}(B_0 - \lambda - z)^{-1} u\, dz$$

and

$$R_\lambda u = \frac{1}{2\pi i}\int_\gamma [B_0, (A_0 - z)^{-1}](B_0 - z - \lambda)^{-1} u\, dz,$$

as $J_\lambda = S_\lambda (1 + R_\lambda)^{-1}$.

Let γ_λ be the oriented path joining $-\infty e^{i\omega}$ with $+\infty e^{i\omega}$, which is the boundary of the domain $\Sigma \bigcup \{z \in \mathbb{C} : \operatorname{Re} z > \frac{\lambda}{2}\}$. Since for $\lambda > 2k$ we have $\Sigma \bigcup \{z \in \mathbb{C} : \operatorname{Re} z > \frac{\lambda}{2}\} \subset \rho(B_0 - \lambda)$, it is easy to see that in the definition of S_λ we can replace γ by γ_λ, so that we have:

$$\|S_\lambda\|_{\mathscr{L}(X_{k,0})} \leq \frac{1}{2\pi}\int_{\gamma_\lambda} \|(A_0 - z)^{-1}\|_{\mathscr{L}(X_{k,0})} \|(B_0 - z)^{-1}\|_{\mathscr{L}(X_{k,0})} |dz| \leq$$

$$\leq \text{const} \int_{\gamma_\lambda} \frac{1}{|\operatorname{Re} z|} \frac{1}{|z + \lambda|} |dz| = \frac{\text{const}}{\lambda} \int_{\gamma_1} \frac{1}{|\operatorname{Re} z|\,|z+1|} |dz|.$$

Since $\{\Lambda(t) - \lambda\}$ satisfies the assumptions H_k, with constants that can be chosen independently from λ, $\forall\, \varepsilon \geq 0$ it is possible to fix $\delta > 0$ such that $\forall\, u \in X_{k,0}$, $\lambda > 0$, $\|R_\lambda u\|_\delta \leq \varepsilon \|u\|_\delta$, and so the operators $(1 + R_\lambda)^{-1}$ are equicontinuous. So we can conclude that $\|J_\lambda\|_{\mathscr{L}(X_{k,0})} \leq \frac{\text{const}}{\lambda}$ $\forall\, \lambda > 0$.

Now we prove that $A_0 - B_0$ is closable.

The family of operators $\{J_\lambda : \lambda > 0\}$ is a pseudoresolvent (see Da Prato [5], prop. I.3.2 and rem. I.3.4) and $\{\lambda J_\lambda : \lambda > 0\}$ is equicontinuous. This implies (see Yosida [15], VIII.4, lemma 1') that, for a fixed λ,

$$\overline{(\overline{A_0 - B_0} + \lambda)^{-1}(X_{k,0})} \bigcap \operatorname{Ker}(\overline{A_0 - B_0} + \lambda)^{-1} = \{0\}.$$

But $X_{k,0} = \overline{D(A_0) \bigcap D(B_0)} \subset \overline{(\overline{A_0 - B_0} + \lambda)^{-1}(X_{k,0})}$, so that $\operatorname{Ker}(\overline{A_0 - B_0} + \lambda)^{-1} = \{0\}$ for every λ. This implies that $\overline{A_0 - B_0}$ is a single valued operator and $A_0 - B_0$ is closable.

To conclude this section we have:

THEOREM 2.6 - <u>If the hypothesis</u> H_k <u>are satisfied then the equation</u>

$$tu'(t) - \Lambda(t)u(t) = f(t)$$

<u>has a unique strong solution</u> u <u>in</u> $X_{k,0}$ <u>for every</u> $f \in X_{k,0}$. <u>Moreover</u>, u <u>depends continuously on</u> f <u>and</u> $u = 0$ <u>implies</u> $f = 0$

3 - STRONG SOLUTIONS IN THE GENERAL CASE

Now we study the equation $(A - B)u = v$ in the spaces X_k with $k \geq 1$.

Strong solutions are here defined as in section 2.
In this case the operator $\Lambda(0)$ is asked to satisfy the following further condition:

(H_k') $\{0,1,\ldots,k-1\} \subset \rho(\Lambda(0))$

We can make use of the results of section 2 thanks to the following remark: assume $u \in D(A) \cap D(B)$ and $(A-B)u = v$; derivating j times $(j \leq k-1)$ we obtain at $t=0$

$$(j - \Lambda(0))u^{(j)}(0) = \sum_{\ell=0}^{j-1} \binom{j}{\ell} \Lambda^{(j-\ell)}(0)u^{(\ell)}(0) + v^{(j)}(0)$$

and, thanks to hypothesis H_k', we can define

$$D_0 v = -\Lambda(0)^{-1} v(0),\ D_j v = (j - \Lambda(0))^{-1} \{\sum_{\ell=0}^{j-1} \binom{j}{\ell} \Lambda^{(j-\ell)}(0) D_\ell v + v^{(j)}(0)\}$$
$$(1 \leq j \leq k-1)$$

so that we have $u^{(j)}(0) = D_j v$ $(0 \leq j \leq k-1)$.
Clearly $D_j \in \mathscr{L}(X_k, F)$. We pose $\Pi : X_k \to X_k$, $\Pi u(t) = \sum_{j=0}^{k-1} \frac{t^j}{j!} u^{(j)}(0)$ $(0 \leq t \leq T)$, $P: X_k \to C^k([0,T], F)$, $Pv(t) = \sum_{j=0}^{k-1} \frac{t^j}{j!} D_j v$.
In view of the preceeding remark, if $(A-B)u = v$, then $\Pi u = Pv$.
Further we have $\operatorname{Ker} P = X_{k,0}$.

THEOREM 3.1 - $A-B$ <u>is closable and</u> $0 \in \rho(\overline{A-B})$.

PROOF Let (u_n) be a sequence in $D(A) \cap D(B)$, $u_n \to 0$ in X_k, $v_n = (A-B)u_n \to v$ in X_k.
We have $(1-\Pi)u_n \in D(A_0) \cap D(B_0)$, $(1-\Pi)u_n \to 0$ in $X_{k,0}$ and $(A_0 - B_0)(1-\Pi)u_n = (A-B)u_n - (A-B)\Pi u_n = v_n - (A-B)Pv_n \to v - (A-B)Pv$ (as $(A-B)P \in \mathscr{L}(X_k)$). But $0 = \lim_{n\to\infty} \Pi u_n = Pv$, so that $v \in X_{k,0}$ and $(A_0 - B_0)(1-\Pi)u_n \to v$.
By prop. 2.5, $v = 0$ and so $A-B$ is closable. Consider now the equation $(\overline{A-B})u = v$ $(v \in X_k)$. We put $v_0 = v - (A-B)Pv$. Since $Pv_0 = Pv - P(A-B)Pv = Pv - \Pi Pv = 0$, we have $v_0 \in X_{k,0}$.
Let $u = (\overline{A_0 - B_0})^{-1} v_0 + Pv$. We have $\overline{A-B}\,u = v_0 + (A-B)Pv = v$, so that $\overline{A-B}$ is surjective.
We prove now that it is also injective. Let (u_n) be a sequence in $D(A) \cap D(B)$, converging to $u \in X_k$ and such that $v_n = (A-B)u_n \to 0$. We have $v_n = (A-B)\Pi u_n + (A-B)(1-\Pi)u_n = (A-B)Pv_n + (A_0 - B_0)(1-\Pi)u_n$.
Since $(A-B)Pv_n \to 0$, we have $(A_0 - B_0)(1-\Pi)u_n \to 0$ and so $(1-\Pi)u = 0$.
On the other hand, $\Pi u = \lim_{n\to\infty} \Pi u_n = \lim_{n\to\infty} Pv_n = 0$. So $u = 0$ and $\overline{A-B}$ is injective.

4 - WEAK SOLUTIONS

We call <u>test function</u> every function $\phi \in C^1([0,T],E^*)$ (E^* is the dual space of E), such that:

a) $\forall\ t \in [0,T]\ \ \phi(t) \in D(\Lambda(t)^*)$

b) $t \to \Lambda(t)^*\phi(t)$ is continuous from $[0,T]$ to E^*

c) $\phi(T) = 0$

We say that $u \in C([0,T],E)$ is a <u>weak solution</u> of the equation $tu'(t) - \Lambda(t)u(t) = f(t)$ ($f \in X_0$) if for every test function ϕ

$$\int_0^T \langle u(t), -\frac{d}{dt}(t\phi(t)) - \Lambda(t)^*\phi(t)\rangle\, dt = \int_0^T \langle f(t),\phi(t)\rangle\, dt$$

PROPOSITION 4.1 - <u>Under the hypothesis</u> H_k, <u>if</u> u <u>is a strong solution in</u> X_k <u>of the equation</u> $tu'(t) - \Lambda(t)u(t) = f(t)$ ($f \in X_k$), <u>then</u> u <u>is a weak solution of the same equation</u>.

PROOF The proposition is easily proved by integration by parts in the case $u \in D(A) \cap D(B)$, and by approximation in the general case.

We are now going to study under what assumptions a weak solution is a strong one.

We suppose that the conditions H_k are satisfied and that:

a) $D(\Lambda(0)^*)$ is dense in E^* (with the strong topology)

(H_k^*) b) $\forall\ s,t,\tau \in [0,T]\ \forall\ y \in Y$

$$\| (\Lambda(s) - k)^{-1}(\Lambda(t) - \Lambda(\tau))y\|_E \le \text{const}\ |t-\tau|^\alpha \| y\|_E .$$

LEMMA 4.2 - <u>Let</u> U, V <u>a couple of densely defined closed operators on a Banach space</u> E, <u>such that</u> $0 \in \rho(U) \cap \rho(V)$. <u>Then</u> $D(U^*) = D(V^*)$ <u>if and only if the operators</u> $U^{-1}V$ <u>and</u> $V^{-1}U$ <u>are bounded</u>.

PROOF Assume $U^{-1}V$ is bounded and let $x^* \in D(U^*)$, $x \in D(V)$. Then $\langle Vx,x^*\rangle = \langle U^{-1}Vx,U^*x^*\rangle$, so that $x^* \in D(V^*)$. On the other hand, assume $D(U^*) \subset D(V^*)$. It follows that V^*U^{*-1} is bounded and this implies easily that $U^{-1}V$ is bounded.

Lemma 4.2 and the hypothesis H_k and H_k^* guarantee that $D(\Lambda(t)^*)$ does not depend on t. We put $F' = D(\Lambda(t)^*) = D(\Lambda(0)^*)$. F' is a Banach space with the norm induced by $\Lambda(0)^*$.

It is also easy to prove that $t \to \Lambda(t)^* \in C^{0,\alpha}([0,T], \mathcal{L}(F',E^*))$, so that the operators $\Lambda(t)^*$ satisfy the conditions H_k if we replace E with E^*, F with F'.

LEMMA 4.3 - <u>Let</u> E <u>be a complex Banach space</u>, $\{\Lambda(t) : t\in[0,T]\}$ <u>a family of operators satisfying the hypothesis</u> H_k.
<u>We pose</u> $Z = L^1([0,T],E)$, $D(N) = \{u\in Z : t\to\frac{d}{dt}(tu(t))\in Z,\ u(T)=0\}$, $Nu(t) = -\frac{d}{dt}(tu(t))$, $D(L) = L^1([0,T],F)$, $Lu(t) = \Lambda(t)u(t)$.
<u>Then the range of the operator</u> $N - L$ <u>is dense in</u> Z.

PROOF It can be easily verified that $\rho(N) = \{z\in\mathbb{C} : \mathrm{Re}\, z < 0\}$ and that $\mathrm{Re}\, z < 0$, $h\in Z$ imply $((N-z)^{-1}h)(t) = t^{-z-1}\int_t^T s^{-z}h(s)\,ds$.
Moreover $\|(N-z)^{-1}\|_{\mathscr{L}(Z)} \le \frac{1}{|\mathrm{Re}\, z|}$.
Since $(N-z)^{-1}(D(L))\subset D(L)$, we can proceed as in the proof of prop.2.2.
We have only to show that also in this case $R\in\mathscr{L}(Z)$ and $\sigma(R) = \{0\}$, with $Ru = \frac{1}{2\pi i}\int_\gamma [L,(N-z)^{-1}](L-z)^{-1}u\,dz$,

$$Ru(t) = \frac{1}{2\pi i}\int_\gamma t^{-z-1}\int_t^T s^z(\Lambda(t)-\Lambda(s))(\Lambda(s)-z)^{-1}u(s)\,ds\,dz.$$

To this end, we introduce a new norm on Z, by putting

$$\|u\|_{\delta,1} = \int_0^T \exp(-\delta(T-t))\,\|u(t)\|_E\,dt\ .$$

Then $\|Ru\|_{\delta,1} \le$

$$\text{const}\int_0^T\exp(-\delta(T-t))\int_\gamma t^{-\mathrm{Re}\, z-1}\int_t^T s^{\mathrm{Re}\, z}(s-t)^\alpha\|u(s)\|_E\,ds\,|dz|\,dt \le$$

$$\text{const}\int_\gamma\int_0^T s^{\mathrm{Re}\, z}\|u(s)\|_E\exp(-\delta(T-s))\int_0^s\exp(-\delta(s-t))t^{-\mathrm{Re}\, z-1}(s-t)^\alpha\,dt\,ds\,|dz| =$$

$$\text{const}\int_\gamma\int_0^T\|u(s)\|_E\exp(-\delta(T-s))\int_0^1 s^\alpha t^{-\mathrm{Re}\, z-1}\exp(-\delta s(1-t))(1-t)^\alpha dt\,ds\,|dz| \le$$

$$\text{const}\,\|u\|_{\delta,1}\sup_{s\in[0,T]} s^\alpha\int_\gamma\int_0^1\exp(-\delta s(1-t))t^{-\mathrm{Re}\, z-1}(1-t)^\alpha dt\,|dz|$$

One has

$$s^\alpha\int_\gamma\int_0^1\exp(-\delta s(1-t))\,t^{-\mathrm{Re}\, z-1}(1-t)^\alpha dt\,|dz| \le s^\alpha\int_\gamma B(-\mathrm{Re}\, z,\alpha+1)\,|dz|$$

and the whole is majorized by ε if $s\le\eta$, for a suitable $\eta > 0$.
On the other hand, if $s > \eta$, $s^\alpha\int_\gamma\int_0^1\exp(-\delta s(1-t))\,t^{-\mathrm{Re}\, z-1}(1-t)^\alpha\,dt\,|dz| \le$
$\le T^\alpha\int_\gamma\int_0^1\exp(-\delta\eta(1-t))t^{-\mathrm{Re}\, z-1}(1-t)^\alpha\,dt\,|dz|\to 0$ as $\delta\to+\infty$.
So we have $\|Ru\|_{\delta,1}\le 2\varepsilon\|u\|_{\delta,1}$ for a suitable δ and this implies $\sigma(R) = \{0\}$.

THEOREM 4.4 - <u>Under the assumptions</u> H_0 <u>and</u> H_0^*, <u>every weak solution of the equation</u> $tu'(t) - \Lambda(t)u(t) = f(t)$, <u>with</u> $f\in X_0$, <u>is a strong solution in</u> X_0.

PROOF By prop. 4.1 every strong solution is a weak solution. As the hypothesis H_0 guarantees the uniqueness of the strong solution in X_0, we have to prove only the uniqueness of the weak solution to have the result. If u is a weak solution of $tu'(t) - \Lambda(t)u(t) = 0$ for every test function ϕ we have $\int_0^T \langle u(t), -\frac{d}{dt}(t\phi(t)) - \Lambda(t)^*\phi(t)\rangle\,dt = 0$.
We choose a sequence (ω_n) of mollifiers converging to δ in the sense of distribution and, for $\psi\in D(N)\cap D(L)$, we put:

$$\psi_n(t) = \int_0^T \omega_n(t-s)\psi(s)\,ds - \int_0^T \omega_n(T-s)\psi(s)\,ds.$$

ψ_n is a test function and

$$0 = \lim_{n\to\infty} \int_0^T \langle u(t), -\frac{d}{dt}(t\psi_n(t)) - \Lambda(t)^*\psi_n(t)\rangle dt = \int_0^T \langle u(t), -\frac{d}{dt}(t\psi(t)) + - \Lambda(t)^*\psi(t)\rangle\, dt.$$

From lemma 4.3 it follows that $\int_0^T \langle u(t), h(t)\rangle\, dt = 0 \ \forall\, h \in Z$, which implies $u = 0$.

For $k \geq 1$ we have the following result:

THEOREM 4.5 - Assume that the hypothesis H_k, H_k', H_k^* are satisfied. Then every weak solution which is of class C^k is a strong solution in X_k.

PROOF For $k \in \mathbb{N} \cup \{0\}$, we indicate with A^k and B^k the operators A and B defined on X_k. To prove the theorem we have to show only that, if u is a weak solution of class C^k of the equation $tu'(t) - \Lambda(t)u(t) = 0$, then $u = 0$. In fact, u is a weak solution of the equation

$$tu'(t) - (\Lambda(t) - k)u(t) = ku(t)$$

but the operators $\Lambda(t) - k$ satisfy the conditions H_0, and so $u \in \in D(\overline{A^0 - B^0})$ and $(\overline{A^0 - B^0} + k)u = ku$.

But obviously the operators $\Lambda(t) - k$ satisfy the conditions H_k and H_k' and, as $u \in X_k$, there exists a unique $v \in X_k$, such that $(\overline{A^k - B^k} + k)v = ku$. This clearly implies $v \in D(\overline{A^0 - B^0})$ and $(\overline{A^0 - B^0} + k)v = ku$.

By theorem 2.4, $u = v$. So $u \in D(\overline{A^k - B^k})$ and $\overline{A^k - B^k}u = 0$.

By theorem 3.1, this implies $u = 0$.

THEOREM 4.6 - Assume that the hypothesis H_k and H_k^* are satisfied. Let $f \in X_0$. If u is a weak solution of $tu'(t) - \Lambda(t)u(t) = f(t)$, and u is of class C^1, then $u \in C([0,T], F)$ and satisfies the equation pointwise.

PROOF Let $\psi \in C_0^\infty([0,T], E)$, $x^* \in F'$. Then $\int_0^T \psi(t) \langle tu'(t) - f(t), x^*\rangle dt = = \int_0^T \psi(t)\langle u(t), \Lambda(t)^* x^*\rangle\, dt$, so that $\forall\, t \in [0,T]$, $\langle tu'(t) - f(t), x^*\rangle = \langle u(t), \Lambda(t)^* x^*\rangle$ and this proves that $u(t) \in D(\Lambda(t))$ (see Ball [2], lemma). and $\Lambda(t)u(t) = tu'(t) - f(t)$.

REFERENCES

[1] C. Baiocchi and M.S. Baouendi, "Singular evolution equations", J. Funct. Anal. 25 (1977), 103-120.

[2] J. M. Ball, "Strongly continuous semigroups, weak solutions and the variation of parameter formula", Proc. Am. Math. Soc. 63 (1977), 370-373.

[3] M. L. Bernardi, "Su alcune equazioni d'evoluzione singolari", Boll. Unione Mat. Ital., V. Ser., B 13 (1976), 498-517.

[4] G. Coppoletta, "Abstract singular evolution equations of "hyperbolic" type", J. Funct. Anal. 50 (1983), 50-66.

[5] G. Da Prato, "Applications croissantes et équations d'évolution dans les espaces de Banach",Institutiones Mathematicae vol.II,Academic Press, London-New York, 1976.

[6] G. Da Prato and P. Grisvard, "Sommes d'opérateurs linéaires et équations differentielles opérationelles", J. Math. Pures Appl., IX. Ser., 54 (1975), 305-387.

[7] G. Da Prato and P. Grisvard, "On an abstract singular Cauchy problem", Commun. Partial Differ. Equations 3 (1978), 1077-1082.

[8] G. Dore and A. Venni, "On a singular evolution equation in Banach spaces", J. Funct. Anal. 64 (1985), 227-250.

[9] A. Friedman and Z. Schuss, "Degenerate evolution equations in Hilbert space", Trans. Am. Math. Soc. 161 (1971), 401-427.

[10] V. P. Gluško, "Smoothness of solutions of degenerate differential equations in Banach space" (russian), Dokl. Akad. Nauk SSSR 198 (1971),20-22 (translated in Sov. Math., Dokl.12 (1971), 701-704).

[11] V. P. Gluško and S. G. Kreĭn, "Degenerating linear differential equations in a Banach space" (russian), Dokl. Akad. Nauk SSSR 181 (1968), 784-787 (translated in Sov. Math., Dokl. 9 (1968), 919-922).

[12] J.E. Lewis and C. Parenti, "Abstract singular parabolic equations", Commun. Partial Differ. Equations 7 (1982), 279-324.

[13] Z. Schuss, "Regularity theorems for solutions of a degenerate evolution equation", Arch. Ration. Mech. Anal. 46 (1972), 200-211.

[14] P. E. Sobolevskiĭ, "On degenerate parabolic operators" (russian), Dokl.Akad. Nauk SSSR 196 (1971), 302-304 (translated in Sov. Math., Dokl. 12 (1971), 129-132).

[15] K. Yosida, "Functional Analysis" 6th ed., Springer Verlag, Berlin Heidelberg New York, 1980.

On the spectrum of certain systems of linear evolution equations

by

K.-J. Engel and R. Nagel

Mathematisches Institut der Universität Tübingen
Auf der Morgenstelle 10
D-7400 Tübingen
Federal Republic of Germany

In this paper we consider systems of linear evolution equations of the form

$$(*) \quad \frac{d}{dt} u_i(t) = \Sigma_{j=1}^{n} a_{ij} A u_j(t) + b_{ij} u_j(t) \quad , \quad u_i(0) = v_i$$

for $i = 1,..,n$ and $t \geq 0$, where A is the generator of a strongly continuous semigroup of linear operators on some Banach space E and $M := (a_{ij})_{n\times n}$, $N := (b_{ij})_{n\times n}$ are complex $n\times n$-matrices. Clearly, this system can be written as

$$(**) \quad \frac{d}{dt} \mathcal{u}(t) = \mathcal{A}\mathcal{u}(t) \ , \ \mathcal{u}(0) = \mathcal{v} \ ,$$

where $\mathcal{A} := (a_{ij}A + b_{ij})_{n\times n}$ is canonically defined on the product space $\mathcal{E} := E^n$.

It is well known that the abstract Cauchy problem (*), resp. (**), is well-posed if and only if $\mathcal{A}$ is the generator of a strongly continuous semigroup on $\mathcal{E}$. Since $(b_{ij}\cdot Id_E)_{n\times n}$ is a bounded perturbation of $(a_{ij}A)_{n\times n}$ theorem 2.3 of [5] (see also [6] or [2]) yields the following characterisation.

Let A be a generator on some Banach space E and consider complex matrices $M := (a_{ij})_{n\times n}$ and $N := (b_{ij})_{n\times n}$.

<u>Theorem 1</u>. For the operator $\mathcal{A} = (a_{ij}A + b_{ij})_{n\times n}$ on $\mathcal{E} = E^n$ the following assertions are equivalent.

(a) $\mathcal{A}$ is the generator of a strongly continuous semigroup on $\mathcal{E}$.

(b) For each eigenvalue α of M the following holds:

(i) αA is the generator of a strongly continuous semigroup on E.

(ii) If α is a multiple zero of the minimal polynomial of M, then αA is the generator of an analytic semigroup on E.

(iii) If $\alpha = 0$ is a multiple zero of the minimal polynomial of M, then A is bounded.

Once the well-posedness of (*) is guaranteed we proceed by studying the qualitative behavior of its solutions. One of the most useful objects in this respect is the spectrum $\sigma(\mathcal{A})$ of the generator $\mathcal{A}$. It is our aim to characterize $\sigma(\mathcal{A})$ as the set of eigenvalues of certain matrices thereby opening the way to the application of classical matrix techniques in order to investigate $\mathcal{A}$.

The main tool in order to determine $\sigma(\mathcal{A})$ will be the use of the operator-valued "determinant" of the matrix $\mathcal{A}$. While this is well known for bounded A we have to overcome certain difficulties as soon as A is unbounded.

In the following let A be a densely defined unbounded operator on E and consider

$$\mathcal{A} = (a_{ij}A + b_{ij})_{n\times n} = A \otimes M + Id \otimes N \quad \text{on} \quad \mathcal{E} = E^n = E \otimes \mathbb{C}^n .$$

We assume that A and $\mathcal{A}$ have non empty resolvent sets (e.g., A and $\mathcal{A}$ are generators) and introduce the following notation.

<u>Definition 2</u>. Let $p_{\mathcal{A}}(x)$ denote the determinant of the matrix $(xM + N)$. Then $p_{\mathcal{A}}(x)$ is an element of the polynomial ring $\mathbb{C}[x]$ and the well defined operator

$$\Delta(\mathcal{A}) := p_{\mathcal{A}}(A)$$

(use [1], p.602) is called the <u>determinant</u> of $\mathcal{A}$.

If A is bounded it is easy to show that $\mathcal{A}$ is invertible in $\mathcal{E}$ if and only if $\Delta(\mathcal{A})$ is invertible in E. For unbounded A the situation is more complicated.

Lemma 3. Under the above assumptions the following assertions are equivalent.

(a) $\mathcal{A}$ is invertible.

(b) (i) $\Delta(\mathcal{A})$ is invertible in E and

(ii) $\lim_{\alpha\to\infty}(\alpha M + N)^{-1}$ exists in $M_n(\mathbb{C})$.

Proof. We start by showing the implication (a)=>(b.i). As a first step we show that $p_{\mathcal{A}}(.)$ is not identically zero. Suppose that $p_{\mathcal{A}}(x) \equiv 0$. This implies that the polynomial matrix $\mathcal{M} := (a_{ij}x + b_{ij})_{n\times n}$ is singular over the quotient field $\mathbb{C}(x)$, i.e. there exists $0 \neq z \in (\mathbb{C}(x))^n$ such that $\mathcal{M}z = 0$. By multiplying z by an appropriate polynomial we may suppose that $z \in (\mathbb{C}[x])^n$, hence $z = x^m a_m + \ldots + x a_1 + a_0$ for non trivial coefficients $a_0, \ldots, a_m \in \mathbb{C}^n$. Applying $\mathcal{M}$ to z yields

$$\begin{aligned} 0 &= \mathcal{M}z \\ &= (xM + N)\,(x^m a_m + \ldots + x a_1 + a_0) \\ &= x^{m+1} M a_m + x^m (N a_m + M a_{m-1}) + \ldots + x(N a_1 + M a_0) + N a_0 \end{aligned}$$

and

$$(\#) \quad 0 = M a_m = N a_0 = N a_k + M a_{k-1} \quad \text{for } k = 1,\ldots,m .$$

For $f \in D(A^{m+1})$ we consider $\mathfrak{f} := \Sigma_{k=0}^{m} A^k f \otimes a_k \in \mathcal{E} = E \otimes \mathbb{C}^n$, which can be identified with the linear operator

$L_{\mathfrak{f}} : \mathbb{C}^n \to E$ where $L_{\mathfrak{f}} a' := \Sigma_{k=0}^{m} \langle a_k, a' \rangle \cdot A^k f$ for $a' \in \mathbb{C}^n$.

Then $\mathfrak{f} \neq 0$ if and only if $L_{\mathfrak{f}} \neq 0$ and in particular if $U(f) := \{A^k f : k = 0,\ldots,m\}$ is linearly independent. Suppose that $U(f)$ is linearly dependent for each $0 \neq f \in D(A^{m+1})$. Consequently for each such f there exist $\nu \in \mathbb{N}$, $\nu \leq m$, and $\alpha_k \in \mathbb{C}$, $k = 0,\ldots,\nu$ such that $A^\nu f = \Sigma_{k=0}^{\nu-1} \alpha_k A^k f$. Since $\rho(A)$ is non empty we may assume that A is invertible and obtain $f = \Sigma_{k=0}^{\nu-1} \alpha_k A^{k-\nu} f$. In particular this implies $f \in D(A^{m+2})$ and therefore $D(A^{m+1}) = D(A^{m+2})$. Since (the restriction of) A is an isomorphism from $D(A^k)$ onto $D(A^{k-1})$, $k \in \mathbb{N}$, we conclude that $D(A) = D(A^0) = E$ in contradiction to the assumption that A is unbounded.

By these considerations there exists $0 \neq f \in D(A^{m+1})$ such that $\mathfrak{f} \neq 0$. Then (#) implies

$$\begin{aligned} \mathcal{A}\mathfrak{f} &= (\, A \otimes M + Id \otimes N \,)(\, \Sigma_{k=0}^{m} A^k f \otimes a_k \,) \\ &= \Sigma_{k=0}^{m} A^{k+1} f \otimes M a_k \;+\; \Sigma_{k=0}^{m} A^k f \otimes N a_k \\ &= -\Sigma_{k=0}^{m-1} A^{k+1} f \otimes N a_{k+1} \;+\; \Sigma_{k=0}^{m-1} A^{k+1} f \otimes N a_{k+1} \\ &= 0 , \end{aligned}$$

i.e. A is not injective. This contradicts the assumption (a) and we showed that $p_A(c) \neq 0$ for some $c \in \mathbb{C}$.
In the next step this will be used to transform M and N into canonical form. In fact, $p_A(c) \neq 0$ implies that

$$S := (cM + N)^{-1}$$

exists and we obtain

$$SN = S(cM + N) - cSM = Id - cSM .$$

In addition there exists an invertible matrix T such that $T^{-1}SMT$ has Jordan form while the above identity implies

$$T^{-1}SNT = Id - cT^{-1}SMT .$$

Since the assertions (a) and (b) remain unchanged if we multiply the matrices M and N by (the same) invertible matrices and since it suffices to consider the restriction of A to invariant subspaces of $E \otimes \mathbb{C}^n$ we showed that M and N may be assumed to be of the form

$$(\#\#) \qquad M = \begin{pmatrix} \mu & 1 & 0 & . & 0 \\ 0 & \mu & 1 & . & 0 \\ . & & . & . & . \\ . & & & \mu & 1 \\ 0 & . & . & 0 & \mu \end{pmatrix}, \quad N = \begin{pmatrix} 1-c\mu & -c & 0 & . & 0 \\ 0 & 1-c\mu & -c & . & 0 \\ . & & . & . & . \\ . & & & 1-c\mu & -c \\ 0 & . & . & 0 & 1-c\mu \end{pmatrix} .$$

After this preparations we prove the assertion. By assumption A^{-1} exists in $L(E)$ and can be represented as a matrix $(R_{ij})_{n\times n}$ where $R_{ij} \in L(E)$. This implies

$$(\mu A + (1 - c\mu)Id)\cdot R_{nn} = Id_E \quad \text{and} \quad R_{11}\cdot(\mu A + (1 - c\mu)Id) = Id_{D(A)}$$

(in the case $\mu \neq 0$, otherwise $R_{11}\cdot(\mu A + (1 - c\mu)Id) = Id_E$). Hence $\mu A + (1 - c\mu)Id$ and therefore $\Delta(A) = (\mu A + (1 - c\mu)Id)^n$ is invertible.
(a)=>(b.ii). As above we may assume M and N to have the normal form (##). If M is invertible we obtain

$$\lim_{\alpha\to\infty}(\alpha M + N)^{-1} = \lim_{\alpha\to\infty}(\alpha + M^{-1}N)^{-1}M^{-1} = 0 .$$

If M is not invertible, i.e. $\mu = 0$, we have

$$A = \begin{pmatrix} Id & A-cId & 0 & . & 0 \\ 0 & Id & A-cId & . & 0 \\ . & & . & . & . \\ . & & & Id & A-cId \\ 0 & . & . & 0 & Id \end{pmatrix}_{m\times m}$$

for some $1 \leq m \leq n$. But for matrices of this form $\rho(A) \neq \emptyset$ only if $m = 1$. In particular $M = 0$ and $\lim_{\alpha\to\infty}(\alpha M + N)^{-1} = N^{-1} = Id$.

(b)=>(a). The condition (i) implies $p_A(.) \not\equiv 0$. This shows that the polynomial matrix $(xM + N)$ is invertible in $M_n(\mathbb{C}(x))$ with inverse $p_A^{-1}(x)\cdot \mathrm{Adj}(xM + N)$. Here the entries are rational functions in x where the degree of the denominator polynomial dominates the degree of the numerator polynomial since (ii) holds. By [1], chap. VII.9 thm.8 the operator matrix $\mathcal{R} := p_A^{-1}(A)\cdot \mathrm{Adj}(a_{ij}A + b_{ij})$ is continuous on E. Elementary calculations show $\mathcal{A}^{-1} = \mathcal{R}$. □

Remarks 1. It is easy to see that condition (b.ii) is equivalent to the property that $\lim_{k\to\infty}(\mu_k M + N)^{-1}$ exists for every unbounded sequence $(\mu_k)_{k\in\mathbb{N}} \subset \mathbb{C}$.
In this case $\lim_{\alpha\to\infty}(\alpha M + N)^{-1} = \lim_{k\to\infty}(\mu_k M + N)^{-1}$ holds.
2. The above lemma yields $\lambda_0 \in \rho(\mathcal{A})$ if and only if (i) $\Delta(\lambda_0 - \mathcal{A})$ is invertible in E and (ii) $\lim_{\alpha\to\infty} R(\lambda_0, \alpha M + N)$ exists in $M_n(\mathbb{C})$.

Unfortunately condition (ii) is necessary: Take

$$\mathcal{A} = \begin{pmatrix} A & 1 \\ 1 & 0 \end{pmatrix}.$$

Then $\Delta(\mathcal{A}) = -\mathrm{Id}$ is invertible while $\mathcal{A}$ is not (for A unbounded).
In order to handle condition (ii) we give the following characterization.

Lemma 4. Take $\mu_0 \in \rho(\mathcal{A})$ and consider $Q := \lim_{\alpha\to\infty} R(\mu_0, \alpha M + N)$. For $\lambda_0 \in \mathbb{C}$ the following assertions are equivalent.
(a) $\lim_{\alpha\to\infty} R(\lambda_0, \alpha M + N)$ exists in $M_n(\mathbb{C})$.
(b) $\lambda_0 \notin \mu_0 - (\sigma(Q) \setminus \{0\})^{-1}$.

Proof. The lemma holds for $\lambda_0 = \mu_0$, hence assume $\lambda_0 \neq \mu_0$.

(a)=>(b). We show $(\mu_0 - \lambda_0)^{-1} \notin \sigma(Q)$. The resolvent equation yields that the inverse of $(\mu_0 - \lambda_0)^{-1} - R(\mu_0, \alpha M + N)$ is given by $(\mu_0 - \lambda_0) + (\mu_0 - \lambda_0)^2 R(\lambda_0, \alpha M + N)$. Hence the assumption (a) implies

$$\begin{aligned}(\mu_0 - \lambda_0) + (\mu_0 - \lambda_0)^2 \lim_{\alpha\to\infty} R(\lambda_0, \alpha M + N) &= \\ = \lim_{\alpha\to\infty} ((\mu_0 - \lambda_0)^{-1} - R(\mu_0, \alpha M + N))^{-1} &= \\ = R((\mu_0 - \lambda_0)^{-1}, Q), \text{ i.e. } (\mu_0 - \lambda_0)^{-1} \notin \sigma(Q).\end{aligned}$$

(b)=>(a). By hypothesis $(\mu_0 - \lambda_0)^{-1} - Q$ is invertible, hence $(\mu_0 - \lambda_0)^{-1} - R(\mu_0, \alpha M + N)$ is invertible for α sufficiently large. Then the identity

$$R(\lambda_0, \alpha M + N) = (\mu_0 - \lambda_0)^{-2}((\mu_0 - \lambda_0)^{-1} - R(\mu_0, \alpha M + N))^{-1} - (\mu_0 - \lambda_0)^{-1}$$

and the continuity of the inversion imply (a). □

The above lemmas yield the desired characterization of $\sigma(\mathcal{A})$.

Theorem 5. Let A be a densely defined operator on some Banach space E satisfying $\rho(A) \neq \emptyset$. For complex matrices $M = (a_{ij})_{n\times n}$, $N = (b_{ij})_{n\times n}$ consider the operator $\mathcal{A} = (a_{ij}A + b_{ij})_{n\times n} = A \otimes M + Id \otimes N$ on $\mathcal{E} = E^n = E \otimes \mathbb{C}^n$ and assume $\mu_0 \in \rho(\mathcal{A})$. Then

$$Q := \lim_{\alpha\to\infty} R(\mu_0, \alpha M + N)$$

exists in $M_n(\mathbb{C})$ and

$$\sigma(\mathcal{A}) = \sigma_1 \cup \sigma_2 ,$$

where
$$\sigma_1 := \cup_{\mu\in\sigma(A)} \sigma(\mu M + N) = \sigma(\sigma(A)\cdot M + N) ,$$

$$\sigma_2 := \mu_0 - (\sigma(Q) \setminus \{0\})^{-1} .$$

Proof. The existence of Q follows from (b.ii) of lemma 3. The second remark in combination with lemma 4 shows that $\lambda \in \sigma(\mathcal{A})$ if and only if $\lambda \in \sigma_1$ or $\Delta(\lambda - \mathcal{A})$ is not invertible. This last property is equivalent to $0 \in \sigma(p_\lambda(A))$ where $p_\lambda(A) = \Delta(\lambda - \mathcal{A})$ is a polynomial in A . By the spectral mapping theorem for polynomials (see [1], p.604) we obtain

$$\sigma(p_\lambda(A)) = p_\lambda(\sigma(A)) = \{\det(\lambda - (\mu M + N)) : \mu \in \sigma(A)\} .$$

But $0 \in \cup_{\mu\in\sigma(A)} \sigma(\lambda - (\mu M + N))$ is equivalent to $\lambda \in \cup_{\mu\in\sigma(A)} \sigma(\mu M + N) = \sigma_1$. □

As a first observation we state that $\sigma(\mathcal{A})$ consists of the "natural" set σ_1 and at most finitely many exceptional points in σ_2 . In fact the proof of lemma 3 shows $|\sigma_2| \leq n-1$ as soon as $M \neq 0$.
In the following corollaries we give even more precise information on the location of σ_2 in $\sigma(\mathcal{A})$.

Corollary 6. If M is invertible then $\sigma_2 = \emptyset$ and therefore

$$\sigma(\mathcal{A}) = \sigma(\sigma(A)\cdot M + N) .$$

Proof. In the proof of lemma 3, implication (a)=>(b.ii) we showed that $Q = 0$ whenever M is invertible. □

Corollary 7. If $\sigma(A)$ is bounded then $\sigma(\mathcal{A})$ is bounded and $\sigma_2 \setminus \sigma_1$ consists of (finitely many) isolated points in $\sigma(\mathcal{A})$.

Proof. This follows from the observation that σ_1 is bounded and closed whenever $\sigma(A)$ is bounded. □

Corollary 8. If $\sigma(A)$ is unbounded then

$$\sigma(\mathcal{A}) = \overline{\sigma_1} = \overline{\cup_{\mu\in\sigma(A)}\sigma(\mu M + N)} .$$

Proof. Take $\lambda_0 \in \sigma_2$, i.e. $(\mu_0 - \lambda_0)^{-1} \in \sigma(Q)$ for some $\mu_0 \in \rho(\mathcal{A})$ and $Q = \lim_{\alpha\to\infty} R(\mu_0, \alpha M + N)$. Since $\sigma(A)$ is unbounded there exists an unbounded sequence $(\lambda_k)_{k\in\mathbb{N}} \subset \sigma(A)$. Then $Q_k := R(\mu_0, \lambda_k M + N)$ exists and converges to Q. By [3], chap.II, thm.5.1 we find a sequence $(\mu_k)_{k\in\mathbb{N}}$ such that $\mu_k \in \sigma(Q_k) = \sigma(\mu_0 - \sigma(\lambda_k M + N))^{-1}$ and $\mu_k \to (\mu_0 - \lambda_0)^{-1}$. Therefore $(\mu_0 - \mu_k^{-1}) \in \sigma(\lambda_k M + N) \subset \sigma_1$ and $(\mu_0 - \mu_k^{-1}) \to \lambda_0$. □

This last corollary is particularly useful since it allows to determine the spectrum of $\mathcal{A}$ by calculating the eigenvalues of the matrices $\mu M + N$, $\mu \in \sigma(A)$, alone without worrying about the "exceptional" spectral values in σ_2.

For applications to stability theory the <u>spectral bound</u>

$$s(\mathcal{A}) := \sup\{\mathrm{Re}\,\lambda : \lambda \in \sigma(\mathcal{A})\}$$

is of special interest (see [7], A-III). Elementary matrix techniques in combination with the above corollary yield interesting and simple estimates for $s(\mathcal{A})$. We present one example only.

Corollary 9. Let $\sigma(A)$ be unbounded and assume that M and N are normal matrices. Then

$$s(\mathcal{A}) \leq \sup_{\alpha\in\sigma(M)} s(\alpha A) + s(N) .$$

Proof. For the numerical range $V(S) := \{(Sx,x) : \|x\| = 1\}$ of a matrix S the following holds: $\sigma(S) \subset V(S)$ and, for normal S, $\mathrm{co}(\sigma(S)) = V(S)$ (see [4]). From these facts we obtain $\sigma(\mu M + N) \subset V(\mu M + N) \subset \mu V(M) + V(N)$ and

$$\begin{aligned}
\sup_{\mu\in\sigma(A)} \mathrm{Re}(\sigma(\mu M + N)) &\leq \sup_{\mu\in\sigma(A)} (\mathrm{Re}(\mu V(M)) + V(N)) \\
&\leq \sup_{\mu\in\sigma(A)} (\mathrm{Re}(\mu\,\mathrm{co}(\sigma(M)) + \mathrm{co}(\sigma(N)))) \\
&= \sup_{\mu\in\sigma(A)} (\mathrm{Re}(\mu\sigma(M)) + \sigma(N)) \\
&\leq \sup_{\mu\in\sigma(A)} \mathrm{Re}(\mu\sigma(M)) + s(N) \\
&= \sup_{\alpha\in\sigma(M)} s(\alpha A) + s(N) .
\end{aligned}$$
□

Remark. If only M is assumed to be normal we obtain

$$s(\mathcal{A}) \leq \sup_{\alpha\in\sigma(M)} s(\alpha A) + \tfrac{1}{2}\, s(N + N^*) .$$

References.

[1] Dunford, N.; Schwartz, J.T.
Linear Operators, Vol.I.
Interscience Publishers, New York 1957.

[2] Engel, K.-J.
Operatormatrizen und einparametrige Halbgruppen.
Diplomarbeit, Tübingen 1985.

[3] Kato, T.
Perturbation Theory of Linear Operators.
Springer, Berlin-Heidelberg-New York, 1966.

[4] Marcus, M.; Minc, H.
A Survey of Matrix Theory and Matrix Inequalities.
Allyn and Bacon, Inc., Boston 1964.

[5] Nagel, R.
Well-posedness and Positivity for Systems of Linear Evolution Equation.
Conf. Sem. Mat. Univ. Bari, 203 (1985).

[6] Nagel, R.
On Operator Matrices with Unbounded Entries.
Semesterbericht Funktionalanalysis
Tübingen, Wintersemester 1984/85.

[7] Nagel, R.
One-parameter Semigroups of Positive Operators.
To appear in Lecture Notes Math.,Springer Verlag.

SOME EXTENSIONS OF THOMAS-FERMI THEORY

Jerome A. Goldstein and Gisèle Ruiz Rieder
Department of Mathematics
and Quantum Theory Group
Tulane University
New Orleans, LA 70118, USA

0. OVERVIEW

We shall formulate and solve a variant of the Thomas-Fermi problem for ground state electron densities. Two novelties are present in this study: We work in $\mathbb{R}^d$ with $d \geq 3$ rather than $d = 3$, and we use weight functions in the (approximate) kinetic energy term. Our presentation is patterned in outline form after the papers of Brezis [3], [4].

1. INTRODUCTION; THE ELECTRON DENSITY

The quantum mechanical Hamiltonian (or energy operator) for a system of N electrons in $\mathbb{R}^d$ is

$$H = T + V_{ne} + V_{ee} + V_{nn}$$

where

$$T = -\frac{1}{2}\Delta = -\frac{1}{2}\sum_{j=1}^{N} \Delta_j \quad \text{is the kinetic energy,}$$

$$V_{ne} = \sum_{j=1}^{N} V(x_j) \quad \text{is the nuclear-electron potential energy,}$$

$$V_{ee} = \frac{1}{2}\sum_{j\neq k} \frac{1}{|x_j - x_k|^{d-2}} \quad \text{is the electron-electron repulsion energy, and}$$

V_{nn}, a constant, is the nuclear-nuclear repulsion energy. Here $d \geq 3$, $x = (x_1,\ldots,x_N) \in \mathbb{R}^{dN}$, where $x_j \in \mathbb{R}^d$ is the position of the jth electron, and most of the physical constants have been set equal to unity. H acts on the space $L_a^2(\mathbb{R}^{dN})$ of antisymmetric square integrable functions on $\mathbb{R}^{dN}$, and, in the most important special case,

$$V(y) = -\sum_{j=1}^{M} \frac{Z_j}{|y-R_j|^{d-2}} \quad (y \in \mathbb{R}^d), \tag{1}$$

where for $j = 1,\ldots,M$, a nucleus with Z_j protons is at the (fixed) point R_j in $\mathbb{R}^d$. In this case

$$V_{nn} = \frac{1}{2} \sum_{j \neq k} \frac{Z_j Z_k}{|R_j - R_k|^{d-2}} .$$

T is a differential operator while V_{ne}, V_{ee}, and V_{nn} are multiplication operators. Since V_{nn} is a constant, it can be ignored in the sequel. Introducing spin causes some minor complications in the wave function formulation below but not in the electron density formulation; thus we ignore spin.

The ground state wave function of H is a unit vector Ψ in $L_a^2(\mathbb{R}^{dN})$ satisfying $H\Psi = E\Psi$ where $E = \inf \sigma(H)$ is the *ground state energy*, i.e. the bottom of the spectrum of H. The *ground state (electron) density* is given by

$$\rho(x_1) = N\int_{\mathbb{R}^d} \cdots \int_{\mathbb{R}^d} |\Psi(x_1,x_2,\ldots,x_N)|^2 \, dx_2 \ldots dx_N; \tag{2}$$

we have $0 \le \rho \in L^1(\mathbb{R}^d)$, $\int_{\mathbb{R}^d} \rho(y)dy = N$ and $\int_\Gamma \rho(y)dy$ is the expected number of electrons one will find in the Borel set Γ in $\mathbb{R}^d$ at any instant of time.

When one is dealing with bulk matter, N is of the order of magnitude of 10^{26}, so even if the ground state wave function Ψ were given, it would be impossible to use it to compute properties of the system since so many variables would be involved. Thus one wishes to find the ground state density ρ and study its properties.

2. MINIMIZING THE ENERGY

Write $E_{QM}(\phi) = \langle H\phi,\phi\rangle$ for ϕ a unit vector in $L_a^2(\mathbb{R}^{dN})$. The ground state energy is

$$E = \inf \{E_{QM}(\phi) : \phi \in \mathrm{Dom}(H), \|\phi\| = 1\};$$

the Euler-Lagrange equation for this minimization problem is the eigenvalue problem $H\phi = \lambda\phi$ (where λ is a Lagrange multiplier). We are concerned with densities, i.e. nonnegative functions ρ in $L^1(\mathbb{R}^d)$ which integrate to N. Write

$$\Psi \longmapsto \rho$$

if one obtains the density ρ from the unit vector Ψ by means of equation (2). Our goal is to view the energy $E_{QM}(\phi)$ as a function of ρ where $\phi \longmapsto \rho$, and then to minimize the energy. But first we want to find, at least approximately, a formula for $E_{QM}(\phi)$ as a function of ρ.

For $S \in \{T, V_{ne}, V_{ee}, V_{nn}\}$, write $\hat{S}(\phi) = \langle S\phi, \phi \rangle$ for ϕ a unit vector in Dom(S). We want to find, in an approximate way, $\hat{S}(\phi)$ as a function of ρ; we shall denote our approximation of this by $\tilde{S}(\rho)$.

First, it is easy to see that

$$\hat{V}_{ne}(\phi) = \int_{\mathbb{R}^d} \rho(y)V(y)dy =: \tilde{V}_{ne}(\rho). \tag{3}$$

Also,

$$\tilde{V}_{nn}(\rho) = \hat{V}_{nn}(\phi) \text{ is the constant } V_{nn}. \tag{4}$$

For $\tilde{V}_{ee}(\rho)$ we take the Hartree approximation

$$\tilde{V}_{ee}(\rho) = \frac{c}{2} \int_{\mathbb{R}^d}\int_{\mathbb{R}^d} \frac{\rho(x)\rho(y)}{|x-y|^{d-2}} \, dxdy. \tag{5}$$

Usually one takes $c = 1$. The Fermi-Amaldi approximation is $c = (N-1)/N$; this essentially agrees with $c = 1$ for N large and vanishes for $N = 1$, when there is no electron-electron repulsion.

It is not obvious how to choose $\tilde{T}(\rho)$; this is where Thomas [17] and Fermi [7] made their contributions. We shall "derive" $\tilde{T}(\rho)$ from scaling considerations.

3. SCALING

For ϕ a unit vector in $L^2_a(\mathbb{R}^{dN})$ and $\lambda > 0$, define the unit vector ϕ_λ by $\phi_\lambda(x) = \lambda^{dN/2}\phi(\lambda x)$. Then an elementary calculation shows

$$\hat{V}_{ee}(\phi_\lambda) = \lambda^{d-2}\,\hat{V}_{ee}(\phi).$$

The corresponding $\tilde{V}_{ee}$ scales the same way. More precisely, if ρ is a density of total mass N and $\rho_\lambda(y) = \lambda^d \rho(\lambda y)$ for $\lambda > 0$ and $y \in \mathbb{R}^d$, then $0 \leq \rho_\lambda$ and $\int_{\mathbb{R}^d} \rho_\lambda(y)dy = N$. Moreover,

$$\tilde{V}_{ee}(\rho_\lambda) = \lambda^{d-2}\tilde{V}_{ee}(\rho).$$

The scaling property for the kinetic energy is easily seen to be

$$\hat{T}(\phi_\lambda) = \frac{1}{2}\int_{\mathbb{R}^{dN}} |\nabla\phi_\lambda(x)|^2\, dx = \lambda^2\hat{T}(\phi).$$

We set

$$\tilde{T}(\rho) = c_{TF}\int_{\mathbb{R}^d} \rho(y)^p\, dy$$

where $c_{TF} > 0$ and $p > 0$. The corresponding scaling equation is

$$\begin{aligned}\tilde{T}(\rho_\lambda) &= c_{TF}\int_{\mathbb{R}^d} [\lambda^d \rho(\lambda y)]^p\, dy \qquad (x = \lambda y)\\ &= c_{TF}\lambda^{dp-d}\int_{\mathbb{R}^d} \rho(x)^p\, dx = \lambda^{d(p-1)}\tilde{T}(\rho).\end{aligned}$$

In order that $\tilde{T}$ scales in the same way as $\hat{T}$ we must choose p so that $d(p-1) = 2$. This entails

$$p = 1 + \frac{2}{d}\,.$$

This becomes $p = 5/3$ when $d = 3$; in this case $\tilde{T}(\rho)$ becomes the classical approximation of Thomas and Fermi. (We omit a discussion of how to choose c_{TF} on physical grounds.)

We will work with the more general kinetic energy term

$$\tilde{T}(\rho) = \int_{\mathbb{R}^d} J(\rho(y))w(y)dy. \tag{6}$$

Here J is a convex function on $[0,\infty)$ satisfying $J(0) = J'(0) = 0$, $J'' \geq 0$, $J(r) > 0$ for $r > 0$. Regarding the *weight function* w we assume the following condition.

(W) $w : \mathbb{R} \to \mathbb{R}$ *is measurable and satisfies* $\varepsilon \leq w(x) \leq \varepsilon^{-1}$ *for all* $x \in \mathbb{R}^d$ *and some* $\varepsilon > 0$.

A more general condition is the following one.

(W') *There are points* $X_1, \ldots, X_Z, Y_1, \ldots, Y_P$ *in* $\mathbb{R}^d$ $(Z, P \in \{0,1,2,\ldots\})$ *and* $\varepsilon > 0$, $\delta > 0$ *such that outside of a* δ*-neighborhood of* $\{X_1, \ldots, X_Z, Y_1, \ldots, Y_P\}$, $\varepsilon \leq w(x) \leq \varepsilon^{-1}$, *and* $w(x) \to 0$ [*resp.* $w(x) \to \infty$] *as* $x \to X_j$ [*resp.* $x \to Y_k$] *at the rate* $|x - X_j|^{\alpha_j}$ [*resp.* $|x - Y_k|^{-\beta_k}$].

For analogues of the results presented below in the context of (W') see Rieder [16].

4. THE MINIMIZATION PROBLEM

Let

$$\tilde{E}(\rho) = \tilde{T}(\rho) + \tilde{V}_{ne}(\rho) + \tilde{V}_{ee}(\rho) + \tilde{V}_{nn}(\rho)$$

(see (3), (4), (5), (6)). We allow for a general potential V. The domain of $\tilde{E}$ is

$$D_N := \{\rho \geq 0 : \rho \in L^1(\mathbb{R}^d),\ \int_{\mathbb{R}^d} \rho(y)dy = N,$$
$$\rho \in \mathrm{Dom}(\tilde{T}) \cap \mathrm{Dom}(\tilde{V}_{ne}) \cap \mathrm{Dom}(\tilde{V}_{ee})\}.$$

Here $N > 0$; N need not be an integer. It is easy to show that the operator $\tilde{E}$ on D_N is convex.

PROBLEM 1. *Find* $\rho_0 \in D_N$ *such that*

$$\tilde{E}(\rho_0) \leq \tilde{E}(\rho)$$

for all $\rho \in D_N$.

Note that $\tilde{V}_{nn}(\rho)$ is a constant, so as far as Problem 1 is concerned, it is irrelevant.

If $0 \leq V \in L^1_{loc}(\mathbb{R}^d)$ and $V(x) \to 0$ as $|x| \to \infty$, then by taking $\rho_R = k_R \chi_{B(R)}$ where $\chi_{B(R)}$ is the characteristic function of the ball $B(R)$ of radius R centered at the origin and $k_R = N(\int_{B(R)} dx)^{-1}$, we compute that $\rho_R \in D_N$ and

$$\tilde{E}(\rho_R) \to V_{nn} = \inf\{\tilde{E}(\rho) : \rho \in D_N\}$$

as $R \to \infty$. (Cf. Brezis [3, p.82].) Thus the unique solution of

$$\begin{aligned}\tilde{E}(\rho_0) &= \min\{\tilde{E}(\rho) : \rho \in \cup\{D_M : 0 \leq M \leq N\}\} \\ &= \inf\{\tilde{E}(\rho) : \rho \in D_N\}\end{aligned}$$

is $\rho_0 \equiv 0$, which does not belong to D_N. Thus a necessary condition for the existence of a solution is:

(V) $V \in L^1_{loc}(\mathbb{R}^d)$ *and* $V < 0$ *on a set of positive measure.*

Define the operator B by $B\rho = \dfrac{C_d}{|\cdot|^{d-2}} * \rho$, i.e.

$$B\rho(x) = C_d \int_{\mathbb{R}^d} \frac{\rho(y)}{|x-y|^{d-2}}\, dy;$$

here $C_d = [(d-2)\sigma_d]^{-1}$ with

$$\sigma_d = \begin{cases} \dfrac{d\pi^{d/2}}{(d/2)!} & \text{for } d \text{ even} \\ \dfrac{d2^{(d+1)/2}\pi^{(d-1)/2}}{d!!} & \text{for } d \text{ odd,} \end{cases}$$

so that $B = (-\Delta)^{-1}$.

If $\tilde{E}(\rho)$ is minimized at ρ_0, then, formally,

$$\begin{aligned}\tilde{E}(\rho) &= \int_{\mathbb{R}^d} J(\rho) w\, dx + \int_{\mathbb{R}^d} V\rho\, dx + \frac{k}{2}\int_{\mathbb{R}^d} \rho B\rho\, dx + V_{nn} \\ &= \tilde{E}(\rho_0) + \langle \tilde{E}'(\rho_0), \rho - \rho_0\rangle + o(\rho - \rho_0) \qquad (7)\end{aligned}$$

where $k = cC_d^{-1}$. The derivative $\tilde{E}'(\rho_0)$ acts on $\rho-\rho_0 \in L^1(\mathbb{R}^d)$ to produce a real number, hence it is a bounded linear functional, that is, a function in $L^\infty(\mathbb{R}^d)$, and the bracketed term in (7) means $\int_{\mathbb{R}^d} \tilde{E}'(\rho_0)(\rho-\rho_0)dx$. Since $\tilde{E}$ is a convex functional of ρ, we expect the solution of Problem 1 to be the unique ρ_0 for which $\tilde{E}'(\rho_0) = 0$.

In computing $\tilde{E}'(\rho)$ we must take into account that $\int \rho = N$; to handle this we add to $\tilde{E}(\rho)$ the term $\lambda(\int_{\mathbb{R}^d} \rho(x)dx - N)$ (where λ is a Lagrange multiplier). Then, formally,

$$\tilde{E}'(\rho) = wJ'(\rho) + V + kB\rho + \lambda.$$

There is one other constraint to take into account: $\rho \geq 0$. Thus if ρ_0 is the minimizing density, in the formal expression (7) we must have $\rho = \rho-\rho_0 \geq 0$ on the set where $\rho_0 = 0$. This leads to the following Euler-Lagrange problem associated with Problem 1.

PROBLEM 2. *Find* $\rho_0 \in D_N$ *and* $\lambda \in \mathbb{R}$ *such that*

$wJ'(\rho) + V + kB\rho + \lambda = 0$ *for a.e.* x *for which* $\rho(x) > 0$,

$wJ'(\rho) + V + kB\rho + \lambda \geq 0$ *for a.e.* x *for which* $\rho(x) = 0$.

PROPOSITION. *If* ρ_0 *is a solution of Problem* 1, *then there is a* $\lambda \in \mathbb{R}$ *such that* (ρ_0,λ) *is a solution of Problem* 2. *Conversely, if* (ρ_0,λ) *is a solution of Problem* 2, *then* ρ_0 *is a solution of Problem* 1 *provided*

$$x \to w(x)J^*((M-V(x))_+/w(x)) \in L^1(\mathbb{R}^d) \tag{8}$$

for some real number M.

Here J^* is the convex conjugate of J:

$$J^*(r) = \sup\{rs - J(s) : s \geq 0\};$$

and for $r \in \mathbb{R}$, r_+ is the positive part of r. The condition (8) involves J and V and w and ensures that $\inf \tilde{E}(\rho) > -\infty$. If

$\liminf_{|x|\to\infty} V(x) > -\infty$, then by taking M small enough we see that (8) is a condition on J and the local singularities of V and the zeroes of w.

Problem 1 was solved by Lieb and Simon for $d = 3$, $J(r) = cr^{5/3}$, $w \equiv 1$ in a pioneering work [13], [14]. Problem 2 was solved by Benilan and Brezis [1], [3], [4] for $w \equiv 1$, $d = 3$. Our work is patterned after [3], [4]. We shall solve Problem 2 with the aid of the theory of nonlinear elliptic partial differential equations in $L^1(\mathbb{R}^d)$.

When $J(r) = c_1 r^p$ $(1 < p < \infty)$ and when V is the molecular Coulomb potential (1), then $J^*(r) = c_2 r^q$ when $p^{-1} + q^{-1} = 1$, and (8) holds (assuming (W)) if and only if $p > d/2$. We shall see below that Problem 2 has a solution whenever $p > 2 - 2/d$.

5. REDUCTION TO A NONLINEAR ELLIPTIC EQUATION

Let (ρ,λ) be a solution of Problem 2. Let

$$u(x) = (-V(x) - kB\rho(x))/w(x).$$

On the set where $\rho > 0$, $J'(\rho) = u - \lambda/w$. Define

$$\gamma(r) = \begin{cases} (J')^{-1}(r) & \text{for } 0 < r < \infty \\ 0 & \text{for } -\infty < r \leq 0. \end{cases} \tag{9}$$

Then Problem 2 formally takes the form

$$\rho = \gamma(u - \lambda/w) \quad \text{a.e.}$$

If $v = wu$, this becomes (after applying $-\Delta = B^{-1}$)

$$\Delta v - k\gamma\left(\frac{v-\lambda}{w}\right) = \Delta V,$$

together with $v(x) \to 0$ as $|x| \to \infty$ in a weak sense and $\int\rho = N(= \int\gamma((v-\lambda)/w))$. More precisely, Problem 2 is equivalent to

PROBLEM 3. *Find a real function* v *in the Marcinkiewicz space (or weak* L^p *space)* $M^{d/(d-2)}(\mathbb{R}^d)$ *(cf.* [2]*) and a number* λ *such that*

$$\Delta v - k\gamma\left(\frac{v-\lambda}{w}\right) = \Delta V \quad \text{a.e.},$$

$$\int_{\mathbb{R}^d} \gamma((v(x) - \lambda)/w(x))dx = N.$$

The detailed proof of this equivalence is omitted.

6. SOLUTION OF PROBLEM 2

We assume (W) and (V). Regarding J we assume (see (9)) that

$$\int_{|x|>1} \gamma(|x|^{2-d}(w(x))^{-1})dx = \infty, \tag{10}$$

and that for some $K \in \mathbb{R}$,

$$\begin{aligned} x &\to \gamma((K-V(x))/w(x)) \in L^1(\mathbb{R}^d), \\ x &\to \gamma((t-V(x))/w(x)) \in L^1_{loc}(\mathbb{R}^d) \end{aligned} \tag{11}$$

for all real t. (Note that if $V(x) = -Z/|x|^{d-2}$ and $K < 0$, assuming (W), then (10) and (11) hold for $J(r) = cr^p$ if and only if $p > 2 - 2/d$. Conditions (10), (11) could have been stated without mention of w. But we want to give an indication of the sort of hypotheses involved when (W') holds.) We also assume that ΔV is either a finite signed measure or else in $L^1(\mathbb{R}^d)$; in either case the integral $\int_{\mathbb{R}^d} \Delta V$ makes sense.

THEOREM. *Under the above hypotheses, there exists an* $N_0 > 0$ *such that Problem* 2 *has a unique solution for* $0 < N \leq N_0$ *and no solution for* $N > N_0$. *If* $V(x) \to 0$ *as* $|x| \to \infty$ *and* $0 < N < N_0$, *then the density* ρ *which solves Problem* 2 *has compact support in* $\mathbb{R}^d$. *Furthermore,*

$$\int_{\mathbb{R}^d} (-\Delta V) \leq N_0 \leq \int_{\mathbb{R}^d} (-\Delta V)_+. \tag{12}$$

For V given by (1) it follows from (12) that

$$N_0 = \sum_{j=1}^{M} Z_j .$$

The proof of the theorem is based on the proofs of Benilan and Brezis [3], [4]. We solve Problem 2 by solving Problem 3. The idea is first to freeze λ, set

$$\beta(x,v) = k\gamma\left(\frac{v - \lambda}{w(x)}\right),$$

and solve

$$-\Delta v + \beta(x,v(x)) = -\Delta V \tag{13}$$

subject to v vanishing at ∞ in a weak sense. Benilan and Brezis used the results of Benilan, Brezis, and Crandall [2] for this purpose. But the results of [2] are too special to apply to (13); we rely on the more general results of Gallouët and Morel [8], [9].

Having solved (13), call the solution v_λ. Then set

$$I(\lambda) = \int_{\mathbb{R}^d} \gamma\left(\frac{v_\lambda(x) - \lambda}{w(x)}\right) dx.$$

One shows that I is continuous on $[0,\infty)$, $I(0) > 0$, I is strictly decreasing where it is positive, and $\lim_{\lambda\to\infty} I(\lambda) = 0$. Much of the theorem follows from this with $N_0 = I(0)$. For full details and greater generality, see Rieder [16].

7. STABILITY OF MATTER

Consider now the molecular Coulomb potential V given by (1) (and do not ignore the V_{nn} terms in the definition of $\mathcal{E}_{QM}(\phi) = \langle H\phi, \phi\rangle$). We assume $d = 3$. Write $E_N = \inf \sigma(H)$ for the (true) ground state energy of the system. The *stability of matter theorem* of Dyson and Lenard [5], [10] and of Lieb and Thirring [15] says that $E_N > -CN$ for some absolute constant C and all $N = 1,2,3,\ldots$. Moreover, there are positive constants c_0, c_1 such that if Ψ is the ground state wave function and if $\Psi \mapsto \rho$, ρ being the true electron density, then

$$c_0\tilde{T}(\rho) = c_1\int_{\mathbb{R}^3}\rho(x)^{5/3}\,dx \leq \langle\nabla\Psi,\nabla\Psi\rangle = \hat{T}(\Psi).$$

The fact that $\rho \in L^1 \cap L^{5/3}(\mathbb{R}^3)$ implies that matter cannot be too heavily concentrated. In particular, if $\Omega \subset \mathbb{R}^3$ is a region containing at least half the electrons in the sense that $\int_\Omega \rho(x)dx \geq N/2$, then one can show that $\mathrm{Vol}(\Omega) \geq c_2 N$ for some absolute constant $c_2 > 0$ and all N. Beautiful proofs of these and related results were given by Lieb and Thirring [15], [11] in the mid 1970s. Their proofs were partly based on Thomas-Fermi theory (see also [12]). Deep and important extensions were discovered in 1983 by Fefferman and Phong (cf. [6, pp.159-179, especially Theorems 9 and 10]). We suspect that various stability of matter theorems are valid in d (≥ 3) dimensions, but that is a subject for further investigations.

This work was partially supported by an NSF grant, which is gratefully acknowledged.

REFERENCES

[1] Benilan, Ph. and H. Brezis, The Thomas-Fermi problem, in preparation.

[2] Benilan, Ph., H. Brezis, and M. G. Crandall, A semilinear elliptic equation in $L^1(\mathbb{R}^N)$, Ann. Scuola Norm. Sup. Pisa 2(1975), 523-555.

[3] Brezis, H., Nonlinear problems related to the Thomas-Fermi equation, in Contemporary Developments in Continuum Mechanics and Partial Differential Equations (ed. by G. M. de la Penha and L. A. Medieros), North-Holland, Amsterdam (1978), 81-89.

[4] Brezis, H., Some variational problems of Thomas-Fermi type, in Variational Inequalities and Complementary Problems: Theory and Applications (ed. by R. W. Cottle, F. Giannessi, and J.-L. Lions), Wiley, New York (1980), 53-73.

[5] Dyson, F. J. and A. Lenard, Stability of Matter, I, J. Math. Phys. 8 (1967), 423-434.

[6] Fefferman, C. L., The uncertainty principle, Bull. Amer. Math. Soc. 9 (1983), 129-206.

[7] Fermi, E., Un metodo statistico per la determinazione di alcune prioretà dell'atome, Rend. Acad. Naz. Lincei 6 (1927), 602-607.

[8] Gallouët, Th. and J.-M. Morel, Resolution of a semilinear equation in L^1, Proc. Roy. Soc. Edinburgh 96A (1984), 275-288 and 99A (1985), 399.

[9] Gallouët, Th. and J.-M. Morel, On some semilinear problems in L^1, Boll. Un. Mat. Ital. 4A (1985), 123-131.

[10] Lenard, A. and F. J. Dyson, Stability of matter, II, J. Math. Phys. 9 (1968), 698-711.

[11] Lieb, E. H., The stability of matter, Rev. Mod. Phys. 48 (1976), 553-569.

[12] Lieb, E. H., Thomas-Fermi theory and related theories of atoms and molecules, Rev. Mod. Phys. 53 (1981), 603-641.

[13] Lieb, E. H. and B. Simon, Thomas-Fermi theory revisited, Phys. Rev. Lett. 33 (1973), 681-683.

[14] Lieb, E. H. and B. Simon, The Thomas-Fermi theory of atoms, molecules and solids, Adv. Math. 23 (1977), 22-116.

[15] Lieb, E. H. and W. E. Thirring, A bound for the kinetic energy of fermions which proves the stability of matter, Phys. Rev. Lett. 35 (1975), 687-689 and 1116.

[16] Rieder, G. R., Mathematical Contributions to Thomas-Fermi Theory, Ph.D. Thesis, Tulane University, New Orleans, 1986.

[17] Thomas, L. H., The calculation of atomic fields, Proc. Camb. Phil. Soc. 23 (1927), 542-548.

THE EXTENT OF SPATIAL REGULARITY FOR PARABOLIC INTEGRODIFFERENTIAL EQUATIONS

Ronald Grimmer
Department of Mathematics
Southern Illinois University
Carbondale, Illinois 62901
U.S.A.

Eugenio Sinestrari
Dipartimento di Matematica
Università di Roma
Piazzale Aldo Moro 2
00785 ROMA, Italy

1. Introduction.

In this note we are concerned with the "smoothing" properties of the resolvent operator for the linear integrodifferential equation

$$\begin{aligned} x'(t) &= Ax(t) + \int_0^t B(t-s)x(s)ds \\ x(0) &= x_0 \end{aligned} \tag{1.1}$$

in a Banach space X with norm $\|\cdot\|$. We shall assume throughout this paper that A generates an analytic semigroup T(t) on X and that B(t)x is Bochner integrable for each x in the domain of A.

In the case $B \equiv 0$ the solution of (1.1) is given by $T(t)x_0$ and it is well known that $T(t)x_0 \in D(A^n)$ $n = 1,2,3,\ldots$, $t > 0$. That is, T(t) is infinitely smoothing. In this note we examine the solution operator or resolvent operator of (1.1) and show that this is not the case in general for the resolvent operator for (1.1).

The resolvent operator for (1.1) is obtained as an integral over an appropriate contour. This technique for obtaining the resolvent operator has been used by a number of authors. In particular, see DaPrato and Iannelli [3], Friedman and Shinbrot [5], Grimmer and Kappel [6] and Grimmer and Pritchard [7].

As A generates an analytic semigroup on X, the domain of A, D(A), together with the graph norm forms a Banach space we denote by Y. We will use a family Y^α $(0 < \alpha < 1)$ of intermediate spaces between Y and X with norm $\|\cdot\|_\alpha$ such that

(i) $0 < \alpha < \beta < 1 \Rightarrow Y \hookrightarrow Y^\beta \hookrightarrow Y^\alpha \hookrightarrow X$

(ii) if $x \in Y^\alpha$ and $t > 0$ we have $\|AT(t)x\| \le ct^{\alpha-1}\|x\|_\alpha$ with We c independent of x and t.

We also set $Y^0 = X$ and $Y^1 = Y$ and if $n \le \alpha < n+1$, $n \in \mathbb{N}$, $Y^\alpha = \{x \in D(A^n) : A^n x \in Y^{\alpha-n}\}$. Such a family can be obtained in various ways: by using the interpolation spaces of Butzer and Berens [1] (see (iii) in example 2.3 later) or the domains of the α-power of

$-A$. The space of bounded linear operators from a Banach space V to another Banach space W we denote by $\mathcal{B}(V,W)$ unless V = W in which case we write simply $\mathcal{B}(V)$. The norm on the space $\mathcal{B}(Y^{\alpha},Y^{\beta})$ shall be written $\|\cdot\|_{\alpha,\beta}$. The Laplace transform is central to our work and the Laplace transform of a function h(t) shall be written $\hat{h}(\lambda)$.

We first note that under appropriate conditions the resolvent or solution operator R(t) maps X into Y for t > 0 and that if $x_0 \in Y^{\alpha}$, $\alpha > 0$, $R(t)x_0$ is a "solution" of (1.1) for t > 0. We shall then obtain conditions on B(t) which ensure that $R(t): X \to Y^{\alpha}$, $2 \geq \alpha > 1$, for t > 0. We show by example that our results cannot be strengthened to $R(t) : X \to Y^{\alpha+\varepsilon}$ where $\varepsilon > 0$. Interpreting $R(t)x_0$ as a solution of a partial differential integral equation, we are presented with a situation where the solution is infinitely differentiable with respect to t but only has a finite degree of spatial regularity, a situation different from that arising in semigroup theory for partial differential equations.

Related work concerning the mapping properties of R(t) can be found in DaPrato and Iannelli [2], Friedman and Shinbrot [5], Grimmer and Kappel [6], Grimmer and Pritchard [7] and Hannsgen and Wheeler [9].

Finally, we apply our results to a problem suggested in the study of viscoelastic fluids. The problem involves a viscoelastic fluid filling a semi-infinite space between two flat plates. The fluid is initially assumed quiescent and then a step jump in the velocity of the lower plate is implemented. If there is a Newtonian contribution in the equation for the velocity of the fluid, the fluid is not really considered viscoelastic, [10], and the discontinuity in the velocity is not expected to be propagated into the interior of the fluid. This is precisely the result obtained in the example.

Our basic hypotheses are those of [6] restated for a sector in the complex plane.

(H1) A generates an analytic semigroup T(t) and satisfies the estimate $\|(\lambda I - A)^{-1}\| \leq M/|\lambda|$, $M \geq 1$, $\lambda \in \Lambda$ where $\Lambda = \{\lambda \in \mathbb{C}: |\arg \lambda| < \frac{\pi}{2} + \delta\}$, $0 < \delta < \frac{\pi}{2}$, and, in addition, $A^{-1} \in \mathcal{B}(X)$.

(H2) For $t \geq 0$, $B(t) \in \mathcal{B}(Y,X)$, $\|B(t)\|_{1,0} \leq b(t)$ for some $b \in L^1_{loc}(0,\infty)$ and B(t)x is strongly measurable for each $x \in D(A)$.

(H3) $\hat{B}(\lambda)$ has an analytic extension to Λ and for $\lambda \in \Lambda$, $\|\hat{B}(\lambda)\|_{1,0} \leq N/|\lambda|^{\beta}$, $N \geq 1$ for some $\beta > 0$.

Our results shall be stated in terms of the resolvent operator or solution operator of (1.1).

<u>Definition 1.1</u>. A solution of (1.1) is a function $x \in C([0,\infty),Y) \cap C^1([0,\infty),X)$ with $x(0) = x_0$ which satisfies (1.1) for $t \geq 0$.

<u>Definition 1.2</u>. $R(t)$ is a resolvent operator for (1.1) if $R(t) \in \mathcal{B}(X)$, $0 \leq t < \infty$, and if it satisfies:

(a) $R(t)$ is strongly continuous for $t \geq 0$ with $R(0) = I$.

(b) For each $t \geq 0$, $R(t) \in \mathcal{B}(Y)$ and $R(t)$ is strongly continuous, $t \geq 0$, on Y.

(c) For each $x \in Y$, $R(t)x$ is continuously differentiable $t \geq 0$, with

$$R'(t)x = AR(t)x + \int_0^t B(t-u)R(u)x\,du$$

and

$$R'(t)x = R(t)Ax + \int_0^t R(t-u)B(u)x\,du.$$

It is clear from the definition of a resolvent operator that if $x_0 \in Y$, then $R(t)x_0$ is the solution of (1.1). In addition, we remark that if a resolvent operator exists, it is unique. For a general discussion of resolvent operators see, for example, Desch, Grimmer and Schappacher [4] and Grimmer and Prüss [8] which also contain further references.

2. <u>Smoothing</u>.

Our interest here is to consider $R(t)x$ when $x \notin Y$. In particular, we wish to determine when $R(t)x$ satisfies (1.1) for $t > 0$ and how smooth $R(t)x$ is as measured by the space Y^α so that $R(t)x \in Y^\alpha$ for $t > 0$. With reference to the physical examples of (1.1) (see e.g. at the end of this paper) this smoothness is a regularity of the solution with respect to the spatial variables.

<u>Theorem 2.1</u>. Assume (H1) - (H3) are valid. Then (1.1) has a resolvent operator $R(t)$ which is strongly continuously differentiable for $t > 0$ and satisfies $R(t) : X \to Y$, $t > 0$. If, in addition, $b(t)$ is bounded on bounded intervals of the form $0 < T_1 \leq t \leq T_2 < \infty$ and $\alpha > 0$, then

$$R'(t)x = AR(t)x + \int_0^t B(t-s)R(s)xds, \quad t > 0, \ x \in Y^\alpha.$$

<u>Proof</u>. Define R(0) = I and for t > 0 define R(t) as in [6] by

$$R(t)x = T(t)x + (2\pi i)^{-1} \int_\Gamma e^{\lambda t} \sum_{j=1}^{\infty} ((\lambda I - A)^{-1}\hat{B}(\lambda))^j (\lambda I - A)^{-1} x d\lambda = T(t)x + R_1(t)x. \tag{(2.1)}$$

Here $\Gamma = \{re^{i\theta}\} \cup \{re^{-i\theta}\} \cup \{e^{i\omega}\}$, $1 < r < \infty$, $\frac{\pi}{2} < \theta < \frac{\pi}{2} + \delta$, $-\theta < \omega < \theta$, oriented so Im λ increases along Γ.

It has been previously shown in [6] that the operator R(t) defined in this manner is a resolvent operator. To obtain the remainder of the theorem one argues as in [6] using Lemma 2.1 of [6].

We now turn to the problem of determining when $R(t) : X \to Y^\alpha$, $\alpha > 1$. As we shall see, this occurs when B(t) is of "lower order than A". In particular, our meaning of "lower order than A" is embodied in our next hypothesis.

(H4) For each $\lambda \in \Lambda$, $\hat{B}(\lambda) \in \mathfrak{B}(Y, Y^\alpha)$ $(0 < \alpha < 1)$ and there exist positive c, σ, L such that

$$\|\hat{B}(\lambda)x\|_\alpha \le c|\lambda|^{-\sigma}\|x\|_1$$

for $\lambda \in \Lambda$, $|\lambda| \ge L$ and $x \in Y$.

<u>Theorem 2.2</u>. Assume (H1) - (H4). Then $R(t) : X \to Y^{1+\alpha}$.

<u>Proof</u>. An examination of (2.1) shows that R(t)x can be written

$$R(t)x = (2\pi i)^{-1} \int_\Gamma e^{\lambda t}(\lambda I - A - \hat{B}(\lambda))^{-1} x d\lambda$$

which shall be more convenient for us here. Further, as $R(t) : X \to Y$, $t > 0$, we may consider AR(t)x for t > 0 and

$$(2\pi i)AR(t)x = \int_\Gamma e^{\lambda t} A(\lambda I - A - \hat{B}(\lambda))^{-1} x d\lambda$$
$$= \int_\Gamma e^{\lambda t}[-I + (\lambda I - \hat{B}(\lambda))(\lambda I - \hat{B}(\lambda) - A)^{-1}] x d\lambda$$

$$= \int_{\Gamma} e^{\lambda t}(\lambda I-\hat{B}(\lambda))(\lambda I-\hat{B}(\lambda)-A)^{-1}x d\lambda.$$

Now

$$(\lambda I-\hat{B}(\lambda)-A)^{-1} = \sum_{j=0}^{\infty} [(\lambda I-A)^{-1}\hat{B}(\lambda)]^{j}(\lambda I-A)^{-1}$$

and an application of Lemma 2.1 of [6] yields that this is a bounded operator from X into Y. Hypothesis (H4) thus implies that this integral converges in Y^{α}. This completes the proof.

Now we give an example showing that the results of Theorem 2.1 and 2.2 cannot be improved in general; in this example we prove that if x(t) is a solution of (1.1) and $\hat{B}(\lambda) \in \mathcal{B}(Y,Y^{\alpha})$, $0 \le \alpha \le 1$ it follows that if $x(0) \in Y^{\beta}$, $0 \le \beta \le 1$ then $Ax(t) \in Y^{\alpha+\beta}$ for $t > 0$.

<u>Example 2.3</u>. Let X be the Banach space of all the sequences of complex numbers converging to zero $x = (x_n)$ with norm $\|x\| = \sup_{n\in\mathbb{N}} |x_n|$. Given $\alpha \in [0,1]$ define the Banach space

$$Y^{\alpha} = \{x \in X : \lim_{n\to\infty} n^{\alpha}x_n = 0\} \tag{2.2}$$

with the norm

$$\|x\|_{\alpha} = \sup_{n\in\mathbb{N}} |n^{\alpha}x_n|.$$

Obviously we have $Y^0 = X$. Let us define the linear operator

$$A : D(A) \subset X \to X$$

as follows

$$\begin{cases} D(A) = Y^1 \\ Ax = (-nx_n). \end{cases} \tag{2.3}$$

It can be easily checked that

(i) A generates the analytic semigroup $T(t)x = (e^{-tn}x_n)$ of negative type;

(ii) $Y^1 \simeq D(A)$ where D(A) is given the norm of the graph of A;

(iii) $Y^{\alpha} = \{x \in X : \lim_{t\to 0} t^{1-\alpha}AT(t)x = 0\}$

(iv) for each $x \in Y^{\alpha}$ and $t > 0$, $\|AT(t)x\| \le t^{\alpha-1}\|x\|_{\alpha}$.

From these properties it follows that the Y^{α} can be chosen as intermediate spaces and A satisfies property (H1). For each $\alpha \in [0,1]$ let us define the linear operator

$$B : Y \to Y^{\alpha}$$

as follows

$$Bx = (-n^{1-\alpha}x_n).$$

Obviously, $B \in \mathcal{B}(Y,Y^{\alpha}) \cap \mathcal{B}(Y,X)$. Given $\omega > 0$ and $t > 0$ set

$$B(t) = e^{-\omega t}B.$$

As for $\mathrm{Re}\,\lambda > -\omega$, we have $\hat{B}(\lambda) = (\lambda+\omega)^{-1}B$ and it can be easily seen that (H2) - (H4) are satisfied. Thus, theorems 2.1 and 2.2 apply. With the preceding definitions, problem (1.1) can be written as

$$\begin{cases} x'(t) = Ax(t) + \int_0^t e^{-\omega(t-s)}Bx(s)ds, & t > 0 \\ x(0) = x_0. \end{cases}$$

Suppose that this problem has a solution for $t > 0$. We deduce the existence of $x_n : \mathbb{R}_+ \to C$ such that

$$\begin{cases} x_n'(t) = -nx_n(t) - n^{1-\alpha}\int_0^t e^{-\omega(t-s)}x_n(s)ds, & t > 0 \\ x_n(0) = x_{on} \end{cases} \tag{2.4}$$

where $x_0 = (x_{on})$. We shall prove the following result

$$x_0 \in Y^{\beta},\ 0 \le \beta \le 1 \Rightarrow Ax(t) \in Y^{\alpha+\beta},\ t > 0. \tag{2.5}$$

Differentiating (2.4) we get

$$\begin{cases} x_n''(t) + (\omega+n)x_n'(t) + (n^{1-\alpha}+\omega n)x_n(t) = 0 \\ x_n(0) = x_{on} \\ x_n'(0) = -nx_{on}. \end{cases} \tag{2.6}$$

We shall take n sufficiently great so that the characteristic equation associated with (2.6) has two distinct real roots

$$r_{1,2} = -\frac{\omega}{2} - \frac{n}{2}(1 \pm a_n)$$

where

$$a_n = \sqrt{(\frac{\omega}{n} - 1)^2 - \frac{4}{n^{1+\alpha}}}$$

and the solution of (2.6) is given by $x_n(t) = y_n(t) + z_n(t)$ where

$$y_n(t) = \frac{-x_{on}}{2na_n}(\omega-n(1+a_n))\exp[-(\omega+n(1+a_n))\frac{t}{2}]$$

$$z_n(t) = \frac{x_{on}}{2na_n}(\omega-n(1-a_n))\exp[-(\omega+n(1-a_n))\frac{t}{2}].$$

By using the following elementary properties

$$\lim_{n\to\infty} a_n = 1$$

$$\lim_{n\to\infty} n^{\alpha}[\omega-n(1-a_n)] = -2$$

$$\lim_{n\to\infty} \omega + n(1-a_n) = \begin{cases} 2\omega & \text{if } 0 < \alpha \le 1 \\ 2\omega+2 & \text{if } \alpha = 0 \end{cases}$$

we see that if $x_0 \in X$ and $\gamma > 0$ we have

$$\lim_{n\to\infty} n^{\gamma} y_n(t) = 0.$$

On the other hand, for each $\alpha, \beta \in [0,1]$ we can write

$$n^{\alpha+\beta+1} z_n(t) = n^{\beta} x_{on} \frac{n^{\alpha}[\omega-n(1-a_n)]}{2a_n} \exp\,[-(\omega+n(1-a_n))\frac{t}{2}]$$

and conclude that $\lim_{n\to\infty} n^{\alpha+\beta+1} z_n(t) = 0$ if and only if $\lim_{n\to\infty} n^{\beta} x_{on} = 0$.

From this (2.5) follows.

As a final example we consider a problem motivated by studies in viscoelasticity.

<u>Example 2.4</u>. Suppose a viscoelastic fluid fills a semi-infinite space between two flat plates

$$\Omega = \{(x,y,z) : 0 < x < 1,\ -\infty < y < \infty,\ -\infty < z < \infty\}$$

and one has a shear flow with the velocity given by $V = e_y v(x,t)$. Assuming the fluid is initially quiescent, a step jump in the velocity of the bottom plate yields the scalar problem

$$\rho \frac{\partial v}{\partial t}(x,t) = \mu \frac{\partial^2 v}{\partial x^2}(x,t) + \int_0^t G(t-s) \frac{\partial^2 v}{\partial x^2}(x,s)ds, \quad \mu > 0,$$

$$v(x,0) = 0 \qquad 0 < x < 1$$

$$v(1,t) = 0 \qquad t > 0$$

$$v(0,t) = 1 \qquad t > 0,$$

(cf. [10] or [11] for example as well as [12] concerning singular kernels G(t)).

For our purposes here it is convenient to consider the difference between v(x,t) and linear shear. Define w(x,t) by

$$w(x,t) = (1-x) - v(x,t)$$

so that w(x,t) satisfies

$$\rho \frac{\partial w}{\partial t}(x,t) = \mu \frac{\partial^2 w}{\partial x^2}(x,t) + \int_0^t G(t-s) \frac{\partial^2 w}{\partial x^2}(x,s)ds$$

$$w(x,0) = (1-x), \qquad 0 < x < 1$$

$$w(1,t) = w(0,t) = 0, \qquad t > 0.$$

Assuming $\mu > 0$, take $X = L^2(0,1)$, $D(A) = H^2(0,1) \cap H_0^1(0,1)$ and $Aw = (\mu/\rho)\dfrac{\partial^2 w}{\partial x^2}$. Then A generates an analytic semigroup on X and we may choose any sector Λ. If $\hat{G}(\lambda)$ can be extended into Λ with the estimate $|\hat{G}(\lambda)| \leq N/|\lambda|^{\beta}$, $N \geq 1$, $\lambda \in \Lambda$, $\beta > 0$, we see that (H1) - (H3) are valid. Thus, we can apply Theorem 2.1 to obtain that for $t > 0$, $w(x,t) \in H^2(0,1) \cap H_0^1(0,1)$. That is,

$$x \to v(x,t) - (1-x) \in H^2(0,1) \cap H_0^1(0,1)$$

for $t > 0$ and the discontinuity of the boundary and initial data does not propagate into the interior of the fluid.

3. Acknowledgment.

The first author gratefully acknowledges the support of C.N.R. of Italy for a visiting professorship at The University of Rome during the fall of 1984.

References

1. P. L. Butzer and H. Berens, Semi-Groups of Operators and Approximation, Springer-Verlag, New York, 1967.
2. G. DaPrato and M. Iannelli, Existence and regularity for a class of integrodifferential equations of parabolic type, to appear J. Math. Anal. Appl.

3. G. DaPrato and M. Iannelli, Linear integrodifferential equations in Banach spaces, Rend. Sem. Mat. Padova, 62(1980), 207-219.

4. W. Desch, R. Grimmer and W. Schappacher, Some considerations for linear integrodifferential equations, J. Math. Anal. Appl., 104(1984), 219-234.

5. A. Friedman and M. Shinbrot, Volterra integral equations in Banach space, Trans. Amer. Math. Soc., 126(1967), 131-179.

6. R. Grimmer and F. Kappel, Series expansions for resolvents of Volterra integrodifferential equations in Banach spaces, SIAM J. Math. Anal., 15(1984), 595-604.

7. R. Grimmer and A. J. Pritchard, Analytic resolvent operators for integral equations in Banach space, J. Diff. Eq., 50(1983), 234-259.

8. R. Grimmer and J. Prüss, On linear Volterra equations in Banach spaces, Special Issue on Hyperbolic Partial Differential Equations, Comp. Math. Appl. to appear.

9. K. B. Hannsgen and R. L. Wheeler, Behavior of the solution of a Volterra equation as a parameter tends to infinity, J. Integral Eq., 7(1984), 229-238.

10. D. D. Joseph, M. Renardy and J. Saut, Hyperbolicity and change of type in the flow of viscoelastic fluids, MRC Technical Summary Report #2657, Madison, Wisconsin, 1984.

11. A. Narain and D. D. Joseph, Remarks about the interpretation of impulse experiments in shear flows of viscoelastic liquids, Rheol. Acta 22(1983), 528-538.

12. M. Renardy, Some remarks on the propagation and non-propagation of discontinuities in linearly viscoelastic liquids, Rheol. Acta 21(1982), 251-254.

An approach to the singular solutions of elliptic problems via the theory of differential equations in Banach spaces

P. Grisvard (Nice - France)

Contents

1. – Introduction

. We focus on a very simple kind of first order ordinary differential equations with operator coefficients. The solutions of such equations exhibit a very peculiar singular asymptotic behavior. We believe that most corner problems for elliptic boundary value problems fall into that pattern after changing variables and reducing the order.

. To be more precise we consider u solution of the equation

$$(1,1) \qquad u'(t) + A\,u(t) = f(t)$$

on the half line $t \geqslant 0$ where f is a given function and A a closed operator having a compact resolvent with minimal growth on the imaginary axis at least near infinity. We call such equations <u>elliptic</u> since the operator valued symbol

$$\tau \longmapsto (i\tau + A)$$

is invertible for large values of $|\tau|$.

. The relation with corner problems is seen by performing the change of variable $r = e^{-t}$ which produces the new equation

$$(1,2) \qquad -r\,u'(r) + A\,u(r) = f(r)$$

in the interval $r \in]0,1[$. The asymptotic behavior when $t \longrightarrow +\infty$ corresponds to behavior when $r \longrightarrow 0$. Now A may stand for a matrix of ordinary first oder differential operators in an auxiliary variable θ varying in an interval $I =]0,\omega[$ with $0 < \omega \leqslant 2\pi$. The definition of the domain of A may include boundary conditions. This is a way of considering many boundary value problems written as first order systems in polar coordinates in the corner domain

$$(1,3) \qquad \Omega = \{(r\cos\theta, r\sin\theta)\ ;\ 0 < r < 1,\ 0 < \theta < \omega\}$$

It is beyond the size of this paper to show the extent of the application of our abstract results to concrete boundary value problems. This is why we shall restrict our illustration to the very simple minded example of § 10.

. Also a much handier abstract statement in view of the applications is likely to be a similar description of the asymptotic behavior of solutions of higher order abstract differential equations generalizing the above.

. Going back to equation (1,1) let us outline the technique we want to apply. The assumptions allow the spectrum of A to be unbounded in the both half planes $\mathrm{Re}\,\lambda > 0$ and $\mathrm{Re}\,\lambda < 0$. We wish to perform the corresponding separation of the spectrum. As it turns out we have not been able to find the suitable statement in the litterature. Here we perform the needed spectral separation with the help of interpolation theory under some additional assumptions in § 3. The general problem of separating two unbounded components of the spectrum remains open.

. Once separation of the spectrum is achieved equation (1,1) is split into three decoupled equations. One is a finite system of ordinary scalar differential equations. The coefficient of the second one is the infinitesimal generator of an analytic semi group. The third one behaves as transposed of the second. See §4. The analysis of those three problems relies on classical tools and is carried out in §§ 5 and 6. This is only for equations in Hilbert spaces. The case of the equations in Banach spaces is briefly described in § 9 and appeals much more to interpolation theory as in Da Prato - Grisvard 1975.

2. - Review of interpolation theory

. Let D and E be a pair of Banach spaces with D continously imbedded in E. Following Lions 1959 we denote by $W(p,\alpha\,;D,E)$ the space of all functions defined in $\mathbb{R}_+ =]0,\infty[$ with values in D such that

$$t \longmapsto t^{\alpha} u(t)$$

belongs to $L^p(\mathbb{R}_+, D)$ while

$$t \longmapsto t^{\alpha} u'(t)$$

belongs to $L^p(R_+, E)$. Here we consider the distribution derivative and we assume $p \in [1,\infty]$, $\theta = \alpha + \frac{1}{p} \in]0,1[$. Under those assumptions u coincides a.e. with a continuous function in $[0,\infty[$ with values in E. This makes it possible to define $u(0)$ and we denote by $(D,E)_{\theta,p}$ the image of the mapping $u \longmapsto u(0)$. Accordingly we have

$$(D,E)_{\theta,p} = \{u(o)\,;\; u \in W(p,\alpha\,;D,E)\}$$

Banach norms on such spaces are obvious.

. Of particular interest is the case when D is the domain D_B of a closed linear operator B in E equipped with the graph norm. We assume that the resolvent set ρ_B contains the negative half-line $\mathbb{R}_-$ and we assume the existence of a constant M such that

$$(2,1) \qquad \|(B+tI)^{-1}\|_{E \to E} \leqslant M/t$$

for every $t > 0$. The space $(D_B, E)_{\theta,p}$ has been characterized in Grisvard 1966 as the subspace of all $x \in E$ such that

$$t \longmapsto t^{1-\theta} B(B+tI)^{-1} x$$

belongs to $L^p(\mathbb{R}_+, E, dt/t)$. This space will be denoted $D_B(1-\theta, p)$ in what follows. The reason for this change of notation appears in the particular case when B is a non negative self adjoint operator in a Hilbert space E. Indeed under such assumption $D_B(1-\theta, p)$ coincides with $D_{B^{1-\theta}}$ the domain of the power $1-\theta$ of B.

. The main reason for introducing such spaces is the property of interpolation of continuous linear operators. Let π be a linear continuous operator in E whose restriction to D is also linear continuous in D, then the restriction of π to $(D, E)_{\theta,p}$ turns out to be continuous in $(D, E)_{\theta,p}$.

Several alternative definitions of $(D, E)_{\theta,p}$ may be found in Lions - Peetre 1964. One of them allows to derive a very powerful interpolation result which we describe now. Let H be a Hilbert space and V a Banach space continuously imbedded in H. The transposed injection makes H a subspace of V'. In such a framework one has

$$(2,2) \qquad (V, V')_{1/2,2} = H$$

. Finally let us emphasize that due to the interpolation property $(D_B, E)_{\theta,p}$ does not really depend on B but only on the subspace D_B. Details are to be found in Lions - Peetre 1964.

. Further properties that we shall take advantage of, are duality and reiteration. Consider again a pair of Banach spaces D and E as above. The reiteration property means that

$$(D, E)_{\theta,p} = ((D, E)_{\theta_1,p_1}, (D, E)_{\theta_2,p_2})_{\alpha,p}$$

where $\theta_1, \theta_2, \alpha \in]0,1[$ and

$$\theta = (1-\alpha)\,\theta_1 + \alpha\theta_2$$

whatever p_1 and p_2 are.

. If in addition we assume that D is dense in E the dual E' is continuously imbedded in D'. This makes it possible to interpolate between D' and E'. The duality property claims that the dual space of $(D,E)_{\theta,p}$ is just $(E',D')_{1-\theta,p'}$.

3. – Spectral separation

. As already emphasized in the Introduction we are going to consider a closed linear operator A with domain D_A whose resolvent set ρ_A contains the imaginary axis and such that there exists a constant M with

$$\text{(3,1)} \qquad \|(A-\lambda I)^{-1}\|_{E \to E}\ (M / |\lambda| + 1)$$

for every $\lambda \in i\mathbb{R}$.

. Accordingly the spectrum σ_A is the union of two disjoint closed subsets σ_+ and σ_- of the complex plane such that

$$\sigma_+ \subseteq \{\lambda \in \mathbb{C} \,;\, \mathrm{Re}\,\lambda > 0\}$$

$$\sigma_- \subseteq \{\lambda \in \mathbb{C} \,;\, \mathrm{Re}\,\lambda < 0\}\ .$$

. It is well known that if σ_+, for instance, were bounded it would be possible to perform the spectral separation as follows. Consider

$$\text{(3,2)} \qquad P_+ = (1/2i\pi) \int_{\gamma_+} (\lambda I - A)^{-1}\, d\lambda$$

where γ_+ is a simple closed curve around σ_+ oriented directly. Then P_+ is a projection which commutes with A. Denote by E_+ and E_- the ranges of P_+ and $I - P_+$ respectively. Then E is the direct sum of E_+ and E_-. Furthermore it turns out that σ_+ and σ_- are the spectra of the restrictions A_+ and A_- of A to E_+ and E_-. This procedure is usually called separation of the spectrum, see Dunford - Schwartz 1958 for instance.

. Also if E were a Hilbert space and A a selfadjoint operator a similar separation of the spectrum could be worked out defining P_+ suitably with the help of the spectral resolution of A. This would work without any assumption of boundedness on σ_+ or σ_-.

. Here we would like to perform a similar separation of the spectrum without assuming the self-adjointness of A or the boundedness of either σ_+ or σ_-. This can be achieved with the help of the theory of distribution semi-groups according to Nikolaenko 1977. However we shall present here a different approach relying on the theory of interpolation.

. First of all let us observe that there exists $\rho > 0$ and $\delta > 0$ such that

$$\rho_A \supset R = \{\lambda \in \mathbb{C}, |\lambda| \leq \rho\} \cup \{\lambda \in \mathbb{C} ; |\arg \lambda \pm \pi/2| \leq \delta\}$$

and

$$\|(A - \lambda I)^{-1}\|_{E \to E} \leq (2M / |\lambda| + 1) \tag{3,3}$$

for every $\lambda \in R$. This follows from (3, 1).

. Let us denote by Γ_+ and Γ_- the right and left boundaries of R respectively. The orientation is from above to below for Γ_+ and from below to above for Γ_- according to the figure.

. The spectrum σ_A is the union of two, possibly unbounded, components σ_+ and σ_-. Γ_+ and Γ_- are oriented positively with respect to σ_+ and σ_- respectively. We define

$$\begin{aligned} P_+ &= (1/2i\pi) \int_{\Gamma_+} A(\lambda I - A)^{-1} (d\lambda / \lambda) \\ P_- &= (1/2i\pi) \int_{\Gamma_-} A(\lambda I - A)^{-1} (d\lambda / \lambda) \end{aligned} \tag{3,4}$$

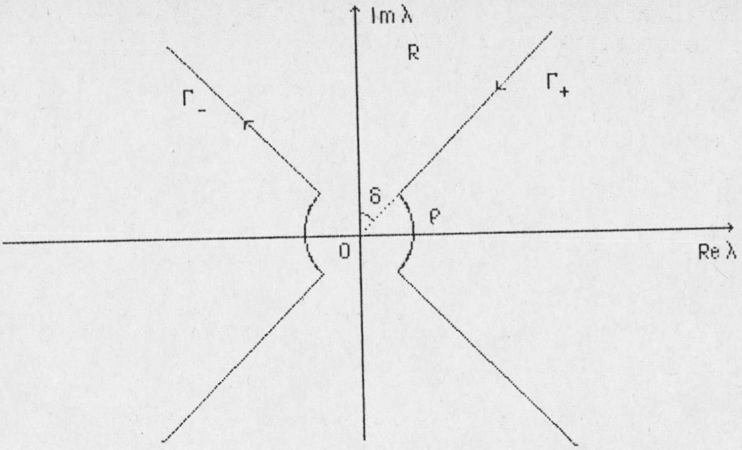

These are linear continuous operators from D_A into E. We would like to be able to prove that P_+ and P_- are actually linear continuous in E and are projectors. A first step is to check continuity in the interpolation spaces.

Proposition 3.1 : Under assumption (3,1), P_+ and P_- are linear continuous operators in $D_A(\theta, p)$ for every $\theta \in]0,1[$ and every $p \in [1,\infty]$. Furthermore one has

$$P^2_+ = P_+ , \; P^2_- = P_- , \; P_+ + P_- = I , \tag{3,5}$$

in the algebra of linear continuous operators in $D_A(\theta, p)$.

Proof : Let us assume in a first step that $p = +\infty$. We apply the characterisation of the interpolation space that we have recalled in § 2 to the operators $\pm \, i \, e^{\pm i\delta} A$ whose domains coincide with D_A. Accordingly if $x \in D_A(\theta, \infty)$ there exists a constant C such that

$$\|A(\lambda I - A)^{-1} x\|_E \leqslant C |\lambda|^{-\theta} \|x\|_{D_A(\theta,\infty)} \tag{3,6}$$

on $\Gamma_+ \cup \Gamma_-$.

. Next in order to show that $P_\pm x$ belongs to $D_A(\theta,\infty) = D_{iA}(\theta,\infty)$, one has to estimate

$$t^\theta A(iA + tI)^{-1} P_\pm x = (1/2i\pi) \int_{\Gamma_\pm} (t^\theta / t+i\lambda) \; A(\lambda I - A)^{-1} x \, d\lambda$$

for every $t > 0$. Note that the last equality is derived applying the resolvent equation as in the usual functional calculus of Dunford.

. It follows that

$$\|t^\theta A(iA+tI)^{-1} P_\pm x\| \leq (1/2\pi) \int_{\Gamma_\pm} (t^\theta / |t+i\lambda|)\,(|d\lambda| / |\lambda|^\theta)\ \|x\|_{D_A(\theta,\infty)}$$

The above integral is bounded uniformly in t and this shows the continuity of $P_\pm$ in the space $D_A(\theta,\infty)$. The continuity in $D_A(\theta,p)$ for every p follows by applying the reiteration property recalled in § 2.

. The derivation of (3.5) is straightforward with the help of the resolvent equation and of the decay given by (3.6) ∎

. Now we restrict our purpose to very special operators in a Hilbert space.

Theorem 3.2 : We assume that E is a Hilbert space and A is a closed densily defined linear operator in E fulfilling (3,1). In addition we assume that there exists $\theta \in]0,1[$ such that

(3,7) $$D_A(\theta,2) = D_{A^*}(\theta,2)\ .$$

Then P_+ and P_- are linear continuous in E and the identities (3,5) hold in the algebra of linear continuous operators in E.

Proof : From proposition 3,1 we know that $P_\pm$ are continuous in $D_A(\theta,2)$. Since assumption (3,1) is also clearly fulfilled by the adjoint operator A^* we know the $Q_\pm$ are continuous in $D_{A^*}(\theta,2)$, where

(3,8) $$Q_\pm = (1/2i\pi) \int_{\Gamma_\pm} A^*(\lambda I - A^*)^{-1}\,(d\lambda/\lambda)$$

. On the other hand it is easy to check that

$$(P_\pm x, y) = (x, Q_\pm y)$$

for every pair $x \in D_A$ and $y \in D_{A^*}$. This implies that $P_\pm$ admit extensions as linear continuous operators $P_\pm$ in the dual space of $D_{A^*}(\theta,2)$.

. Now set $V = D_A(\theta,2)$. Taking into account (3,7) we have shown that $P_\pm$ are continous linear operators in V and V' . By interpolation $P_\pm$ are continuous in E by identity (2,2).

. The validity of (3,5) on E follows from the density of V in E ■

. Theorem 3.2 makes it possible to define

(3,9) $$E_+ = P_+E\ ,\ E_- = P_-E$$

also A^{-1} maps E_+ and E_- into themselves. We define two unbounded operators A_+ and A_- in E_+ and E_- respectively by considering the inverse operators of the restrictions of A^{-1} to E_+ and E_- . Since A^{-1} is injective continuous, A_+ and A_- are unbounded closed in E_+ and E_- respectively. It will be no surprise that $\sigma_{A+} = \sigma_+$, $\sigma_{A_-} = \sigma_-$. The following says much more

<u>Corollary</u> 3.3 : <u>Under the assumptions of Thm</u> 3.2, A_- <u>and</u> $-A_+$ <u>generate bounded analytic semi-groups in</u> E_- <u>and</u> E_+ <u>respectively.</u>

<u>Proof</u> : Let us consider A_- only. The proof for $-A_+$ will be quite similar.

. We start from the following representation formula for the resolvent of A_- :

(3,10) $$(\lambda_0 I - A_-)^{-1} = (1/2i\pi) \int_{\Gamma_-} (\lambda I - A)^{-1} (d\lambda/(\lambda_0 - \lambda))$$

for λ_0 on the "right" of Γ_- . Indeed it is easy to check that

$$(\lambda_0 I - A)\,(1/2i\pi) \int_{\Gamma_-} (\lambda I - A)^{-1} x\,(d\lambda/(\lambda_0 - \lambda)) = P_- x$$

for every $x \in D_A$. Identity (3.7) follows by density.

. On the other hand, for $\lambda \in R$, $(\lambda_0 I - A_-)^{-1}$ is the restriction to E_- of $(\lambda_0 I - A)^{-1}$.

. Altogether this shows that the spectrum of A_- lies in the part of the complex plane which is "left" to Γ_-. Therefore σ_{A-} is just σ_-. In addition one has

$$\|(\lambda_0 I - A_-)^{-1}\|_{E- \to E-} \leqslant 2M/(|\lambda_0| + 1)$$

for $\lambda_0 \in R$, while (3,10) shows only a decay in $O((\mathrm{Log}|\lambda_0|)/|\lambda_0|)$ when λ_0 is right of Γ_+. This is enough to apply Phragmen - Lindelöf principle. See Titchmarsh 1932 for instance. The conclusion is a decay in $O(1/|\lambda_0|)$ right of Γ_- ∎

4. - Splitting an elliptic equation in a Hilbert space

. From now on we assume that A is a closed densely defined linear operator in E a Hilbert space which fulfills (3,7). We release the assumption (3,1) as follows. We assume the existence of two constants M and r such that

$$(4,1) \qquad \|(A - \lambda I)^{-1}\|_{E \to E} \leqslant M/|\lambda|$$

for every $\lambda \in i\mathbb{R}$ and $|\lambda| \geqslant r$. To compensate for the weakening of (3,1) we assume in addition that

$$(4,2) \qquad (A - \lambda I)^{-1} \ \underline{\text{is compact}}$$

for $\lambda \in \rho_A$ or, in other words, that the injection of D_A into E is compact.

. Again here there exists $\delta > 0$ such that

$$\rho_A \supset R_{2r} = \{\lambda \in \mathbb{C}\ ; |\lambda| \geqslant 2r, |\arg \lambda \pm \pi/2| \leqslant \delta\}$$

and

$$(4,3) \qquad \|(A - \lambda I)^{-1}\|_{E \to E} \leqslant 2M/|\lambda|$$

for every $\lambda \in R_{2r}$.

. Also σ_A is a discrete set with no finite cluster point. Each spectral value is an eigenvalue with finite multiplicity. Therefore for any given pair of real numbers a, b with $a < b$ there is only a finite number of eigenvalues in the strip

$$\{\lambda \in \mathbb{C} \ ; \ a \leqslant \operatorname{Re}\lambda \leqslant b\}.$$

. This allows one to perform separation of the spectrum corresponding to the three subsets

$$(4,4) \qquad \begin{aligned} \sigma_0 &= \{\lambda \in \sigma_A ; a \leqslant \operatorname{Re}\lambda \leqslant b\} \\ \sigma_+ &= \{\lambda \in \sigma_A ; \operatorname{Re}\lambda > b\} \\ \sigma_- &= \{\lambda \in \sigma_A ; \operatorname{Re}\lambda < a\} \end{aligned}$$

To be more precise we shall derive the following

Theorem 4.1 : Under assumptions (3,7), (4,1) and (4,2) E is the direct sum of three closed subspaces E_0, E_+ and E_- all invariant by $(A-\lambda I)^{-1}$ for every $\lambda \in \rho_A$. In addition if we denote by A_0, A_+ and A_- the restrictions of A to E_0, E_+ and E_- respectively, the following properties hold :

(i) E_0 is finite dimensional, $\sigma_{A_0} = \sigma_0$.

(ii) $A_- + aI$ is the infinitesimal generator of an analytic semi-group with exponential decay in E_-, $\sigma_{A_-} = \sigma_-$.

(iii) $-A_+ + bI$ is the infinitesimal generator of an anlytic semi-group with exponential decay in E_+, $\sigma_{A+} = \sigma_+$.

. In order to avoid any misunderstanding let us emphasize that restrictions are defined as follows

$$D_{A_0} = D_A \cap E_0, D_{A\pm} = D_A \cap E_\pm$$

$$A_0 x = Ax \ \text{ for } \ x \in D_{A_0}$$

$$A_\pm x = Ax \ \text{ for } \ x \in D_{A\pm}$$

Equivalently, for every $\lambda \in \rho_A$, $(A_0 - \lambda I)^{-1}$, $(A_\pm - \lambda I)^{-1}$ are the restrictions to E_0 and $E_\pm$ respectively of the bounded operator $(A-\lambda I)^{-1}$.

. Also the exponential decays of the semi-groups mean the existence of $\varepsilon > 0$ and $M_\varepsilon > 0$ such that

$$(4,5)\qquad \begin{aligned} &\|e^{t(A_- + aI)}\|_{E_- \to E_-} \leqslant M_\varepsilon e^{-\varepsilon t} \\ &\|A_- e^{t(A_- + aI)}\|_{E_- \to E_-} \leqslant M_\varepsilon (e^{-\varepsilon t}/t) \\ &\|e^{t(-A_+ + bI)}\|_{E+ \to E+} \leqslant M_\varepsilon e^{-\varepsilon t} \\ &\|A_+ e^{t(-A_+ + bI)}\|_{E+ \to E+} \leqslant M_\varepsilon (e^{-\varepsilon t}/t) \end{aligned}$$

for every $t \geqslant 0$.

. The <u>proof</u> is two-step. One first separates the compact component σ_0 of the spectrum as usual. See (3,2). This reduces the general situation to the one solved by Corollary 3.3. The estimates (4,5) follow from general results in the theory of analytic semi-groups. See <u>Kato</u> 1966 ■

. Now let us consider the equation

$$(4,6)\qquad u'(t) + Au(t) = f(t)$$

for $t \geqslant 0$, where f is a given function with values in E and u is an unknown function with values in D_A. Applying Theorem 4.1 one can split (4,6) into a system of three independant equations

$$(4,7)\qquad u'_0(t) + A_0 u_0(t) = f_0(t)$$

$$(4,8)\qquad u'_-(t) + A_- u_-(t) = f_-(t)$$

$$(4,9)\qquad u'_+(t) + A_+ u_+(t) = f_+(t).$$

The notation is self-explanatory.

. Since E_0 is finite dimensional, and consequently A_0 is bounded, equation (4,7) is just a finite system of ordinary first order linear differential equations. We shall investigate the properties of such a system in § 5.

We shall call (4,9) the forward equation and (4,8) the bakward equation. They will be investigated in § 6.

5. - Asymptotic behavior for a system of ordinary differential equations.

. We are going to consider equation (4,7). In order to make out the basic ideas we begin with the very simple-minded particular case when E_0 is one dimensional. Thus problem (4,7) reduces to the single scalar equation

$$u' + \lambda u = f \tag{5,1}$$

All the solutions are known :

$$u(t) = e^{-\lambda t}u(0) + \int_0^t e^{-\lambda(t-s)}f(s)ds \tag{5,2}$$

. We shall assume that f is given such that

$$e^{-\beta t}f \in L^2(\mathbb{R}_+) . \tag{5,3}$$

We do not care with the existence of u . Instead we assume that u is such that

$$e^{-\alpha t}u \in L^2(\mathbb{R}_+) \tag{5,4}$$

The numbers α and β are given such that $\beta < \alpha$ and we ask whether the faster decay of f implies a similar decay of u . The answer is given by

Proposition 5.1 : Assume (5,4)(5,2) and (5,3), then one has

$$e^{-\beta t}u \in L^2(\mathbb{R}_+) \tag{5,5}$$

when $\mathrm{Re}\,\lambda \leqslant -\alpha$ or when $\mathrm{Re}\,\lambda > -\beta$. Otherwise there exists a constant c such that

$$e^{-\beta t}(u - c\,e^{-\lambda t}) \in L^2(\mathbb{R}_+) , \tag{5,6}$$

when $-\alpha < \mathrm{Re}\,\lambda < -\beta$.

. We are not able to provide any conclusion in the limit case when $\operatorname{Re}\lambda = -\beta$.

The proof relies on a very elementary lemma.

Lemma 5.2 : Assume $e^{-\gamma t} f \in L^2(\mathbb{R}_+)$ then

$$e^{-\gamma t}\int_0^t e^{-\lambda(t-s)} f(s)\, ds \in L^2(\mathbb{R}_+)$$

when $\operatorname{Re}\lambda > -\gamma$, while

$$e^{-\gamma t}\int_t^\infty e^{-\lambda(t-s)} f(s)\, ds \in L^2(\mathbb{R}_+)$$

when $\operatorname{Re}\lambda < -\gamma$.

. This lemma is an easy consequence of Young's inequality on convolutions. Then the proof of Proposition 5.1 proceeds by considering the several different cases.

. First we assume $\operatorname{Re}\lambda \leqslant -\alpha$ and consequently $\operatorname{Re}\lambda < -\beta$. We deduce from (5,2) that

$$\begin{cases} u(t) = c\, e^{-\lambda t} + g(t) \quad \text{where} \\ \\ c = u(0) + \displaystyle\int_0^\infty e^{\lambda s} f(s)\, ds \text{ and } g(t) = -\int_t^\infty e^{-\lambda(t-s)} f(s)\, ds . \end{cases}$$

Lemma 5.2 implies that $e^{-\beta t} g \in L^2(\mathbb{R}_+)$. Since β is less than α it follows from (5,4) that

$$c\, e^{-\alpha t} e^{-\lambda t} = e^{-\alpha t}(u - g) \in L^2(\mathbb{R}_+) .$$

This forces c to vanish. Thus $u = g$ and (5,5) holds.

. Next we assume $-\alpha < \operatorname{Re}\lambda < -\beta$. Then with the same notation as before we have

$$e^{-\beta t}(u(t) - c\, e^{-\lambda t}) = e^{-\beta t} g \in L^2(\mathbb{R}_+) .$$

This shows (5,6).

. Finally we assume $\operatorname{Re}\lambda > -\beta$. We can no longer define g but it is obvious on (5,2) that (5,5) holds ∎

. The above statement is easily extended to the case when E_0 has arbitrary finite dimension. We need some more notation. We denote by $\lambda_1, \ldots \lambda_k$ the eigenvalues of A_0 and by $E_1, \ldots E_k$ the corresponding generalized eigenspaces. This means for each j, E_j is a minimal subspace invariant by A_0 in which A_0 has λ_j as only eigenvalue. Thus the numbers λ_j are not necessarily all distinct. We denote by m_j the dimension of E_j. There exists a basis of E_j in which the matrix of A_0 is the Jordan block of dimension m_j :

$$\begin{bmatrix} \lambda_j & 1 & & \\ & \ddots & \ddots & \\ & & \ddots & 1 \\ & & & \lambda_j \end{bmatrix}$$

Writing the projection of equation (4,7) on E_j in such a basis one easily derives the following :

Theorem 5.3 : Assume u_0 is a solution of (4,7) with $e^{-\alpha t} u_0 \in L^2(\mathbb{R}_+, E_0)$. Assume in addition that $e^{-\beta t} f_0 \in L^2(\mathbb{R}_+, E_0)$ with $\beta < \alpha$. Assume also that A_0 has no eigenvalue with real part equal to $-\beta$. Then there exist polynomials p_j of order $m_j - 1$ with values in E_j such that

$$e^{-\beta t}\Big(u_0(t) - \sum_{-\alpha < \mathrm{Re}\,\lambda j < -\beta} e^{-\lambda j t} p_j(t)\Big) \in L^2(\mathbb{R}_+, E_0) \ . \tag{5,7}$$

. In what follows we shall refer to the sum in (5,7) as the singular part of u. Terms corresponding to semi-simple eigenvalues have the particular form

$$e^{-\lambda j t} \phi_j$$

where ϕ_j is an eigenvector of A_0 corresponding to the eigenvalue λ_j.

6. – Asymptotic behavior of the forward and backward equations.

. We turn now to the study of equation (4,9) under the assumptions that

(6,1) $$u_+ \in L^2([0,T]; D_{A_+}) \text{ for every } T < \infty \text{ and}$$

(6,2) $$e^{-\beta t} f_+ \in L^2(\mathbb{R}_+ ; E_+) .$$

The question is whether u_- has a growth property similar to that of f_-.

Theorem 6.1 : Assume $-A_+$ is the infinitesimal generator of an anlytic semi-group fulfilling (4,5). Assume that u_+ is a solution of equation (4,9) satisfying (6,1). Finally assume $b \geq -\beta$ than (6,2) implies

(6,3) $$e^{-\beta t} u_+ \in L^2(\mathbb{R}_+ ; D_{A_+})$$

Proof : We observe that (6,1) and (6,2) imply

$$u'_+ \in L^2([0,T]; E_+)$$

for every $T < \infty$. Taking into account the results recalled in § 2 we conclude that

(6,4) $$u_+(0) = \xi_+ \in D_{A_+}(1/2, 2) .$$

Then we write

$$u_+(t) = e^{-tA_+} \xi_+ + \int_0^t e^{-(t-s)A_+} f_+(s)\, ds$$

hence

$$e^{-\beta t} u_+(t) = e^{-t(A_+ + \beta I)} \xi_+ + \int_0^t e^{-(t-s)(A_+ + \beta I)} e^{-\beta s} f_+(s)\, ds$$

The conclusion follows from the following "classical" result.

Lemma 6.2 : Let L be the infinitesimal generator of an analytic semi-group e^{tL} in a Hilbert space H . Assume there exist $\varepsilon > 0$ and $M > 0$ such that

$$\|e^{tL}\|_{H \to H} + \|tLe^{tL}\|_{H \to H} \leq M e^{-\varepsilon t}$$

for every $t \geq 0$. Then for every $x \in D_L(1/2, 2)$ and every $g \in L^2(\mathbb{R}_+ ; H)$ the unique solution of

$$(6,5) \qquad \begin{cases} v'(t) = Lv(t) + g(t) \\ v(0) = x \end{cases}$$

fulfills

$$v \in L^2(\mathbb{R}_+ ; D_L) .$$

. This is easily derived via Fourier transform when $x = 0$. The non homogeneous case appeals to the interpolation results recalled in § 2. See also De Simon 1964 or Da Prato - Grisvard 1975.

. In a very similar fashion we can investigate the equation (4,8) under the assumptions that

$$(6,6) \qquad e^{-\alpha t} u_- \in L^2(\mathbb{R}_+, E_-)$$

$$(6,7) \qquad u_- \in L^2([0,T] ; D_{A_-}) \text{ for every } T < \infty .$$

and that

$$(6,8) \qquad e^{-\beta t} f_- \in L^2(\mathbb{R}_+, E_-)$$

with $\beta < \alpha$. We seek a similar behavior for u_- .

Theorem 6.3 : Assume A_- is the infinitesimal generator of an anlytic semi-group fulfilling (4,5). Assume that u_- is a solution of equation (4,8) satisfying (6,6) (6,7) Then when $\alpha \leq a$ (6,8) implies

$$(6,9) \qquad e^{-\beta t} u_- \in L^2(R_+, D_{A_-}) .$$

. The underlying idea in the proof is that (4,9) behaves as transposed of (4,8). Accordingly we shall not need to consider the initial value $u_-(0)$ but rather the value of u_- at infinity which has to be zero by (6,6). The preliminary result corresponding to Lemma 6.2 is given by

Lemma 6.4 : Let L be the infinitesimal generator of an analytic semi-group e^{tL} in a Hilbert space H. Assume there exist $\varepsilon > 0$ and $M > 0$ such that

$$\|e^{tL}\|_{H \to H} + \|t L e^{tL}\|_{H \to H} \leq M e^{-\varepsilon t}$$

for every $t \geq 0$. Then for $g \in L^2(\mathbb{R}_+, H)$ given the problem

(6,10) $$v'(t) = -Lv(t) + g(t)$$

has a unique solution such that

(6,11) $$v \in L^2(\mathbb{R}_+ ; H)$$

(6,12) $$v \in L^2([0,T] ; D_L)$$

for every $T < \infty$. This solution is given by

(6,13) $$v(t) = -\int_t^\infty e^{(s-t)L} g(s)\, ds$$

and furthermore

(6,14) $$v \in L^2(\mathbb{R}_+ ; D_L) .$$

Proof : Everything is based on (6,13) which is derived by integrating equation (6,10) against $e^{(s-t)L}$ in $[t,T]$ and then letting $T \to \infty$:

$$\int_t^T e^{(s-t)L} g(s)\, ds = \int_t^T e^{(s-t)L} v'(s)\, ds + \int_t^T e^{(s-t)L} Lv(s)\, ds =$$

$$= e^{(s-t)L} v(s) \Big|_t^T = e^{(T-t)L} v(T) - v(t) .$$

Now v is continuous with values in H and therefore (6,11) implies the existence of a sequence $T_n \nearrow \infty$ such that

$$v(T_n) \to 0$$

hence (6,13).

. This shows the uniqueness we have claimed. Next (6,14) follows by Fourier transform. ∎

. The proof of theorem 6.3 follows by applying lemma 6.4 to $L = (A_- + \alpha I)$. This shows

$$e^{-\alpha t} u(t) = - \int_t^\infty e^{(s-t)(A_- + \alpha I)} e^{-\alpha s} g(s)\, ds$$

hence

$$e^{-\beta t} u(t) = - \int_t^\infty e^{(s-t)(A_- + \beta I)} e^{-\beta s} g(s)\, ds .$$

A second application of lemma 6.4 to $L = (A_- + \beta I)$ implies (6,14) ∎

7. – Summing up

. We are going to glue together the results of Theorems 4.1, 5.3, 6.1 and 6.3 assuming $b = -\beta$ and $\alpha = a$. The rsult reads as follows :

Theorem 7.1 Assume (3,7) (4,1) and (4,2). Let u be a solution of (4,6) such that

(7,1) $$e^{-\alpha t} u \in L^2(\mathbb{R}_+ ; E)$$

(7,2) $$u \in L^2([0,T] ; D_A) \quad \text{for every } T < \infty .$$

Assume in addition that

(7,3) $$e^{-\beta t} f \in L^2(\mathbb{R}_+ ; E)$$

with $\beta < \alpha$ such that A has no eigenvalue on the line $\mathrm{Re}\,\lambda = -\beta$. Then there exist polynomials p_j with values in E such that

(7,4) $$e^{-\beta t}\Big(u(t) - \sum_{-\alpha < \mathrm{Re}\,\lambda_j < -\beta} e^{-\lambda_j t} p_j(t)\Big) \in L^2(\mathbb{R}_+ ; D_A) .$$

· Go back to theorem 5.3 for more information about the polynomials p_j.

8. – Back to corner problems

. We now perform the change of variable

$$r = e^{-t}$$

that maps the half-line $\mathbb{R}_+$ onto $]0,1[$.

Theorem 8.1 : Assume (3,7) (4,1) and (4,2). Let u be a solution of

such that

(8,2) $$r^{\alpha} u \in L^2([0,1]; E)$$

(8,3) $$u \in L^2([\varepsilon,1]; D_A) \quad \text{for every } \varepsilon > 0 .$$

Assume in addition that

(8,4) $$r^{\beta} f \in L^2([0,1]; E)$$

with $\beta < \alpha$ such that A has no eigenvalue on the line $\mathrm{Re}\,\lambda = -\beta - 1/2$. Then there exist polynomials p_j with values in E such that

(8,5) $$r^{\beta}\Big(u(r) - \sum_{-\alpha-1/2 < \mathrm{Re}\,\lambda_j < -\beta-1/2} r^{\lambda_j} p_j(\mathrm{Log}\, r)\Big) \in L^2([0,1]; D_A) .$$

. When λ_j is a semi-simple eigenvalue then

$$p_j(\mathrm{Log}\, r) \equiv \phi_j$$

an eigenvector of A corresponding to the eigenvalue λ_j .

9. – Equations in Banach spaces

. We consider again equation (4,6) but we no longer assume that E is a Hilbert space. Nevertheless we assume (4,1) and (4,2). We forget (3,7) which is meaningless in the Banach space setting. The continuity of the projectors P_+ and P_- in the interpolation spaces suggest that the same splitting of the elliptic equation is workable out in such interpolation spaces.

. We observe that $B = i(A - irI)$ fulfills (2,1) and therefore $D_A(\theta, p)$ is well defined. The statement corresponding to Theorem 7.1 is as follows.

Theorem 9.1 : Assume (4,1) and (4,2) (in a B space E). Let u be a solution of (4,6) such that

$$e^{-\alpha t} u \in L^p(\mathbb{R}_+ ; E) \tag{9,1}$$

$$u \in L^p([0,T] ; D_A) \quad \text{for every } T < \infty . \tag{9,2}$$

Assume in addition that

$$e^{-\beta t} f \in L^p(\mathbb{R}_+ ; D_A(\theta, p)) \tag{9,3}$$

with $\theta \in]0,1[$ and $1 < p < +\infty$ and $\beta < \alpha$. Finally assume A has no eigenvalue on the line $\operatorname{Re} \lambda = \beta - {}^1/_2$. Then there exist polynomials p_j with values in E such that

$$e^{-\beta t}\Big(u(t) - \sum_{-\alpha < \operatorname{Re} \lambda_j < -\beta} e^{-\lambda_j t} p_j(t)\Big) \in L^p(\mathbb{R}_+ ; D_A(\theta, p)) . \tag{9,4}$$

The proof is by spectral separation in the space $D_A(\theta, p)$ as in the Hilbert case. One then applies results on maximal regularity for backward or forward evolution problems which are consequences of the general Theorems in Da Prato – Grisvard 1975

. The same change of variable as in § 8 leads to

Theorem 9.2 : Assume (4,1) and (4,2) (in a B space E). Let u be a solution of (8,1) such that

(9,5) $$r^{\alpha} u \in L^p([0,1];E)$$

(9,6) $$u \in L^p([\varepsilon,1];D_A) \quad \text{for every } \varepsilon > 0.$$

Assume in addition that

(9,7) $$r^{\beta} f \in L^p([0,1];D_A(\theta,p))$$

with $\beta < \alpha$, $1 < p < +\infty$, $0 < \theta < 1$. Finally assume A has no eigenvalue on the line $\operatorname{Re}\lambda = -\beta - 1/p$. Then there exist polynomials p_j with values in E such that

(9,8) $$r^{\beta}\Big(u(r) - \sum_{-\alpha-1/2<\operatorname{Re}\lambda_j<-\beta-1/p} r^{\lambda_j} p_j(\operatorname{Log} r)\Big) \in L^p([0,1];D_A(\theta,p)).$$

10. – Corner singularities of the Dirichlet problem

. Let Ω be defined in polar coordinates by $0 < \theta < \omega$ and $0 < r < 1$ with $0 < \omega \leqslant 2\pi$. Consider

$$w \in W^{1,p}(\Omega)$$

solution of

(10,1) $$\Delta w = g$$

with g given in $L^p(\Omega)$ and with the boundary conditions

(10,2) $$\begin{cases} w = 0 & \text{on } \theta = 0 \text{ and } \theta = \omega \\ w = \phi & \text{on } r = 1 \end{cases}$$

We assume ϕ is given in $W_p^{2-1/p,p}(I) \cap \mathring{W}^{1,p}(I)$ where $I =]0,\omega[$ and ask whether $w \in W^{2,p}(\Omega)$. The assumptions on g and ϕ are obvious necessary conditions.

We reduce (10,1) to the form (8,1) by setting

$$\vec{u} = \{ r(\partial w/\partial r) , \partial w/\partial \theta \}$$

$$A = \begin{bmatrix} 0 & -\partial/\partial\theta \\ +\partial/\partial\theta & 0 \end{bmatrix}$$

$$\vec{f} = \{r^2 g , 0\}$$

We work in the framework of $E = L^p(I)^2$ and define the domain of A as follows :

$$D_A = \overset{\circ}{W}{}^{1,p}(I) \times W^{1,p}(I) .$$

. From the assumptions that $w \in W^{1,p}(\Omega)$ we deduce

$$\int_\Omega [|\partial w/\partial r|^p + |1/r (\partial w/\partial \theta)|^p] \, r \, dr \, d\theta < \infty ,$$

hence $r^{-1+1/p} \vec{u} \in L^p([0,1] ; E)$. This is (8,2) or (9,5) with $\alpha = -1 + 1/p$. The regularity of w is no problem at a right angle and therefore we know that $w \in W^{2,p}(\Omega_\varepsilon)$ This implies (8,3) or (9,6).

. When $p = 2$ we have assumed $g \in L^2(\Omega)$ hence $r^{-3/2} \vec{f} \in L^2([0,1] ; E)$. This is (8,4) with $\beta = -3/2$.

. When $p \neq 2$ we make the extra assumption that $g \in W^{2\theta,p}(\Omega)$ for some $\theta > 0$. This is more than enough to imply

$$r^{-2+1/p} \vec{f} \in L^2([0,1] ; D_A(\theta, p))$$

since by <u>Grisvard</u> 1969 we have

$$D_A(\theta,p) = W^{2\theta,p}(I)^2$$

when $0 < 2\theta < 1/p$. This shows (9,7) with $\beta = -2 + 1/p$.

. The application of Theorem 8.1 requires the knowledge of σ_A . When $p = 2$ it is easily seen that A is selfadjoint. Hence (3.7) and (4,1) are obvious. In any case A has a compact resolvent.

Simple calculations show that the eigenvalues of A are the numbers $k(\pi/\omega)$, $k \in \mathbb{Z}$, $k \neq 0$ while the related eigenfunctions are the functions $\{\sin(k\pi\theta/\omega), \cos(k\pi\theta/\omega)\}$ up to a multiplication by a scalar coefficient.

. We are left with checking (4,1) when $p \neq 2$. Explicitly we have to derive the estimate

(10,3) $$\|\phi_1\|_{L^p(I)} + \|\phi_2\|_{L^p(I)} \leqslant (M / |\lambda|) \{\|\psi_1\|_{L^p(I)} + \|\psi_2\|_{L^p(I)}\}$$

when $\phi_1 \in \overset{\circ}{W}{}^{1,p}(I)$, $\phi_2 \in W^{1,p}(I)$ and

$$\begin{cases} -\lambda\phi_1 - \phi'_2 = \psi_1 \\ +\phi'_1 - \lambda\phi_2 = \psi_2 \end{cases}$$

or equivalently

$$\begin{cases} -\phi''_1 - \lambda^2\phi_1 = \lambda\psi_1 - \psi'_2 \\ -\phi''_2 - \lambda^2\phi_2 = \psi'_1 + \lambda\psi_2 \end{cases}$$

This makes (10,3) classical for every $\lambda \in i\mathbb{R}$.

. Application of Theorem 8.1 yields (assuming $g \in L^2(\Omega)$)

$$r^{-3/2}(\vec{u} - \sum_{0 < k\pi/\omega < 1} c_k r^{k\pi/\omega}\{\sin k\pi\theta/\omega, \cos k\pi\theta/\omega\}) \in L^2([0,1]; H^1(I)^2)$$

Going back to w this means

$$\begin{cases} r^{-1/2}(\partial w/\partial r - \sum_{0<k<\omega/\pi} c_k r^{k\pi/\omega - 1} \sin k\pi\theta/\omega) \in L^2([0,1]; H^1(I)) \\ r^{-1/2}(1/r(\partial w/\partial\theta) - \sum_{0<k<\omega/\pi} c_k r^{k\pi/\omega - 1} \cos k\pi\theta/\omega) \in L^2([0,1]; H^1(I)) \end{cases}$$

Integration shows

(10,4) $$w - \sum_{0<k<\omega/\pi} c_k r^{k\pi/\omega} \sin k\pi\theta/\omega \in H^2(\Omega)$$

. The condition that there is no eigenvalue on the line $\operatorname{Re}\lambda = -\beta - 1/2$ rules out $\omega = \pi$ and $\omega = 2\pi$. The meaning of (10,4) is the well known result that $w \in H^2(\Omega)$ when $\omega < \pi$ a convex angle while $w \in \operatorname{Span}(H^2(\Omega); r^{\pi/\omega}\sin\pi\theta/\omega)$ when $\omega > \pi$ a reentrant corner.

. Observe that this example is sort of trivial since A is self adjoint a situation when separation of the spectrum is well known.

. Finally application of Theorem 9.2 yields (assuming $g \in W^{\varepsilon,p}(\Omega)$).

$$w - \sum_{0<k<\omega/\pi\,(2-2/p)} c_k\, r^{k\pi/\omega} \sin k\pi\theta/\omega \in W^{2,p}(\Omega) \tag{10,5}$$

provided $\omega/\pi\,(2 - 2/p)$ is not an integer. This agrees with results derived in Grisvard 1985.

11. - References

Da Prato - Grisvard : Sommes d'opérateurs... J. Maths. Pures et Appl. 54, 1975, p. 305-387.

De Simon : Un' applicazione della teoria degli integrali singolari, Rendiconti di Padova, 34, 1964, p. 205-223.

Dunford - Schwartz : Linear operators, Interscience Publishers, 1958 (book).

Grisvard : Commutativité de deux foncteurs... J. Maths. Pures et Appl. 45, 1966, p. 143-290.

Grisvard : Equations différentielles abstraites... Annales Sc. Ec. Norm. Sup., 1969, p. 311-395.

Grisvard : Elliptic problems in non-smooth domains. Pitman Pub., 1985 (book)

Kato : Perturbation Theory for linear operators, Springer Verlag Vol. 132, 1966 (book)

Lions : Théorèmes de trace et d'interpolation, Annali S.N.S. di Pisa, 13, 1959, p. 389-403.

Lions-Peetre : Sur une classe d'espaces d'interpolation. Publications de l'I.H.E.S., 19, 1964, p. 5-68.

Nikolaenko : Technical report, Los Alamos, 1977.

Titchmarsh : The theory of functions, Oxford University Press, 1932 (book).

A TWO POINT PROBLEM FOR A SECOND ORDER ABSTRACT DIFFERENTIAL EQUATION

Rabah Labbas
Département de Mathematiques - I.M.S.Ph.
Université de Nice Parc Valrose
06034 NICE

I - Statement of the problem and assumptions. Let E be a complex Banach space and let A be a closed linear operator with domain D_A which is not necessarily dense in E; we consider the following abstract elliptic problem:

$$(1) \qquad \begin{cases} u''(t) + Au(t) = f(t), & t \in [0,1] \\ u(0) = x \\ u(1) = y \end{cases}$$

where x, y $\in$ E and $f \in C([0,1];E)$ (or, more concisely, $f \in C(E)$).

We suppose that A satisfies the following assumption (H):

$$(H) \qquad \begin{cases} \exists\, c_0 \in \mathbb{R},\ M \in \mathbb{R}^+_*:\ \rho(A) \supset [-c_0^2, +\infty[\text{ and} \\ \|(A-\lambda)^{-1}\|_{\mathcal{L}(E)} \le \dfrac{M}{1+|\lambda|} \quad \forall \lambda \in [-c_0^2, +\infty[. \end{cases}$$

If x=y=0 it is known that problem (1) is a special case of the theory of sums of linear operators developed by Da Prato and Grisvard [3].

Our purpose is the study of existence, uniqueness and maximal regularity of strict solutions of (1) when f is sufficiently smooth and x, y satisfy certain compatibility conditions which are naturally connected with the differential equation.

We remark that Hypothesis (H) does not imply that A is the infinitesimal generator of an analytic semigroup, but it does imply that $(-A)^{1/2}$ is. In particular, it follows by (H) that $\rho(A)$ in fact contains a sector $\{\lambda \in \mathbb{C} : |\arg \lambda| \leq \theta_0\}$ where $\theta_0 \in]0, \pi/2[$.

Our method is based essentially on the explicit construction of the solution by means of a Dunford integral as in Labbas-Terreni [5], and on an abstract characterization of the interpolation spaces $D_A(\delta, \infty)$ proved by Grisvard [4].

II - Construction of the solution. In order to obtain the solution we can use the following (just formal) argument: consider the problem

$$(2) \qquad \begin{cases} v''(t) + zv(t) = f(t), \ t \in [0,1], \\ v(0) = x \\ v(1) = y \end{cases}$$

where $x, y \in E$, $f \in C(E)$ and $z \in \mathbb{C} - \mathbb{R}$.

A direct computation shows that the solution of (2) is given by

$$v_z(t) = \frac{\sinh(\sqrt{-z}(1-t))}{\sinh\sqrt{-z}} x + \frac{\sinh(\sqrt{-z}t)}{\sinh\sqrt{-z}} y - \int_0^1 K_z(t,s) f(s) ds,$$

where $\mathrm{Re}\sqrt{-z} > 0$ and K_z is defined by

$$K_z(t,s) = \begin{cases} \dfrac{\sinh(\sqrt{-z}(1-t)) \cdot \sinh(\sqrt{-z}s)}{\sqrt{-z}\ \sinh\sqrt{-z}} & \text{if } 0 \leq s \leq t \\[2em] \dfrac{\sinh(\sqrt{-z}(1-s)) \cdot \sinh(\sqrt{-z}t)}{\sqrt{-z}\ \sinh\sqrt{-z}} & \text{if } t \leq s \leq 1 . \end{cases}$$

Now let γ be a simple curve joining $+\infty e^{-i\theta}$ and $+\infty e^{i\theta}$ ($\theta \in]0, \theta_0[$), and such that $\gamma \subset \rho(A) - \overline{\mathbb{R}^+}$. Then the desired solution of (1) is given by the following Dunford integral:

(3) $$u(t) = \frac{1}{2\pi i}\int_\gamma v_z(t)(A-z)^{-1}dz, \qquad t\in]0,1[,$$

which is absolutely convergent for $t\in]0,1[$.

III - Main results. Assume that $x,y \in D_A$ and $f\in C^\delta(E)$ ($\delta\in]0,1[$).

Theorem. The representation formula (3) yields the unique solution of (1) and satisfies moreover:

(i) $Au(.)\in C(E)$ if and only if $f(0)-Ax$, $f(1)-Ay\in \overline{D_A}$;

(ii) $Au(.)\in C^\delta(E)$ if and only if $f(0)-Ax$, $f(1)-Ay\in D_A(\delta/2,\infty)$;

(iii) $Au(.)-f(.)\in B(D_A(\delta/2,\infty))$ if and only if $f(0)-Ax$, $f(1)-Ay\in D_A(\delta/2,\infty)$.

Here $B(D_A(\delta/2,\infty))$ is the space of bounded, strongly measurable functions with values in $D_A(\delta/2,\infty)$.

Sketch of the proof: The fact that the function (3) solves the equation (1) follows by direct computation, noting that all Dunford integrals involved are convergent. The check of the boundary conditions and the proof of statements (i), (ii), (iii) follow by the same arguments used for parabolic abstract equations by Sinestrari [6] and, in the non-autonomous case, by Acquistapace-Terreni [1].

Similar results hold if we assume instead $x, y \in D_A$ and $f \in C(E)\cap B(D_A(\delta/2,\infty))$ ($\delta\in]0,1[$).

IV - An example. The results of the preceding section can be applied to the following Dirichlet problem on the square $[0,1]\times[0,1]$:

(4) $$\begin{cases} \frac{\partial^2 u}{\partial x^2}(x,y) + \frac{\partial^2 v}{\partial y^2}(x,y) = f(x,y) \\ \\ u(0,y) = g_0(y),\ u(1,y) = g_1(y) \\ u(x,0) = \varphi_0(x),\ u(x,1) = \varphi_1(x)\ . \end{cases}$$

Indeed, we can decompose (4) as the "sum" of two problems of this kind:

$$\begin{cases} \frac{\partial^2 v}{\partial \xi^2}(\xi,\eta) + A(\eta,D)v(\xi,\eta) = F(\xi,\eta) \\ \\ v(0,\eta) = \psi_0(\eta),\ v(1,\eta) = \psi_1(\eta) \\ v(\xi,0) = v(\xi,1) = 0; \end{cases}$$

here $A(\eta,D)$ is the operator $\frac{\partial^2}{\partial \eta^2}$ with homogeneous Dirichlet boundary conditions and the abstract theorem is applicable. However our results cover more generally the case of problems like (5) in a cylinder $[0,1] \times \overline{\Omega}$ (Ω open set of $\mathbb{R}^n$), when $A(\eta,D)$ is a differential operator with variable coefficients, elliptic in the sense of Agmon-Douglis-Nirenberg [2], provided the boundary conditions are homogeneous.

References

[1] ACQUISTAPACE P.- TERRENI B., Une méthode unifiée pour l'étude des équations linéaires non autonomes paraboliques dans les espaces de Banach. C.R. Acad. Sci. Paris (1) 301 (1985) 107-110.

[2] AGMON S.-DOUGLIS A.-NIRENBERG L., Estimates near the boundary for solutions of elliptic partial differential equations satisfying general boundary conditions, I, Comm. Pure Appl. Math. 12 (1959) 623-727.

[3] DA PRATO G.- GRISVARD P., Sommes d'opérateurs linéaires et équations différentielles opérationnelles, J. Math. Pures Appl. 54 (1975) 305-387.

[4] GRISVARD P., Commutativitè de deux foncteurs d'interpolation et applications, J. Math. Pures Appl. 45 (1966) 143-290.

[5] LABBAS R.- TERRENI B., Sommes d'opérateurs de type parabolique et elliptique, C.R. Acad. Sci. Paris (1) 301 (1985) 169-172.

[6] SINESTRARI E., On the abstract Cauchy problem of parabolic type in spaces of continuous functions, J. Math. Anal. Appl. 66 (1985) 16-66.

"Sharp" regularity results for mixed hyperbolic problems of second order

Irena Lasiecka
Mathematics Department
University of Florida
Gainesville, Florida 32611

1. Introduction

In this note we wish to discuss the question of optimal regularity of solutions to second order hyperbolic mixed problems with L_2-nonhomogeneous terms in the Neumann B.C. While we refer to the joint work [L-T.3] in progress, for a more extnesive and complete account on this problem, we choose here to provide - within the limits of allowed space - only a few results which, though far less general than it would be possible [L-T.3], have the virtues of allowing for a quicker and simpler proof than in the general case, while at the same time still providing a sufficient insight into the problem.

Let Ω be an open, bounded domain in R^n with smooth boundary Γ. Let $A(x,\partial)$ be a second order strongly elliptic operator with smooth coefficients whose principal part A_0 is given by

$$A_0(x) = \sum_{ij=1}^{n} \frac{\partial}{\partial x_j}(a_{ij}(x)\frac{\partial}{\partial x_i})$$

Consider

$$\text{(1.1)}\quad \begin{aligned} &a)\quad u_{tt} = A(x,\partial)u && \text{in } Q \equiv \Omega \times [0,T] \\ &b)\quad u(0) = u_t(0) = 0 && \text{in } \Omega \\ &c)\quad \frac{\partial u}{\partial \eta_A}\Big|_{\Gamma} = g && \text{on } \Sigma \equiv \Gamma \times [0,T] \end{aligned}$$

where $\partial/\partial_{\eta_A}$ is the co-normal derivative with respect to A_o. As mentioned, we shall discuss optimal regularity results for the solution u of (1.1) corresponding to the Neumann boundary data $g \in L_2(\Sigma)$.

Regularity of the solutions to hyperbolic equations with nonhomogeneous Neumann boundary data were studied also in [L-M] and [M-1] and, for special geometries, in [L-T]. A brief account thereof will be given below. Let L^N stands for the solution map corresponding to (1.1) to be defined as

$$L^N g \equiv u \qquad \text{where } g,\ u \text{ satisfy (1.1).}$$

The results given in [L-M] yield that:

$$\text{(1.2)}\qquad L^N \in \mathscr{L}(L_2(\Sigma) \to H^{1/2,\,1/2}(Q))$$

where $H^{r,s}(Q)$ stands for the usual Sobolev space of order r,s (see [L-M]).

In [M-1] boundary data smoother than L_2 were considered. Namely, the results of [M-1] say that

$$\text{(1.3)}\qquad L^N \in \mathscr{L}(H^{1/2,\,1/2}(\Sigma) \to H^{1,1}(Q) \cap C[OT;H^1(\Omega)])$$

In both cases we notice that the "gain" of the regularity of the solution u with respect to the boundary data g is " $1/2$

derivative." On the other hand, optimal regularity results in the Dirichlet case (i.e.: with boundary conditions $u|_{\Gamma} = g$) proved recently in [L-T-1], [L-T-2], [L-L-T.1] give that the solution map corresponding to Dirichlet boundary condition is bounded from $L_2(\Sigma)$ into $C[0T; L_2(\Omega)]$. More precisely with L^D defined as $L^D g = u$ where g, u satisfy (1.1) a, b and $u|_{\Gamma} = g$ on Σ, we have

Theorem 1 (L-T-1), (L-T-2), (L-L-T),

$$L^D \in \mathscr{L}(L_2(\Sigma) \to C[0T; \ L_2(\Omega)])$$

$$L_t^D \in \mathscr{L}(L_2(\Sigma), \to C[0T; \ (H_0^{+1}(\Omega))'])\text{and}$$

$$\frac{\partial}{\partial \eta} L^D \in \mathscr{L}(L_2(\Sigma) \to L^2[0T; H^{-1}(\Gamma)])$$

where $H^{-s} \equiv (H^s)'$ - dual of H^s.

We now recall that the difference of regularity of solutions between Neumann and Dirichlet case in elliptic and parabolic theory is "one derivative." Thus, one may then ask on the basis of Theorem 1 whether the optimal regularity of the map L^N should likewise provide a "gain" of "one derivative" over that of L^D! Indeed, in one dimensional case, one can readily verify that $L^N \in \mathscr{L}(L_2(\Sigma) \to C[0T; H^1(\Omega)] \cap H^{1,1}(Q))$, thus confirming the above conjecture if $n = 1$. On the other hand, in the case of special geometries like spheres or parallelopipeds for $n \geqslant 2$ and $A(x,\partial) = \Delta$, one can prove that the achieved regularity of the solution u with L_2-boundary data is less than $H^{1,1}(Q)$. More precisely, in [L-T-1] it was shown (by using eigenfunctions expansions) that if $A(x,\partial) = \Delta$, then

(1.4) $\quad L^N \in \mathscr{L}(L_2(\Sigma)) \to H^{2/3,2/3}(\Omega))$ for Ω = sphere

and

(1.5) $\quad L^N \in \mathscr{L}(L_2(\Sigma)) \to H^{3/4-\varepsilon,3/4-\varepsilon}(\Omega))$, for Ω = parallelopiped

The above results are not in line with the stated conjecture in higher dimension and thus raises the question: what is the optimal ("sharp") regularity for the Neumann map L^N. Here below, we shall provide two types of results: on the negative side, a simple counterexample showing that a gain of "1 derivative" is in general not true; on the positive side, a result giving a gain of almost "2/3 derivative": $L^N \in \mathscr{L}(L_2(\Sigma) \to H^{2/3-\varepsilon,2/3-\varepsilon}(Q))$ where ε is arbitrary small number. The latter will be stated under some technical assumption for the purpose of greatly simplifying the proof given here and keeping it self-contained. Removal of this assumption requires a completely different approach based on pseudo-differential operator theory, for which we refer to our forthcoming joint work [L-T.3].

2. Counterexample and statement of main result.

We start by providing a very simple counterexample which confirms that the "gain of one derivative" for Neumann map is not a generic property. To this end, let us consider the following half-space problem

$$(2.1) \quad \begin{cases} u_{tt} = u_{xx} + u_{yy} & \text{in } \Omega \times [0,T] \equiv Q \\ u(0) = u_t(0) = 0 & \\ u_x(x=0) = g & \text{on } \Gamma \times [0, T] \equiv \Sigma \end{cases}$$

where $\Omega = \{(x,y) \in R^2;\ x > 0\}$ and $\Gamma = \{(x,y) \in R^2;\ x = 0\}$.

We shall prove that there exists $g \in L_2(\Sigma)$ such that

$$L^N g \notin H^{1,1}(Q) \tag{2.2}$$

To accomplish this let us introduce Fourier-Laplace's transform $\hat{u}(x,s,w)$ of function $\hat{u}(x,y,t)$ where

$$t \to s = \beta + i\alpha$$

$$y \to iw$$

where β, α, w are real, $\alpha > 0$ and $\hat{u}(y,s,w) = \int_{-\infty}^{+\infty}\int_0^{\infty} e^{ist+iwy}u(x,y,t)dt\,dy$. Problem (2.1) can be now equivalently rewritten as

$$\begin{cases} (-s^2 + w^2)\hat{u}(x,w,s) = \hat{u}_{xx}(x,w,s) \\ \hat{u}_x(0,w,s) = \hat{g}(w,s) \end{cases} \tag{2.3}$$

Solving (2.3) explicitly yields:

$$\hat{u}(x,w,s) = \frac{\hat{g}(w,s)e^{-\sqrt{w^2-s^2}\,x}}{\sqrt{w^2 - s^2}} \tag{2.4}$$

Since

$$\int_0^{\infty} |\hat{u}(x,w,s)|^2 dx = \frac{|\hat{g}(w,s)|^2}{|w^2 - s^2|\,2\mathrm{Re}\sqrt{w^2 - s^2}}$$

We have

$$\mathscr{C}\|u\|^2_{H^{1,1}(Q)} > \int_{-\infty}^{+\infty}\int_{-\infty}^{-\infty} \frac{|\hat{g}(w,s)|^2 w^2 dw\ d\beta}{|w^2 - s^2|\,\mathrm{Re}\sqrt{w^2 - s^2}} \tag{2.5}$$

where $c > 0$

Now let us define region $S_{\alpha,c}$

$$S_{\alpha,c} \equiv \{(\beta,w);\ C_1 \leq \alpha^2 - \beta^2 + w^2 \leq C_2\}$$

It can be readily verified that

$$\frac{w^2}{|w^2 - s^2|\,\mathrm{Re}\sqrt{-s^2 + w^2}} = \mathcal{O}(\sqrt{w}) \quad \text{on } S_{\alpha,c}$$

Taking

$$g(s,w) \equiv \begin{cases} \dfrac{\sqrt{\beta}}{w^{1/2+\varepsilon}} & (\beta,\ w) \in S_{\alpha c} \cap \{|\beta|,\ w > 1\} \equiv S \\ 0 & \text{outside} \end{cases}$$

we compute using (2.5) (with $\varepsilon \leq 1/4$)

$$\|u\|^2_{H^{1,1}(Q)} \geq \int_S \int \frac{\beta\sqrt{w}}{w^{1+2\varepsilon}} d\beta\ dw = \infty$$

thus proving (2.2). ∎

<u>Remark</u> 1

Using representation (2.4) one can directly verify that for (2.1) we have

$$(2.6) \qquad \begin{cases} L^N \in \mathcal{L}(L_2(\Sigma) \to H^{3/4,3/4}(Q)) \\ \\ L^N|_\Gamma \in \mathcal{L}(L_2(\Sigma) \to H^{1/2,1/2}(\Sigma)) \end{cases}$$

Also, above example can be easily generalized to an arbitrary elliptic operator with constant coefficients defined on a half-space of R^n.

In order to state our main result, we need to introduce the following technical assumption.

Assume that there exist vector field

$\vec{h}(x) \subset (h_1(x), h_2(x) \ldots h_n(x)) \in [C^2(R^n)]^n$ such that

(i) $\vec{h}(x)|_\Gamma \parallel \vec{n}(x)$ where $\vec{n}$ is normal direction to Γ

$$*(ii) \quad \sum_{j=1}^{n} \Big[a_{ij}\Big(\frac{\partial}{\partial x_i} h_\ell - \frac{h_\ell}{h_i} \frac{\partial}{\partial x_j} h_i\Big) - h_j\Big(\frac{\partial}{\partial x_j} a_{i\ell}$$

$$- \frac{h_\ell}{h_i} \frac{\partial}{\partial x_j} a_{ii}\Big) + a_{\ell j}\Big(\frac{\partial}{\partial x_j} h_i - \frac{h_i}{h_\ell} \frac{\partial}{\partial x_j} h_\ell\Big)$$

$$-h_j\Big(\frac{\partial}{\partial x_j} a_{\ell i} - \frac{h_i}{h_\ell} \frac{\partial}{\partial x_i} a_{\ell\ell}\Big)\Big] = 0 \text{ for } i, \ell = 1, \ldots n \ i \neq \ell$$

Our main result is

<u>Theorem</u> 2

Let assumption (H) be satisfied. Then

(i) $L^N \in \mathscr{L}(L_2(\Sigma) \to H^{2/3-\varepsilon, 2/3-\varepsilon}(Q))$

(ii) $L^N|_\Gamma \in \mathscr{L}(L_2(\Sigma) \to H^{1/3-\varepsilon, 1/3-\varepsilon}(\Sigma))$

---

* Assumption (H) is equivalent to saying that after partition of unity which transfers the original Ω to the half-space, the principal part of the corresponding elliptic operator is of the form

$$A^o(x,y) = a(x,y)\frac{\partial^2}{\partial x^2} + \sum_{i_j=1}^{n-1} c_{ij}(y)\frac{\partial^2}{\partial y_i \partial y_j} + \sum_{i=1}^{n-1} d_i(x,y)\frac{\partial^2}{\partial x \partial y_i}$$

where $y = (y_1, y_2, \ldots, y_{n-1})$

Corollary 3

(i) $L^N \in \mathscr{L}(L_2(\Sigma) \to C[OT; H^{2/3-\varepsilon}(\Omega)])$

(ii) $L_t^N \in \mathscr{L}(L_2(\Sigma) \to C[OT; H^{-1/3-\varepsilon}(\Omega))$

Remark 2

As mentioned before, removal of the technical assumption H requires a completely different approach (pseudo differential operator theory) for which we refer to the forthcoming joint work [L-T.3]. Here, instead assumption H allows us to keep our presentation self-contained and much simpler while still giving a flavor on the general line of attack to the problem.

Remark 3

Notice that the trace regularity of L^N given by (ii) does not follow from interior regularity (i) via application of Trace Theorem. This is an independent regularity result, which gives more regularity of the solution u on the boundary then Trace Theory combined with interior regularity would imply.

Remark 4

Higher regularity and in particular differentiability of the solutions (for smoother boundary data g satisfying certain compatability conditions) can be obtained directly from Theorem 2 via "lifting techniques" for example [L-T.2] [L-L-T.1] or [M-1]).

Next section will be devoted to the proof of Corollary 3, while the proof of main Theorem 2 is relagated to section 4.

3. Proof of Corollary 3

The proof of the Corollary follows along similar lines to those in [L-T-2].

Let us define the operator $A : L_2(\Omega) \to L_2(\Omega)$ by the following formula

$$Au = -A(x,\partial)u \qquad \text{on } \mathcal{D}(A) \text{ where}$$

$$\mathcal{D}(A) = \{u \in L_2(\Omega); \qquad Au \in L_2(\Omega)\ \frac{\partial u}{\partial \eta_A}\Big|_\Gamma = 0\}.$$

Without loss of generality we can assume that 0 is not an eigenvalue of A. Define $N : L_2(\Gamma) \to L_2(\Omega)$ as

$$\begin{cases} A(x,\partial)Ng = 0 \\ \dfrac{\partial}{\partial \eta_A} Ng\Big|_\Gamma = g \end{cases}$$

It is well known that

$$(3.1) \qquad N \in \mathcal{L}(L_2(\Gamma) \to H^{3/2}(\Omega))$$

The solution to (1.1) can now be written as in [L-T-1]

$$u(t) = A\int_0^t S(t - z)Ng(z)dz \equiv (L^N g)(t)$$

where S(t) stands for sine operator generated by -A.

From Theorem 2 we have

$$L^N \in \mathcal{L}(L_2(\Sigma) \to H^{2/3-\varepsilon,2/3-\varepsilon}(Q))$$

Hence

$$(3.2) \qquad (L^N)^* \in \mathscr{L}(H^{-2/3+\varepsilon,-2/3+\varepsilon}(Q) \to L_2(\Sigma))$$

where adjoint $(L^N)^*$ is computed with respect to $L_2(Q)$ inner product.

Now let us take $f_0 \equiv C^*(t)x$ where $x \in H^{-2/3+\varepsilon}(\Omega)$ and $C(t)$ stands for cosine operator associated with $-A$. Since we have that

$$(3.3) \qquad \mathscr{D}(A^{*1/3-\varepsilon/2}) \sim H^{2/3-\varepsilon}(\Omega)$$

$$(3.4) \qquad f_0 \in C[0T; H^{-2/3+\varepsilon}(\Omega)] \subset H^{-2/3+\varepsilon,-2/3+\varepsilon}(Q)$$

On the other hand, using sine and cosine operator properties we obtain

$$(3.5) \quad (L^{N*}f_0)(t) = N^*A^* \int_t^T S^*(z - t)C^*(z)x\, dz$$

$$= \frac{1}{2} N^*[C^*(2T - t)x - x] + N^*A^*S^*(t)x \cdot (T - t)$$

From (3.1), (3.3) we have that

$$(3.6) \qquad N^*C^*(2T - t)x \in L_2(\Sigma) \qquad \text{for } x \in H^{-2/3+\varepsilon}(\Omega)$$

Thus (3.2), (3.5) and (3.6) yield

$$(T-\cdot)N^*A^*S^*(\ )x \in L_2(\Sigma) \qquad x \in H^{-2/3+\varepsilon}(\Omega)$$

Since the above relation remains true for all T we obtain

$$(3.7) \qquad N^*A^*S^*(\cdot)x \in L_2(\Sigma) \qquad \text{with } x \in H^{-2/3+\varepsilon}(\Omega)$$

Repeating the same procedure with f_0 replaced by f_1 where $f_1(t) \equiv S^*(t)x$ and $x \in H^{-4/3+\varepsilon}(\Omega)$ we obtain

$$(3.7') \qquad N^*A^{*\,1/2}\, C^*(\cdot)x \in L_2(\Sigma) \qquad \text{with } x \in H^{-2/3+\varepsilon}(\Omega)$$

Now let us consider the following closed and densely defined operators from $L_2(\Sigma)$ into $L_2(Q)$

$$(J_1 g)(t) \equiv A \int_0^t S(z)Ng(z)dz$$

$$(J_2 g)(t) \equiv A^{1/2} \int_0^t C(z)Ng(z)dz$$

It can be easily verified that

$$(J_1^* v)(z) = N^*A^*S^*(z) \int_z^T v(t)dt$$

$$(J_2^* v)(z) = N^*A^{*\,1/2}\, C^*(z) \int_z^T v(t)dt$$

Thus from (3.7), (3.7') and closed Graph Theorem we deduce that

$$\begin{cases} J_1 \in \mathscr{L}(L_2(\Sigma) \to L_\infty(OT; \mathscr{D}(A^{*\,1/3-\varepsilon/2})) \\ J_2 \in \mathscr{L}(L_2(\Sigma) \to L_\infty(OT; \mathscr{D}(A^{*\,1/3-\varepsilon/2})) \end{cases}$$

Since $u(t) = A \int_0^t S(t - z)Ng(z)dz$

$$= A^{1/2} S(t) \int_0^t A^{1/2} C(z)Ng(z)dz \qquad -C(t)\int_0^t AS(z)Ng(z)dz$$

$$= A^{1/2} S(t)(J_2 g)(t) - C(t)(J_1 g)(t)$$

and $A^{1/2} S \in \mathscr{L}(L_2(\Omega) \to C[0T;\ L_2(\Omega)])$ we obtain that

$$(3.8) \qquad u \in L_\infty(0T;\ H^{2/3-\varepsilon}(\Omega)) \qquad \text{with } g \in L_2(\Sigma)$$

In order to obtain that

$$(3.9) \qquad u \in C[0T;\ H^{2/3-\varepsilon}(\Omega)]$$

we apply the usual density argument. More precisely we take a sequence $g_n \in H^1[0T;\ L_2(\Gamma)]$ converging to g in $L_2(\Sigma)$, It can be easily verfied that

$$(3.10) \qquad L^N g_n \in C[0T;\ H^1(\Omega)] \subset C[0T;\ H^{2/3-\varepsilon}(\Omega)]$$

(3.8), (3.10) together with the completion of $C[0T;\ H^{2/3-\varepsilon}(\Omega)]$ yields the desired conclusion (3.9) hence part (i) of the Corollary 3. The proof of part (ii) is similar and therefore it is omitted.

4. Proof of Theorem 2

With u the solution corresponding to (1.1) we define variable z by

$$z \equiv \nabla u \cdot \vec{h}$$

It is well known ([L-M], [M-1]) that

$$(4.1) \quad u \in C[0T;\ H^{1/2}(\Omega)] \cap H^{1/2,\,1/2}(Q) \qquad \text{for } g \in L_2(\Sigma)$$

It can be easily verified that z satisfies

$$(4.2)\quad \begin{cases} z_{tt} = A(x,\partial)z + A(x,\partial)(\nabla u \cdot \vec{h}) - \nabla(A(x,\partial)u) \cdot \vec{h} \\ z|_\Gamma = g \end{cases}$$

In view of our assumption H one can check that

$$A(x,\partial)(\nabla u \cdot \vec{h}) - \nabla(A(x,\partial)u) \cdot \vec{h} = I(u) + I(z)$$

where I(u) stands for the first order differential operator defined on Ω

Then (4.1) becomes

$$(4.2')\quad \begin{cases} z_{tt} = A(x,\partial)z + I(u) + I(z) \\ z|_\Gamma = g \\ z(0) = z_t(0) = 0 \end{cases}$$

Using a'priori regularity of u given by (4.1) we deduce that:

$$I(u) \subset C[0T;\ H^{-1/2}(\Omega)] \subset C[0T;\ H^{-1}(\Omega)].$$

Then we are in a position to apply Theorem 1 to the problem (4.2). This gives:

$$(4.3)\qquad z \in C[0T;\ L_2(\Omega)] \qquad \text{for } g \in L_2(\Sigma)$$

or equivalently

$$(4.3')\qquad \nabla u \cdot \vec{h} \in C[0T;\ L_2(\Omega)] \qquad \text{for } g \in L_2(\Sigma)$$

Using the above regularity of the derivative of u in the direction of $\vec{h}$, we will next improve the regularity of $u|_\Gamma$. To accomplish this, we first notice that divergence theorem gives:

$$(4.4) \qquad \delta\langle u\rangle_\Gamma^2 = 2(\nabla u \cdot \vec{h}, u)_\Omega + (|u|^2, \operatorname{div} \vec{h})$$

More generally, using energy methods and interpolation one can show that:

Lemma 1

Let u be the solution of (1.1). Then there exists constant $c > 0$ such that for all $0 < \alpha < 1/2$

$$\langle u\rangle^2_{H^{\alpha,\alpha}(\Sigma)} < C|\nabla u \cdot \vec{h}|_{L_2(Q)}|u|_{H^{2\alpha,2\alpha}(Q)} + C|u|^2_{H^{\alpha,\alpha}(Q)}$$

Now Lemma (1) applied with $\alpha = 1/4$ together with (4.1) and (4.3') yields:

$$(4.5) \qquad \langle u\rangle_{H^{1/4,1/4}(\Sigma)} < C|g|_{L_2(\Sigma)}$$

which is our first "improvement of the regularity of the $u|_\Gamma$. In order to "improve" the regularity in the interior we need another Lemma which again can be easily proved using standard energy methods and interpolation.

Lemma 2

Let u be the solution of (1.1). Then there exists constant $c > 0$ such that for all $0 < \alpha < 1/2$ we have

$$|u|^2_{H^{\alpha+1/2,\alpha+1/2}(Q)} < C|u|_{H^{2\alpha,2\alpha}(\Sigma)}|g|_{L_2(\Sigma)} + C|u|^2_{H^{\alpha,\alpha}(Q)}$$

Applying Lemma (2) with $\alpha = 1/8$ together with (4.5) yields

(4.6) $$|u|^2_{H^{1/2+1/8,1/2+1/8}(Q)} < C|g|^2_{L_2(\Sigma)}$$

Now we can apply boot-strap argument. In fact, using information about improved regularity in the interior of Ω we apply Lemma (1) (taking now $\alpha = \frac{1}{4} + \frac{1}{16}$). This yields

(4.7) $$|u|^2_{H^{1/4+1/16,1/4+1/16}(\Sigma)} < C|g|^2_{L_2(\Sigma)}$$

Again we apply Lemma (2) ($\alpha = \frac{1}{8} + \frac{1}{32}$) to obtain

(4.8) $$|u|^2_{H^{1/2+1/8+1/32,1/2+1/8+1/32}(Q)} < C|g|^2_{L_2(\Sigma)}$$

Repeating the same argument N number of times yields:

$$|u|^2_{H^{r^N,r^N}(\Sigma)} < C_N|g|^2_{L_2(\Sigma)}$$

where $r^N = \sum_{i=1}^{N} \frac{1}{4^i}$ and

$$|u|^2_{H^{s^N,s^N}(Q)} < C_N|g|^2_{L_2(\Sigma)}$$

where $s^N = 1/2 + \frac{1}{8} \sum_{i=0}^{N} \frac{1}{4^i}$

Since

$$r_\infty = \frac{\frac{1}{4}}{1 - \frac{1}{4}} = \frac{\frac{1}{4}}{\frac{3}{4}} = \frac{1}{3}$$

and

$$s_\infty = \frac{\frac{1}{2}}{1 - \frac{1}{4}} = \frac{\frac{1}{2}}{\frac{3}{4}} = \frac{4}{2 \cdot 3} = \frac{2}{3}$$

Hence

$$|u|^2_{H^{1/3-\varepsilon,1/3-\varepsilon}(\Sigma)} \leq C_\varepsilon |g|_{L_2(\Sigma)}$$

and

$$|u|_{H^{2/3-\varepsilon,2/3-\varepsilon}(Q)} \leq C_\varepsilon |p|_{L_2(\Sigma)}$$

which proves the conclusion (i) of Theorem 2. Proof of part (ii) follows along the same lines and is thus omitted. ∎

References

[L-M] J. L. Lions - E. Magenes, Nonhomogeneous boundary value problems and applications I, II Springer-Verlag 1972

[L-T-1] I. Lasiecka, R. Triggiani, A cosine operator approach to modeling $L_2(0T; L_2(\Gamma))$-boundary input hyperbolic equations. Appl. Math. Optim. 7, 35-93 1981

[L-T-2] I. Lasiecka, R. Triggiani, Regularity of hyperbolic equations under $L_2(0T; L_2(\Gamma))$-Dirichlet boundary terms. Appl. Math. Optimiz. 10; 275-286 1983

[L-T.3] I. Lasiecka, R. Triggiani, Regularity of solutions to second order hyperbolic problems with Neumann boundary conditions. Work in progress.

[M-1] S. Miyatake, Mixed problem for hyperbolic equation of second order. J. Math. Kyoto Univ. 13-3 435-487 1973

[L-L-T.1] I. Lasiecka, J. L. Lions, R. Triggiani, Nonhomogeneous boundary value problems for second order hyperbolic operators to appear in Journal de Mathematique Pure et Appliquè,

C^∞ REGULARITY FOR FULLY NONLINEAR ABSTRACT EVOLUTION EQUATIONS

Alessandra Lunardi

Dipartimento di Matematica
Università di Pisa
Via Buonarroti 2
56100 Pisa, Italy

We give here further regularity results for the solution of the initial value problem

$$(*)\quad \begin{cases} u'(t) = f(t,u(t)) \quad , \quad t \geq 0 \\ u(0) = u_0 \end{cases}$$

where $f : [0,+\infty[\times D \to X$ is a smooth function, D,X are Banach spaces with D continuously embedded in X, and $u_0 \in D$. Local and global existence and uniqueness results for the solution of (*) have been proved in [LS1], [LS2] under the assumptions that $f(0,u_0)$ belongs to a suitable intermediate space between D and X, and for each $t \geq 0$ and $y \in D$ the linear operator $f_u(t,y) : D \to X$ generates an analytic semigroup in X. Here we show that, if f belongs to $C^\infty([0,+\infty[\times D;X)$ and $[0,\tau[$ is the maximal interval of definition of the solution u of (*), then u belongs to $C^\infty(]0,\tau[;D)$. Moreover, if certain compatibility conditions hold, u belongs to $C^\infty([0,\tau[;D)$.

1. NOTATIONS AND PRELIMINARIES: C^∞ REGULARITY FOR A LINEAR PROBLEM

Throughout the paper it is assumed that D (with norm $\|\cdot\|_D$) and X (with norm $\|\cdot\|$) are real Banach spaces, D being continuously (not necessarily densely) embedded in X. We denote by $\tilde{D} = \{x+iy;\ x,y \in D\}$ and $\tilde{X} = \{x+iy;\ x,y \in X\}$ the complexification of

D and X respectively, endowed with their natural norms.

Let $A : D \to X$ be a linear operator; then we denote by $\tilde{A} : \tilde{D} \to \tilde{X}$ its complexification: $\tilde{A}(x+iy) = Ax + iAy \quad \forall x,y \in D$.

We assume that

$$(1.1) \quad \begin{cases} \text{the resolvent set } \rho(\tilde{A}) \text{ of } \tilde{A} \text{ contains a sector } S = \{ z \in \mathbb{C}; \\ z \neq 0,\ |\arg z| < \theta_0\} \text{ with } \theta_0 \in \,]\pi/2,\pi[\,\text{; there exists } M_0 > 0 \\ \text{such that} \\ \qquad \|\lambda(\lambda - A)^{-1}\|_{L(\tilde{X})} \le M_0 \quad \text{for } \lambda \in S \end{cases}$$

(1.2) the graph norm of A is equivalent to the norm of D

Then $\tilde{A}$ generates a bounded analytic semigroup $e^{t\tilde{A}}$ in $\tilde{X}$ (which is strongly continuous at $t = 0$ if and only if D is dense in X), whose restriction e^{tA} to X is an analytic semigroup in X (see [S] and the appendix of [LS1]). The interpolation spaces $D_A(\alpha,\infty)$ and $D_A(\alpha)$ $(0 < \alpha < 1)$ are defined by

$$D_A(\alpha,\infty) = \{ x \in X;\ [x]_\alpha = \sup_{t>0} \|t^{1-\alpha} A e^{tA} x\| < +\infty\}$$

$$\|x\|_{D_A(\alpha,\infty)} = \|x\| + [x]_\alpha$$

$$D_A(\alpha) = \{x \in X;\ \lim_{t\to 0} t^{1-\alpha} A e^{tA} x = 0\}$$

Then $D_A(\alpha)$ is the closure of D in the norm of $D_A(\alpha,\infty)$ (see [S]) and, defining analogously $D_{\tilde{A}}(\alpha,\infty)$ and $D_{\tilde{A}}(\alpha)$, we have $D_A(\alpha,\infty) = D_{\tilde{A}}(\alpha,\infty) \cap X$, $D_A(\alpha) = D_{\tilde{A}}(\alpha) \cap X$ (see [LS1]).

We shall use the functional spaces listed below. Let Y be a real or complex Banach space, $a,b \in \mathbb{R}$, $a < b$, $\alpha \in]0,1[$, $k \in \mathbb{N} \cup \{+\infty\}$. $C([a,b];Y)$ (resp. $C^\alpha([a,b];Y)$, $h^\alpha([a,b];Y)$, $C^k([a,b];Y)$) is the Banach space of all continuous (resp. α-Hölder continuous, α-little Hölder continuous, k times continuously differentiable) functions $f : [a,b] \to Y$, endowed with its usual norm. $C(]a,b];Y)$, $C(]a,b[;Y)$, $C^k(]a,b[;Y)$ are similarly defined. $C^{k,\alpha}([a,b];Y)$ (resp. $h^{k,\alpha}([a,b];Y)$) is the subspace of $C^k([a,b];Y)$ consisting of the functions f such that $f^{(k)}$ belongs to

$C^\alpha([a,b];Y)$ (resp. to $h^\alpha([a,b];Y)$).

We shall give now some regularity results for the solution of the linear problem

$$(1.3)\quad \begin{cases} v'(t) = \Lambda(t)v(t) + \phi(t) \ , & t_0 < t \le t_1 \\ v(t_0) = x \end{cases}$$

where $t_1 > t_0$ and the operators $\Lambda(t) : D \to X$ are such that

$$(1.4)\quad t \to \Lambda(t) \in C^\alpha([t_0,t_1];L(D,X))$$

(1.5) there exists $\omega_0 \in \mathbb{R}$ such that for each $t \in [t_0,t_1]$ the operator $A(t) = \Lambda(t) - \omega_0 : D \to X$ satisfies (1.1) and (1.2)

Assumptions (1.4) and (1.5) imply that the interpolation spaces $D_{\Lambda(t)-\omega_0}(\alpha,\infty)$, $t_0 \le t < t_1$, coincide algebraically and topologically: they will be denoted by $D_\Lambda(\alpha,\infty)$.

If ϕ is continuous and $x \in X$, a function $v \in C([t_0,t_1];X) \cap C^1(]t_0,t_1];X) \cap C(]t_0,t_1];D)$ which satisfies (1.3) is said to be a classical solution of (1.3). In the following proposition we give some conditions for the existence of a classical solution of (1.3) together with some properties of the solution itself. The proofs are in [AT], where a complete treatment of problem (1.3) can be found.

PROPOSITION 1.1 *If* $x \in \bar{D}$ *and* $\phi \in C^\alpha([t_0,t_1];X)$, *problem* (1.3) *has a unique classical solution* v, *and there exists* $C_1 > 0$ *such that*

$$(1.6)\quad \|v\|_{C([t_0,t_1];X)} \le C_1(\|x\| + \|\phi\|_{C^\alpha([t_0,t_1];X)})$$

Moreover v *belongs to* $C^{1,\alpha}([t_0+\varepsilon,t_1];X) \cap C^\alpha([t_0+\varepsilon,t_1];D)$ *for any* $\varepsilon \in]0,t_1-t_0[$. *If, in addition,* $x \in D_\Lambda(\alpha,\infty)$, *then* v *belongs to* $C^\varepsilon([t_0,t_1];D_\Lambda(\alpha-\varepsilon,\infty))$ *for each* $\varepsilon \in]0,\alpha[$, *and*

$$(1.7)\quad \sup_{t_0<t\le t_1} \|(t-t_0)^{1-\alpha} v(t)\|_D < +\infty$$

_If_, _moreover_, $x \in D$ _and_ $\Lambda(t_0)x + \phi(t_0) \in D_\Lambda(\alpha,\infty)$, _then_ v _belongs to_ $C^{1,\alpha}([t_0,t_1];X) \cap C^\alpha([t_0,t_1];D)$ _and_ $v'(t)$ _belongs to_ $D_\Lambda(\alpha,\infty)$ _for each_ $t \in [t_0,t_1]$. ■

We shall need also some higher regularity results for the solution of (1.3), assuming more regularity on $\Lambda(\cdot)$ and $\phi(\cdot)$. It is easy to get a heuristic formula for the derivatives of v at $t = t_0$: setting $x_n = v^{(n)}(t_0)$ $(n = 0,1,2,...)$ it should hold

$$(1.8)\quad \begin{cases} x_0 = x \\ x_{n+1} = \sum_{k=0}^{n} \binom{n}{h} \Lambda^{(k)}(t_0) x_{n-k} + \phi^{(n)}(t_0) \end{cases}$$

PROPOSITION 1.2 _Let_ $n \in \mathbb{N}$, $0 < \alpha < 1$. _Assume that the linear operators_ $\Lambda(t) : D \to X$ _satisfy_ (1.5), _and_

$$(1.9)\quad \Lambda(\cdot) \in C^{n,\alpha}([t_0,t_1];L(D,X)) \;,\quad \phi \in C^{n,\alpha}([t_0,t_1];X)$$

Then for each $x \in \bar{D}$ _the classical solution_ v _of_ (1.3) _belongs to_ $C^{n+1,\alpha}([t_0+\varepsilon,t_1];X) \cap C^{n,\alpha}([t_0+\varepsilon,t_1];D)$ _for each_ $\varepsilon \in]0,t_1-t_0[$. _If, in addition_

$$(1.10)\quad \begin{cases} x_k \in D \text{ for } 0 \le k \le n \\ x_{n+1} \in D_\Lambda(\alpha,\infty) \end{cases}$$

then v _belongs to_ $C^{n+1,\alpha}([t_0,t_1];X) \cap C^{n,\alpha}([t_0,t_1];D)$.

Proof: For $n = 0$ the statements follow from proposition 1.1: therefore v belongs to $C^{1,\alpha}([t_0+\varepsilon,t_1];X) \cap C^\alpha([t_0+\varepsilon,t_1]);D)$ for $0 < \varepsilon < t_1-t_0$, and, if $x \in D$, $\Lambda(t_0)x + \phi(t_0) \in D_\Lambda(\alpha,\infty)$, v belongs also to $C^{1,\alpha}([t_0,t_1];X) \cap C^\alpha([t_0,t_1];D)$. Let us prove the proposition for $n = 1$: assume $\Lambda(\cdot) \in C^{1,\alpha}([t_0,t_1];L(D,X))$, $\phi \in C^{1,\alpha}([t_0,t_1];X)$ and consider the problem

$$(1.11)\begin{cases} w'(t) = \Lambda(t)w(t) + \Lambda'(t)v(t) + \phi'(t) & t_0 + \varepsilon < t \le t_1 \\ w(t_0+\varepsilon) = v'(t_0+\varepsilon) \end{cases}$$

with $\varepsilon \in \,]0,\frac{t_1-t_0}{2}[$. Since $v'(t_0+\varepsilon)$ belongs to $D_\Lambda(\alpha,\infty)$, by proposition 1.1 problem (1.11) has a unique classical solution w, which belongs to $C^{1,\alpha}([t_0+2\varepsilon,t_1];X) \cap C^\alpha([t_0+2\varepsilon,t_1];D)$.
We want to show that $w = v'|_{[t_0+\varepsilon,t_1]}$: to this aim, fix $\delta \in \,]0,\frac{t_1-t_0}{2}[$ and set

$$w_h(t) = h^{-1}(v(t+h)-v(t)) \quad , \quad t_0 + \varepsilon \le t \le t_1 - \delta \quad , \quad 0 < h < \delta$$

Then the function $z_h = w_h - w$ satisfies

$$\begin{cases} z_h'(t) = \Lambda(t)z_h(t) + h^{-1}(\Lambda(t+h) - \Lambda(t))v(t+h) - \Lambda'(t)v(t) + \\ \qquad + h^{-1}(\phi(t+h) - \phi(t)) - \phi'(t) \quad , \quad t_0 + \varepsilon < t \le t_1 - \delta \\ z_h(t_0+\varepsilon) = h^{-1}(v(t_0+\varepsilon+h)-v(t_0+\varepsilon)) - v'(t_0+\varepsilon) \end{cases}$$

Since $\psi_h(t) = h^{-1}(A(t+h) - A(t))v(t+h) - A'(t)v(t) + h^{-1}(\phi(t+h) - \phi(t)) - \phi'(t)$ converges to 0 in $C([t_0+\varepsilon,t_1-\delta];X)$ as $h \to 0$, and $h^{-1}(v(t_0+\varepsilon+h) - v(t_0+\varepsilon)) - v'(t_0+\varepsilon)$ converges to 0 in X as $h \to 0$, by (1.6) z_h converges to 0 as $h \to 0$, that is, $w(t) = v'(t)$ for $t_0 + \varepsilon \le t \le t_1 - \delta$. Letting $\delta \to 0$ and using the continuity of w and v' at $t = t_1$ we get $w = v'|_{[t_0+\varepsilon,t_1]}$. Since w belongs to $C^{1,\alpha}([t_0+2\varepsilon,t_1];X) \cap C^\alpha([t_0+2\varepsilon,t_1];D)$ the first part of the proposition is proved. Assume now that also (1.10) holds for $n = 1$; that is, $x \in D$, $\Lambda(t_0)x + \phi(t_0) \in D$, $\Lambda(t_0)(\Lambda(t_0)x + \phi(t_0)) + \Lambda'(t_0)x + \phi'(t_0) \in D_\Lambda(\alpha,\infty)$. By proposition 1.1, v is continuously differentiable up to $t = t_0$. Consider problem (1.11) with $\varepsilon = 0$: by proposition 1.1 there exists a unique solution w, which belongs to $C^{1,\alpha}([t_0,t_1];X) \cap C^\alpha([t_0,t_1];D)$. Arguing as before, we get $w = v'$ and the second part of the proposition holds for $n = 1$. The case $n > 1$ follows easily by induction. ∎

Finally, we shall use an estimate which follows from [L], prop. A1, and [AT], lemmas 3.2-(i), 3.4-(ii), 3.5-(ii).

PROPOSITION 1.3 Let (1.4), (1.5) hold, let $x \in D$ and $\phi :]t_0,t_1] \to X$ be a continuous function such that

$$(1.12) \quad \sup_{t_0<t\le t_1} \|(t-t_0)^\sigma \phi(t)\| < +\infty$$

for some $\sigma \in]0,1[$. Assume that a solution v of (1.3) belongs to $C(]t_0,t_1];D) \cap C^1(]t_0,t_1];X) \cap C^\varepsilon([t_0,t_1];D_\Lambda(\theta,\infty))$ for some $\varepsilon,\theta \in]0,1[$. Then there exists $C_2 > 0$ (depending only on $\|\Lambda(\cdot)\|_{C^\alpha([t_0,t_1];L(D,X))}$, $\omega_0,\theta_0,M_0,\varepsilon,\sigma,\theta,(t_1-t_0)$) such that

$$(1.13) \quad \|v(t)\| \le C_2(\|x\| + \sup_{t_0<s\le t_1} \|(s-t_0)^\sigma \phi(t)\|) \quad \text{for } t_0 \le t \le t_1. \blacksquare$$

2. THE NONLINEAR PROBLEM

Let us recall a result about a class of nonlinear initial value problems which has been proved in [LS1].

THEOREM 2.1 Let $f : [0,+\infty[\times D \to X$ be such that

$$(2.1) \quad f_t, f_u, f_{ut}, f_{uu} \text{ are continuous}$$

(2.2) for each $t_0 \ge 0$ and $y_0 \in D$ there exists $w_0 \in \mathbb{R}$ such that, setting

$$A(t_0,y_0) = f_u(t_0,y_0) - w_0 : D \to X$$

then $A(t_0,y_0)$ satisfies (1.1) and (1.2)

Let $u_0 \in D$ and $\alpha \in]0,1[$ be such that

$$(2.3) \quad f(0,u_0) \in D_{A(0,0)}(\alpha)$$

Then there exists $\tau = \tau(u_0) > 0$ and a unique function $u : [0,\tau[\to D$ such that

(2.4) $u \in h^{\alpha}([0,T];D) \cap h^{1,\alpha}([0,T];X) \cap C([0,T];D_{A(0,0)}(\alpha))$

$$\forall T \in]0,\tau[$$

(2.5) $$\begin{cases} u'(t) = f(t,u(t)) , & 0 \le t < \tau \\ u(0) = u_0 \end{cases}$$

In addition, for each $\tau' > \tau$ it is not possible to define a function $u : [0,\tau'[\to D$ verifying (2.4)-(2.5) with τ replaced by τ'. ■

$u : [0,\tau[\to D$ is said to be the maximal solution of problem (2.5). Concerning existence in the large, see [LS2].

A result of further regularity of the solution is provided by the following theorem.

THEOREM 2.2 Let $f : [0,+\infty[\times D \to X$ and $u_0 \in D$ satisfy (2.2) and (2.3). Assume moreover

(2.6) $f \in C^n([0,+\infty[\times D;X)$, $n \in \mathbb{N}$, $n \ge 2$

Then the solution u of (2.5) given by theorem 2.1 belongs to $C^{n,\alpha}([\varepsilon,\tau-\varepsilon];X) \cap C^{n-1,\alpha}([\varepsilon,\tau-\varepsilon];D)$ for any $\varepsilon \in]0,\tau/2[$.

Proof: First we prove the theorem for $n = 2$. Let $u : [0,\tau] \to D$ be the solution of (2.5) and fix $\varepsilon \in]0,\tau/2[$. Consider the problem

(2.7) $$\begin{cases} v'(t) = \Lambda(t)v(t) + f_t(t,u(t)) , & 0 < t \le \tau - \varepsilon \\ v(0) = f(0,u_0) \end{cases}$$

where $\Lambda(t) = f_u(t,u(t))$ satisfies (1.4) and (1.5), as it is easy to check. Since $t \to f_t(t,u(t))$ belongs to $C^{\alpha}([0,\tau-\varepsilon];X)$ and $f(0,u_0)$ belongs to $D_{A(0,0)}(\alpha,\infty) = D_{\Lambda}(\alpha,\infty)$, by proposition 1.1 problem (2.7) has a unique classical solution v, which belongs to $C^{\alpha}([\varepsilon,\tau-\varepsilon];D) \cap C^{1,\alpha}([\varepsilon,\tau-\varepsilon];X)$, and

$$\sup_{0<t\leq\tau-\varepsilon} \|t^{1-\alpha}v(t)\|_D < +\infty \tag{2.8}$$

We shall show that $v = u'|_{[0,\tau-\varepsilon]}$. To this aim, we set

$$v_h(t) = h^{-1}(u(t+h)-u(t))-v(t) \quad , \quad 0 \leq t \leq \tau - \varepsilon \; , \quad 0 < h < \varepsilon$$

Then v_h satisfies

$$\begin{cases} v_h'(t) = \Lambda_h(t)v_h(t) + \phi_h(t) \quad , \quad 0 < t \leq \tau - \varepsilon \\ v_h(0) = h^{-1}(u(h) - u_0) - u'(0) \end{cases}$$

with

$$\Lambda_h(t) = \int_0^1 f_u(t,\sigma u(t+h) + (1-\sigma)u(t))d\sigma$$

$$\phi_h(t) = [\Lambda_h(t) - f_u(t,u(t))]v(t) + h^{-1}[f(t+h,u(t+h)) - $$
$$- f(t,u(t+h)] - f_t(t,u(t))$$

It is easy to show, by elementary perturbation arguments, that the operators $\Lambda_h(t)$ satisfy (1.4) and (1.5) for h sufficiently small, with $t_0 = 0$, $t_1 = \tau-\varepsilon$, and constants ω_0, θ_0, M_0 , $\|\Lambda_h(\cdot)\|_{C^{\alpha}([0,\tau-\varepsilon];L(D,X))}$ not depending on h. In addition, ϕ_h is continuous in $]0,\tau-\varepsilon]$, and

$$\|\phi_h(t)\| \leq \int_0^1 \|f_u(t,\sigma u(t+h) + (1-\sigma)u(t)) - f_u(t,u(t))\|_{L(D,X)} \|v(t)\|_D +$$
$$+ \int_0^1 \|f_t(t+\sigma h,u(t+h)) - f_t(t,u(t))\| d\sigma$$

Then, by estimate (2.8), $\|t^{1-\alpha}\phi_h(t)\|$ is bounded for each h, and by the Hölder continuity of u we get also

$$\lim_{h\to 0} \sup_{0<t\leq\tau-\varepsilon} \|t^{1-\alpha}\psi_h(t)\| = 0$$

Therefore, by proposition 1.3, we have: $\lim_{h\to 0^+} v_h = 0$ in $C([0,\tau-\varepsilon];X)$, that is $v = u'_{|[0,\tau-\varepsilon]}$ and the proposition is proved for $n = 2$.

Assume now that the statement of the proposition holds for some $\bar{n} \geq 2$ and that $f \in C^{\bar{n}+1}([0,+\infty[\times D;X)$. Fix $\varepsilon \in]0,\tau/3[$. Then u belongs to $C^{\bar{n},\alpha}([\varepsilon,\tau-\varepsilon];X) \cap C^{\bar{n}-1,\alpha}([\varepsilon,\tau-\varepsilon];D)$, hence the function $t \to \Lambda(t) = f_u(t,u(t))$ belongs to $C^{\bar{n}-1,\alpha}([\varepsilon,\tau-\varepsilon];L(D,X))$ and the function $t \to \phi(t) = f_t(t,u(t))$ belongs to $C^{\bar{n}-1,\alpha}([\varepsilon,\tau-\varepsilon];X)$. Since $v = u'$ satisfies

$$\begin{cases} v'(t) = \Lambda(t)v(t) + \phi(t) \quad , \quad \varepsilon \leq t \leq \tau-\varepsilon \\ \\ v(\varepsilon) = u'(\varepsilon) \end{cases}$$

and $u'(\varepsilon)$ belongs obviously to $\bar{D}$, by proposition 1.2 v belongs to $C^{\bar{n},\alpha}([2\varepsilon,\tau-\varepsilon];X) \cap C^{\bar{n}-1,\alpha}([2\varepsilon,\tau-\varepsilon];D)$, so that u belongs to $C^{\bar{n}+1,\alpha}([2\varepsilon,\tau-\varepsilon];X) \cap C^{\bar{n},\alpha}([2\varepsilon,\tau-\varepsilon];D)$. By the arbitrariness of ε, the statement of the proposition follows for $n = \bar{n}+1$. ∎

<u>COROLLARY 2.3</u> <u>Let</u> $f \in C^{\infty}([0,+\infty[\times D;X)$ <u>satisfy</u> (2.2) <u>and let</u> $u_0 \in D$ <u>be such that</u> $f(0,u_0) \in D_{A(0,0)}(\alpha,\infty)$ <u>for some</u> $\alpha \in]0,1[$ ($A(0,0)$ <u>is given in</u> (2.2)). <u>Then the solution</u> u <u>of</u> (2.5) <u>given by theorem</u> 2.1 <u>belongs to</u> $C^{\infty}(]0,\tau[;D)$. <u>Define moreover</u>

$$Y_1 = \{f(0,u_0)\} \quad , \; Y_2 = \{f_t(0,u_0), f_u(0,u_0)f(0,u_0)\}$$

$$Y_{k+1} = \{\frac{\partial^k f}{\partial t^k}(0,u_0), \frac{\partial^k f}{\partial t^h \partial u^{k-h}}(0,u_0)((y_1)^{p_1},\ldots,(y_r)^{p_r}); \quad 0 \leq h \leq k-1 ,$$

$$p_1+\ldots+p_r = k-h \; , \; y_i \in \bigcup_{\alpha=1}^{k} Y_\alpha \; , \; i = 1,\ldots,r\}$$

<u>Then</u>, <u>if</u> $Y_k \subset D$ <u>for each</u> $k \in \mathbb{N}$, u <u>belongs to</u> $C^{\infty}([0,\tau[;D)$.

Proof: The first part of the corollary is an obvious consequence of theorem 2.2. The second part can be shown by induction, arguing as in the proof of theorem 2.2 and using the second part of proposition 1.2. ∎

REFERENCES

[AT] P. ACQUISTAPACE, B. TERRENI, On the abstract nonautonomous parabolic Cauchy problem in the case of constant domains, Ann. Mat. Pura Appl. (IV) 140 (1985), 1-55.

[L] A. LUNARDI, Abstract quasilinear parabolic equations, Math. Ann. 267 (1984), 395-415.

[LS1] A. LUNARDI, E. SINESTRARI, Fully nonlinear integrodifferential equations in general Banach space, Math. Z. 190 (1985), 225-248.

[LS2] A. LUNARDI, E. SINESTRARI, Existence in the large and stability for fully nonlinear integrodifferential equations, J. Int. Eq. (to appear).

[S] E. SINESTRARI, On the abstract Cauchy problem of parabolic type in spaces of continuous functions, J. Math. An. Appl. 107 (1985), 16-66.

SEMILINEAR EVOLUTION EQUATIONS IN FRÉCHET SPACES

Shinnosuke Oharu
Department of Mathematics
Hiroshima University
Hiroshima 730, Japan

This paper is concerned with semilinear evolution equations in Fréchet spaces of the form

$$u'(t) = Au(t) + B(t)u(t), \quad t > 0, \tag{1}$$

where A and B(t) are respectively linear and nonlinear operators in a Fréchet space X and u' stands for the time derivative (perhaps in a generalized sense) of u. We consider equation (1) on the supposition that there is a linear semigroup $\mathfrak{T} = \{T(t):t \geq 0\}$ in X which provides strong solutions of the linear evolution equation

$$u'(t) = Au(t), \quad t > 0, \tag{2}$$

and the nonlinear operators B(t) are continuous in some appropriate sense.

The importance of semilinear equations of the type (1) has constantly been recognized for many years in various branches of mathematical analysis. In particular, for semilinear parabolic equations such as reaction-diffusion systems and several types of semilinear hyperbolic systems it is important to advance existence theories for their solutions which belong to function spaces with Fréchet space structure and enjoy the order-preserving property. However most of the literature dealing with the operator theoretic approach to such equations have been devoted to the studies of semilinear evolution equations in ordered Banach spaces, and it seems that very little is known about evolution equations of the form (1) in ordered Fréchet spaces. In this paper we restrict ourselves to evolution equations (1) in Fréchet lattices and discuss the existence and qualitative properties of mild solutions of (1).

Our discussion involves three problems which are important from the point of view of the application to concrete pde. Firstly, Fréchet spaces are often too large to think of well-posed problems for linear evolution equations of the type (2) even though they are equipped with lattice structure. Hence, according to the nature of the problem for (2) under consideration, it is necessary to choose a suitable subspace of the Fréchet space X on which the problem would be well-posed.

Secondly, it is a new attempt to evolve a theory of linear semigroups in Fréchet spaces in such a way that linear problems as mentioned above can be treated in the framework of the semigroup theory. The third problem is to discuss what kinds of semilinear operators $A + B(t)$ may be involved in the evolution equation (1) and are useful for applications.

Here we consider to define on a vector sublattice X_0 of a given Fréchet lattice X a new topology (denoted γ and called a mixed topology) so that (X_0,γ) forms a "complete locally convex topological vector lattice" which would be a suitable space for treating our problem. In such a new topological vector lattice we formulate an abstraction of so-called supersolution-subsolution method and establish three types of existence theorems for mild solutions of (1). Conditions imposed on the operators A and B(t) are strongly affected by the lattice structure of X, and these existence results imply three different types of generation theorems for order-preserving nonlinear evolution operators in X. Finally, we make brief mention of the application of our results to concrete pde.

1. Mixed topological vector lattices.

In this section we discuss mixed topological vector lattices which are naturally constructed in Fréchet lattices. Let $(X,\leq)$ be a Fréchet lattice. We denote by τ the metric topology of X. (For the definition and basic properties of Fréchet lattices we refer to Yosida's book [12], Chapter XII). Throughout this paper we put the following conditions:

(F1) There is a seminorm system $\{p_i : i = 1,2,\cdots\}$ such that

(i) $\{p_i\}$ defines the topology of X and the sequence $(p_i(x))$ is monotone nondecresing for $x \in X$, and

(ii) if $x, y \in X$ and $|x| \leq |y|$ then $p_i(x) \leq p_i(y)$ for $i = 1,2,\cdots$, where $|x|$ denotes the absolute value of x.

(F2) There is a proper *l.s.c.* seminorm $q : X \to [0,\infty]$ such that

(i)' any q-bounded set is τ-bounded in the sense that a subset C of X is bounded in (X,τ) provided $\sup q(C) < \infty$;

(ii)' if x, y belong to the effective domain $X_0 \equiv D(q)$ and $|x| \leq |y|$, then $q(x) \leq q(y)$; and $q(|x|) = q(x)$ for $x \in X$.

In condition (F2) we have thought of a single seminorm q, but it would be more natural to think of a countable family of *l.s.c.* seminorms $\{q_j, j = 1,2,\cdots\}$ on X. In this case each q_j is assumed to satisfy (ii)', $X_0 \equiv \cap_{j\geq 1} D(q_j)$ and condition (i)' should be changed as follows:

(i)" A subset of X is τ-bounded whenever it is q_j-bounded for each j.

From condition (F2) it follows that X_0 is a linear subspace of X and $\mathrm{Ker}(q) = \{0\}$, and so q is eventually a norm on X_0.

We now introduce the notion of mixed topological vector lattice. For each $\alpha > 0$ we define the level set of q by

$$X_\alpha = \{x \in X : q(x) \leq \alpha\}.$$

Since q is *l.s.c.* over X, each X_α is an absolutely convex τ-closed subset of X (hence each X_α becomes a complete metric space). This implies that $(X_0,\{p_i\},q)$ forms a complete mixed topological space in the sense that $\{X_\alpha : \alpha > 0\}$ defines a bornology of countable type. Namely, let (V_n) be a sequence of absolutely convex τ-neighbourhood of 0 and write

$$\gamma((V_n)) \equiv \cup_{n=1}^{\infty}(V_1 \cap B_1 + \cdots + V_n \cap B_n),$$

where B_n denotes the level set X_α with $\alpha = 2^n$. Then the set of all such $\gamma((V_n))$ forms a basis of neighbourhoods of 0 for a locally convex structure on X_0 and the countable family $\{B_n\}$ gives a basis for the convex bornology on X_0. The topology so defined is denoted by $\gamma \equiv \gamma[\{X_\alpha\},\tau]$ and is called the mixed topology on X_0. We here list some fundamental facts concerning the mixed space (X_0,γ). (For the proofs we refer to Cooper's book [1], Section I.1..)

Proposition 1. (a) (X_0,γ) is a complete locally convex space.
(b) γ is the finest linear topology on X_0 which coincides with τ on the level sets X_α, $\alpha > 0$.
(c) A sequence (x_n) in X_0 converges to x in (X_0,γ) iff all the elements x_n belong to some X_α and (x_n) converges to x in (X,τ).
(d) A subset of X_0 is γ-bounded iff it is q-bounded.
(e) Let $\{T_\lambda : \lambda \in \Lambda\}$ be a family of linear operators from X_0 into itself such that $q(T_\lambda x) \leq Mq(x)$ for $x \in X_0$ and some constant $M \geq 0$. Then $\{T_\lambda\}$ is γ-equicontinuous iff it is τ-equicontinuous on each of X_α, $\alpha > 0$.

Fact (c) states that the convergence as well as the compactness can be characterized in terms of the seminorm system $\{p_i\}$, while Fact (d) means that the γ-boundedness is specified by the seminorm q. Moreover, the lattice structure of X induces a vector lattice structure on X_0; and (F1) together with Fact (b) implies that the space $(X_0,\gamma,\leq)$ forms a topological vector lattice as mentioned in Fremlin's book [3], Chapter 2. (In [3] a vector lattice is called a Riesz space.) We call it the *mixed topological vector lattice* defined by $\{p_i\}$ and q.

In what follows, we write $x_n \overset{\gamma}{\to} x$ if a sequence (x_n) converges to x in (X_0,γ). For the purpose of investigating continuous semigroups of linear operators on (X_0,γ) we employ two notions for linear operators in X_0:

Definition 1. A linear operator A in X_0 is said to be sequentially γ-closed, if $x_n \in D(A)$ for each n, $x_n \overset{\gamma}{\to} x$ and $Ax_n \overset{\gamma}{\to} y$ imply that $x \in D(A)$ and $y = Ax$. A linear operator T from X_0 into itself is said to be sequentially γ-continuous on X_0, if $Tx_n \overset{\gamma}{\to} Tx$ whenever $x_n \overset{\gamma}{\to} x$.

Condition (F1) implies that the mapping $x \to |x|$ is τ-continuous on X. Using this fact and Proposition 1, we obtain the following:

Proposition 2. (a) Under condition (F1) the lattice operations $(x,y) \to x \wedge y$ and $(x,y) \to x \wedge y$ are τ-continuous on $X \times X$. Therefore, given a pair $v, w \in X$ with $v \leqq w$, the order interval $[v,w] \equiv \{x \in X : v \leqq x \leqq w\}$ is a τ-closed convex subset of (X,τ).

(b) Under conditions (F1) and (F2) the lattice operations are sequentially γ-continuous on $X_0 \times X_0$ and, given a pair $v, w \in X_0$ with $v \leqq w$, $[v,w]$ is a γ-closed convex subset of (X_0,γ).

2. Vector integration in mixed topological vector lattices.

In order to discuss differential equations in the mixed topological vector lattice $(X_0,\{p_i\},q)$, we need a few results on the Bochner integration of X_0-valued functions defined on Lebesgue measurable subsets of $\mathbb{R}$. More general and precise results will be discussed in the forthcoming paper [5] by Hashimoto and the author.

Bochner integration theory for Fréchet-space-valued functions can be advanced in a way similar to the Banach space case and a detailed exposition on the theory may be found for instance in the book of Garnir, De Wilde and Schmets [4]. Let E be a fixed measurable subset of $\mathbb{R}$. We write $\mathfrak{M}(E;X)$ for the family of all (strongly) measurable functions on E. $\mathfrak{L}(E;X)$ stands for the space of all X-valued, Bochner integrable functions on E. In particular, $\mathfrak{L}(E;\mathbb{R})$ is simply denoted by $\mathfrak{L}(E)$. By $\mathfrak{L}^\infty(E;X)$ we denote the space of essentially bounded and strongly measurable functions on E. Further, all the integrals are taken in (X,τ) in the sense of Bochner. Basic to our argument is the following result:

Proposition 3. (a) Let $u \in \mathfrak{L}(E;X)$. If $u(t) \in X_0$ for a.e.t $\in$ E, then

$$q\left(\int_E u(t)dt\right) \leqq \int_E q(u(t))dt.$$

(b) Let $u_n \in \mathfrak{L}(E;X)$ for $n = 1,2,\cdots$ and $u : E \to X$. Suppose for each i there is $g_i \in \mathfrak{L}(E)$ such that $p_i(u_n(t)) \leqq g_i(t)$ for a.e.t $\in$ E and $n = 1,2,\cdots$. If $\tau\text{-}\lim u_n(t) = u(t)$ for a.e.t $\in$ E, then $u \in \mathfrak{L}(E;X)$ and

$$\tau\text{-}\lim \int_E u_n(t)dt = \int_E u(t)dt.$$

(c) Let $u_n \in \mathfrak{M}(E;X)$ for $n = 1,2,\cdots$ and $u : E \to X$. Assume that there exist $v, w \in \mathcal{L}(E;X)$ such that $v(t) \leq u_n(t) \leq w(t)$ for $t \in E$ and $n = 1,2,\cdots$. If $\tau\text{-}\lim u_n(t) = u(t)$ for a.e.t $\in E$, then $u \in \mathcal{L}(E;X)$, $v(t) \leq u(t) \leq w(t)$ for a.e.t $\in E$, and

$$\tau\text{-}\lim \int_E u_n(t)dt = \int_E u(t)dt \text{ and } \int_E v(t)dt \leq \int_E u(t)dt \leq \int_E w(t)dt.$$

Assertion (a) is obtained via the approximation of $X \times \mathbb{R}$-valued function $[u(\cdot),q(u(\cdot))]$ on E by $X \times \mathbb{R}$-valued measurable simple functions $[u^n(\cdot),q(u^n(\cdot))]$ with the property [*] below:

[*] There exist a sequence (E_i^n) of measurable subsets of E and a sequence (s_i^n) in E such that $s_i^n \in E_i^n$, $E_i^n \cap E_j^n = \phi$ for $j \neq i$, $\cup_i E_i^n = E$ and $u^n(t) = u(s_i^n)$ for $t \in E_i^n$, $i = 1,2,\cdots, N_n$, $n = 1,2,\cdots$, and

$$\tau\text{-}\lim \sum_{i=1}^{N_n} [u(s_i^n),q(u(s_i^n))]\mu(E_i^n) = [\int_E u(s)ds, \int_E q(u(s))ds]$$

Assertion (b) is known as the dominated convergence theorem and is proved in a way similar to the Banach space case. Assertion (c) may be called the order-dominated convergence theorem and is proved by applying assertion (b) and condition (F1).

The next result is often employed to advance a theory of linear semigroups in the mixed topological vector lattice X_0.

<u>Proposition 4.</u> Let $u : E \to X$. Suppose that $u(t) \in X_0$ for a.e.t $\in E$.

(a) Let A be a sequentially γ-closed operator in X_0. Suppose that $u(\cdot) \in \mathcal{L}(E;X)$ and $q(u(\cdot)) \in \mathcal{L}(E)$. If in addition $Au(\cdot) \in \mathcal{L}(E;X)$ and $q(Au(\cdot)) \in \mathcal{L}(E)$, then we have

$$A\int_E u(s)ds = \int_E Au(s)ds.$$

(b) Let T be a sequentially γ-continuous operator from X_0 into itself such that $q(Tx) \leq Mq(x)$ for $x \in X_0$ and some constant M. Suppose that $u(\cdot)$ is q-bounded and measurable. Then, for every $\varphi \in \mathcal{L}(E)$, $\varphi(\cdot)Tu(\cdot) \in \mathcal{L}(E;X)$ and

$$\int_E \varphi(t)Tu(t)dt = T\int_E \varphi(t)u(t)dt.$$

In case A is closed in the Fréchet space X, assertion (a) is well-known. Hence it is sufficient to find $X \times X$-valued measurable simple functions $[u^n(\cdot),Au^n(\cdot)]$ on E such that $\limsup q(\int_E u^n(t)dt) \leq q(\int_E u(t)dt)$, $\limsup q(A\int_E u^n(t)dt) \leq \int_E q(Au(t))dt$ and

$$\tau\text{-}\lim \int_E [u^n(t),Au^n(t)]dt = \int_E [u(t),Au(t)]dt.$$

But this can be done in a way similar to the proof of Proposition 3 (a). Assertion (b) is a direct consequence of (a).

3. Linear semigroups in mixed topological lattices.

In this section we outline the main points of the theory of linear semigroups on the mixed topological vector lattice $(X_0,\{p_i\},q)$. In what follows we use the same symbols τ and γ as before to denote the topology of X and the mixed topology of X_0, respectively. We begin by introducing a class of linear continuous semigroups on X_0.

The identity operator on X_0 is denoted by I and, given a subinterval J of $\mathbb{R}$, $\mathfrak{C}(J;X)$ stands for the space of X-valued, τ-continuous functions on J. A one-parameter family $\mathfrak{T} \equiv \{T(t) : t \geq 0\}$ of linear operators from X_0 into itself is called a *q*-bounded continuous semigroup of linear operators on X_0, if it has the following properties:

(S1) $T(0) = I$ and $T(s+t) = T(s)T(t)$ for $s, t \geq 0$.

(S2) $T(\cdot)x \in \mathfrak{C}([0,\infty);X)$ for $x \in X_0$.

(S3) There is a number $M \geq 1$ such that

$$q(T(t)x) \leq Mq(x) \quad \text{for } t \geq 0 \text{ and } x \in X_0.$$

(S4) $\mathfrak{T}$ is τ-equicontinuous on each X_α, $\alpha > 0$.

In view of Proposition 1 we see that a *q*-bounded continuous semigroup $\mathfrak{T}$ on X_0 can be regarded as an equicontinuous semigroup of class (C_0) on the locally convex sequentially complete space (X_0,γ) as mentioned in Yosida's book [12], Chapter IX. Hence the infinitesimal generators of *q*-bounded continuous semigroups on X_0 should be defined as follows:

<u>Definition 2.</u> Let $\mathfrak{T} \equiv \{T(t)\}$ be a *q*-bounded continuous semigroup on X_0. The linear oprator A in X_0 defined on the set

$$D(A) = \left\{ x \in X_0 : \begin{array}{l} \tau\text{-}\lim_{h\downarrow 0} h^{-1}(T(h)x - x) \text{ exists} \\ \text{and } \limsup_{h\downarrow 0} q(h^{-1}(T(h)x - x)) < \infty \end{array} \right\}$$

by

$$Ax = \tau\text{-}\lim_{h\downarrow 0} h^{-1}(T(h)x - x) \text{ for } x \in D(A)$$

is called the infinitesimal generator of $\mathfrak{T}$; D(A) is the domain of A. (Henceforth we permit ourselves the abbreviation, the *i.g.* of $\mathfrak{T}$, in referring to the infinitesimal generator of $\mathfrak{T}$.)

In the usual track of the generation theory for linear semigroups in Fréchet spaces, we can obtain a characterization of *q*-bounded continuous semigroups on X_0 in terms of the corresponding *i.g.*'s.

Theorem 5. (I) If A is the *i.g.* of a linear semigroup $\mathfrak{T}$ on X_0 with the properties (S1) - (S4), then we have:

(a) A is sequentially γ-closed and $D(A) \cap X_\alpha$ is τ-dense in X_α for $\alpha > 0$. Hence for each $x \in X_0$ there is a sequence (x_n) in $D(A) \cap X_0$ such that $x_n \overset{\gamma}{\to} x$.

(b) $q((I - hA)^{-n}x) \leq Mq(x)$ for $x \in X_0$, $h > 0$ and $n = 1,2,\cdots$.

(c) The family of linear operators $\{(I - hA)^{-n} : h \in (0,h_0), n = 1,2,\cdots\}$ is τ-equicontinuous on each of the level sets X_α, $\alpha > 0$.

(II) Let A be a linear operator in X_0. If A satisfies conditions (a), (b) and (c) mentioned above, then it is the *i.g.* of a linear semigroup $\mathfrak{T} \equiv \{T(t)\}$ on X_0 with the properties (S1) - (S4) and we have

$$T(t)x = \gamma\text{-lim}\ (I - (t/n)A)^{-n}x \qquad \text{for } x \in X_0 \text{ and } t \geq 0, \tag{3}$$

where the convergence is uniform on every bounded subinterval of $[0,\infty)$;

$$(\lambda - A)^{-1}x = \int_0^\infty e^{-\lambda t}T(t)x\,dt \qquad \text{for } x \in X_0 \text{ and } \lambda > 0; \tag{4}$$

$$T(t)x - x = \int_0^t T(s)Ax\,ds \qquad \text{for } x \in D(A) \text{ and } t > 0. \tag{5}$$

Outline of Proof. (I): We first observe that the integral on the right side of (4) makes sense by Proposition 4 (b). Hence we can show by the usual argument that (4) and (5) are valid for the *i.g.* of $\mathfrak{T}$. It then follows from (5) and Proposition 3 (a) that A is γ-closed, and condition (a) is obtained. Next, the resolvent equation for $(\lambda - A)^{-1}$, the properties (S1) through (S4) and condition (F2) together imply that $(\lambda - A)^{-1}$ is C^∞ with respect to $\lambda > 0$, and that the relations

$$(\lambda - A)^{-n}x = (n-1)!^{-1}\int_0^\infty t^{n-1}e^{-\lambda t}T(t)x\,dt$$

hold for $\lambda > 0$ and $n = 1,2,\cdots$. Thus, applying (S3), (S4) and Proposition 3 to the above relations, we obtain (b) and (c).

(II): Under conditions (a), (b) and (c) we can employ the same argument as in Takahashi-Oharu [11, Section 2] to get the exponential formula (3). Since q is *l.s.c.*, condition (b) implies (3) and the family $\mathfrak{T} \equiv \{T(t): t \geq 0\}$ of linear operators from X_0 into itself which are defined by (3) forms a q-bounded continuous semigroup on X_0. It is now easy to see that A is the *i.g.* of $\mathfrak{T}$, and the proof is complete.

In order to discuss the semilinear evolution equations in the space $(X_0,\{p_i\},q)$, we employ three kinds of q-bounded continuous semigroups on X_0.

Definition 3. A q-bounded continuous semigroup $\mathfrak{T} \equiv \{T(t)\}$ on X_0 is said to be *order-preserving*, if it has the property (S5) below:

(S5) If $x, y \in X_0$ and $x \leq y$, then $T(t)x \leq T(t)y$ for $t \geq 0$.

By Theorem 5 and Proposition 2, we obtain the following.

Corollary 6. A q-bounded continuous semigroup $\mathfrak{T}$ on X_0 is order-preserving iff the *i.g.* A of $\mathfrak{T}$ has the property that $(\lambda - A)^{-1}$ is order-preserving for $\lambda > 0$.

Definition 4. A q-bounded continuous semigroup $\mathfrak{T} \equiv \{T(t)\}$ on X_0 is said to be *analytic*, if it has the three properties below:

(S6) $R(T(t)) \subset D(A)$ for $t > 0$ and $AT(\cdot)x \in \mathfrak{C}((0,\infty);X)$ for $x \in X_0$.

(S7) There exist $C > 0$ and $\omega \in \mathbb{R}$ such that

$$q(AT(t)x) \leq Ct^{-1}e^{\omega t}q(x) \quad \text{for } t > 0 \text{ and } x \in X_0.$$

(S8) The family $\{te^{-\omega t}AT(t)\colon t > 0\}$ of linear operators in X_0 is τ-equicontinuous on each X_α, $\alpha > 0$.

In view of Theorem 5 we can apply the same argument as in Crandall-Pazy-Tartar [2], Theorem 1, to get a characterization of infinitesimal generators of q-bounded continuous semigroups on X_0 which are analytic in the sense of Definition 4. Moreover, the following result is obtained from (S7), (S8) and (5).

Proposition 7. If $\mathfrak{T}$ is analytic, then $T(\cdot)x \in \mathfrak{C}^1((0,\infty);X)$ for $x \in X_0$ and

$$(d/dt)T(t)x = AT(t)x \quad \text{for } t > 0 \text{ and } x \in X_0.$$

Definition 5. A q-bounded continuous semigroup $\mathfrak{T} \equiv \{T(t)\}$ on X_0 is said to be *compact*, if it has the property (S9) below:

(S9) For each $t > 0$, $T(t)$ maps the level sets X_α, $\alpha > 0$, into τ-compact sets in X.

It is seen ([1], Propositions 1.11 and 1.12) that if $\mathfrak{T}$ is compact then each $T(t)$ maps γ-bounded sets in X_0 into relatively γ-compact sets in X_0.

4. Semilinear evolution equations in mixed topological lattices.

In this section we introduce a class of semilinear operators in the mixed topological vector lattice $(X_0,\{p_i\},q)$ and formulate an abstraction of the supersolution-subsolution method for semilinear evolution equations of the type (1) in X_0.

Consider the initial-value problem for (1) in X_0

$$\text{(IVP)} \qquad \begin{aligned} &u'(t) = (A + B(t))u(t), \quad s < t < T, \\ &u(s) = x \in X_0 \end{aligned}$$

where s, x are initial data given in $[0,T) \times X_0$, A is a linear operator in X_0 and $B(t)$, $0 \leq t \leq T$, are possibly nonlinear operators with domains $D(B(t))$ and ranges $R(B(t))$ in X_0. We are concerned with the semilinear operators specified below.

We say that a one-parameter family $\mathfrak{G} = \{A + B(t): 0 \leq t \leq T\}$ of semilinear operators in X is of class $\mathfrak{S}(X_0)$, if it satisfies the following conditions:

(L) A is the *i.g.* of a *q*-bounded continuous semigroup $\mathfrak{T} \equiv \{T(t)\}$ on X_0 which is order-preserving.

(N) (i) For $t \in [0,T]$ and $v, w \in D(B(t))$ with $v \leq w$, the order interval $[v,w]$ is contained in $D(B(t))$ and $B(t)$ is τ-continuous on the γ-closed convex subset $[v,w]$ of X_0.

(ii) Given an X_0-valued, *q*-bounded function *u* on [0,T] such that $u(t) \in D(B(t))$ for $t \in [0,T]$, $B(\cdot)u(\cdot) \in \mathfrak{M}([0,T];X)$ whenever $u \in \mathfrak{M}([0,T];X)$; and if $u \in \mathfrak{C}([0,T];X)$, then so is $B(\cdot)u(\cdot)$.

(iii) For every $\alpha > 0$ there is $\beta > 0$ such that $\cup_{0 \leq t \leq T} B(t)[X_\alpha] \subset X_\beta$.

(iv) Each $B(t)$ is order-preserving in the sense that $B(t)x \leq B(t)y$ whenever $x, y \in D(B(t))$ and $x \leq y$.

By a strong solution of (IVP) is meant an X_0- valued, absolutely continuous function *u* on [0,T] which is γ-differentiable and satisfies (1) for a.e. $t \geq 0$ as well as the initial-condition. However, semilinear problems of the form (IVP) do not necessarily admit strong solutions even if the initial data x belong to $D(A) \cap D(B(s))$. In this paper we employ the notion of mild solution of (1) that is defined as a τ-continuous solution *u* of the variation of parameters formula

$$u(t) = [\Gamma_{s,x} u](t), \quad s \leq t \leq T, \tag{6}$$

where $\Gamma_{s,x}$ denotes the integral operator defined for $u \in \mathfrak{C}([s,T];X)$ with $u(t) \in D(B(t))$, $t \in [0,T]$, by

$$[\Gamma_{s,x} u](t) \equiv T(t-s)x + \int_s^t T(t-\xi)B(\xi)u(\xi)d\xi \text{ for } s \leq t \leq T. \tag{7}$$

A mild solution *u* of (IVP) is regarded as a fixed point of the integral operator $\Gamma_{s,x}$.

We now define super- and subsolutions of the integral equation (6).

<u>Definition 6.</u> By a *q-bounded subsolution* of (6) on [s,T] we means a function $v \in \mathfrak{C}([s,T];X)$ such that $v(t) \in D(B(t)) \cap X_\alpha$ for $t \in [s,T]$ and some $\alpha > 0$, $y \equiv v(s) \leq x$, and

$$v(t) \leq [\Gamma_{s,y} v](t) \quad \text{for } s \leq t \leq T,$$

while a function $w \in \mathfrak{C}([s,T];X)$ is called a *q-bounded supersolution* of

(6) on $[s,T]$, if there is $\alpha > 0$ such that $z \equiv w(s) \geq x$, $w(t) \in D(B(t)) \cap X_\alpha$ and

$$w(t) \geq [\Gamma_{s,z}w](t) \quad \text{for } s \leq t \leq T.$$

The following fact is important:

Lemma 8. If $u \in \mathfrak{m}([s,T];X)$ and $u(t) \in D(B(t)) \cap X_\alpha$ for $t \in [0,T]$ and some $\alpha > 0$, then $[\Gamma_{s,x}u](\cdot) \in \mathfrak{C}([s,T];X)$.

Hence a q-bounded measurable solution of the integral equation (6) is necessarily a mild solution of (6). In this sense it is sufficient to impose in Definition 6 that both v and w are only q-bounded and measurable over $[s,T]$.

We now give an abstraction of the supersolution-subsolution method.

Theorem 9. Let $s \in [0,T]$. Let v, $w \in \mathfrak{C}([s,T];X)$ and suppose that $v(t) \leq w(t)$ and $v(t)$, $w(t) \in D(B(t)) \cap X_\alpha$ for $t \in [s,T]$ and some $\alpha > 0$. Moreover, let $\mathfrak{W}_s$ be the family of all functions $u \in \mathfrak{C}([s,T];X)$ satisfying $v(t) \leq u(t) \leq w(t)$ for $t \in [s,T]$. Then:

(a) For each $x \in [v(s),w(s)]$, the integral operator $\Gamma_{s,x}$ is monotone on its domain with respect to the natural ordering on $\mathfrak{C}([s,T];X)$ defined as follows: $u_1 \leq u_2$ iff u_1, $u_2 \in \mathfrak{C}([s,T];X)$ and $u_1(t) \leq u_2(t)$ for $t \in [s,T]$.

(b) The closed covex subset $\mathfrak{W}_s$ of $\mathfrak{C}([s,T];X)$ is invariant under the integral operator $\Gamma_{s,x}$ for $x \in [v(s),w(s)]$ iff v, w are a subsolution and a supersolution of (6), respectively.

Assertion (a) is a direct consequence of conditions (N), (L) and Proposition 3 (c). It is clear from Proposition 2 that $\mathfrak{W}_s$ is a closed convex subset of $\mathfrak{C}([s,T];X)$. Hence assertion (b) follows from Definition 6 and the first assertion (a).

Thus the problem of finding mild solutions of (IVP) is reduced to that of finding fixed points of $\Gamma_{s,x}$ if there exist a subsolution v and a supersolution w of (6) which are q-bounded and satisfy $v(t) \leq w(t)$ for $t \in [s,T]$. In the following three sections we discuss three types of existence theorems for mild solutions.

5. Monotone successive approximation.

This section discusses the construction of mild solutions of the problem (IVP) as mentioned in the preceding section by means of the monotone method. In addition to our basic hypotheses (L) and (N), we here put two conditions below:

(F3) (i)" The F-lattice $(X,\leq)$ is σ-order complete in the sense that any countable subset $\{x_n\}$ has the supremum $\sup x_n$ whenever it is bounded above with respect to the order "$\leqq$".

(ii)" For any i the seminorm p_i is σ-order countinuous in the sense that if (x_n) is monotone nonincreasing and $\inf x_n = 0$, then (x_n) converges to 0 in (X,γ).

(S) There exist a q-bounded subsolution v and a q-bounded supersolution w of (6) on $[s,T]$ such that $v \leqq w$ on $[s,T]$, where $v \leqq w$ on $[s,T]$ means that $v(t) \leqq w(t)$ for $t \in [s,T]$.

A typical example of Fréchet spaces satisfying (F1) and (F3) is the space $L^1_{\ell oc}(\mathbb{R}^N)$. In case X is $L^1_{\ell oc}(\mathbb{R}^N)$, the functional q on X which assigns to each $x \in X$ the essential supremum of the function $|x(\cdot)|$ over $\mathbb{R}^N$ satisfies (F2).

Let x be any element of X_0 with $v(s) \leqq x \leqq w(s)$. To construct a mild solution u of (6) satisfying the initial condition $u(s) = x$, we define a sequence (u_n) of approximate solutions as follows: First, taking the subsolution v, we define

$$u_0(t) = T(t-s)x + \int_s^t T(t-\xi)B(\xi)v(\xi)d\xi \qquad \text{for } s \leqq t \leqq T. \tag{8}$$

By condition (N) and Definition 6, u_0 is well-defined on $[s,T]$ and belongs to the family $\mathfrak{W}_s$ mentioned in Theorem 9. Inductively, we can define a sequence (u_k) by

$$u_k(t) = T(t-s)x + \int_s^t T(t-\xi)B(\xi)u_{k-1}(\xi)d\xi \tag{9}$$

for $s \leqq t \leqq T$ and $k = 1,2,\cdots$. By the same reason as above, each u_k is well-defined as an element of $\mathfrak{C}([s,T];X)$ and we have $v \leqq u_0 \leqq \cdots \leqq u_k \leqq \cdots \leqq w$ on $[s,T]$. Hence it follows from (F3) that for each $t \in [s,T]$ the order limit $\underline{u}(t) \equiv \text{O-lim}\, u_k(t)$ exists and so $(u_k(t))$ converges to $\underline{u}(t)$ in (X,τ). Apparently, the limit function $\underline{u}$ is in $\mathfrak{M}([s,T];X)$ and satisfies $v \leqq \underline{u} \leqq w$ on $[s,T]$. Thus, by Lemma 8, $\underline{u}$ belongs to $\mathfrak{C}([s,T];X)$ and gives a q-bounded mild solution of (IVP) with $\underline{u}(s) = x$.

We have found this mild solution $\underline{u}$ by starting with the subsolution v and by employing the monotone scheme (8) - (9). Hence if u is any mild solution of (IVP) satisfying $v \leqq u \leqq w$ on $[s,T]$, then it is a q-bounded supersolution of (6) such that $v \leqq u$ on $[s,T]$, and so we see that $v \leqq \underline{u} \leqq u$ on $[s,T]$. This means that $\underline{u}$ is a minimal mild solution of (IVP) with $u(s) = x$.

Likewise, if we start with the supersolution w then we obtain a maximal solution $\overline{u}$ of (IVP) such that $v \leqq \overline{u} \leqq w$. More precisely, we can define u_0 by (8) with v replaced by w and a monotone nonincreasing sequence (u_k) of approximate solutions such that $v \leqq \cdots \leqq u_k \leqq \cdots \leqq u_0 \leqq w$. Therefore there is a measurable function $\overline{u}$ on [s,T] such that $(u_k(t))$ converges to $\overline{u}(t)$ for $t \in [s,T]$. This limit function $\overline{u}$ gives another q-bounded mild solution of (IVP) with $\overline{u}(s) = x$. Clearly, we have $v \leqq \underline{u} \leqq \overline{u} \leqq w$.

Combining the above-mentioned, we obtain

Theorem 10. Suppose conditions (F1), (F2), (F3), (L), (N) and (S) hold. Then for every $x \in [v(s),w(s)]$ the problem (IVP) admits both minimal mild solution $\underline{u}$ and maximal mild solution $\overline{u}$ such that $\underline{u}(s) = \overline{u}(s) = x$ and $v \leqq \underline{u} \leqq \overline{u} \leqq w$ on [s,T].

As is easily seen, $\underline{u}$ (resp. $\overline{u}$) is minimal (resp. maximal) in the sense that for every $s' \in [s,T)$, $\underline{u}$ restricted to [s',T] is minimal (resp. maximal) as a solution on [s',T] of (6) with x replaced by $\underline{u}(s')$ (resp. $\overline{u}(s')$) which takes its value at t in $[v(t),w(t)]$. Further, it is seen from the construction via the scheme (9) that if $v(s) \leqq x_1 \leqq x_2 \leqq w(s)$ then the associated minimal solutions $\underline{u}_1$ and $\underline{u}_2$ (or the associated maximal solutions $\overline{u}_1$, $\overline{u}_2$) satisfy $\underline{u}_1 \leqq \underline{u}_2$ (or $\overline{u}_1 \leqq \overline{u}_2$) on [s,T].

Suppose there exist q-bounded sub- and supersolutions v, w of (6) on [0,T], and define

$$\overline{v}(t) = [\Gamma_{0,v(0)}v](t) \quad \text{and} \quad \underline{w}(t) = [\Gamma_{0,w(0)}w](t) \tag{10}$$

for $t \in [0,T]$. Then $v \leqq \overline{v} \leqq \underline{w} \leqq w$ on [0,T],

$$\overline{v}(t) \leqq [\Gamma_{s,y}\overline{v}](t),\ y = \overline{v}(s) \quad \text{and} \quad \underline{w}(t) \geqq [\Gamma_{s,z}\underline{w}](t),\ z = \underline{w}(s)$$

for $0 \leqq s \leqq t \leqq T$. Hence Theorem 10 implies that for each $s \in [0,T)$ and each $x \in [\overline{v}(s),\underline{w}(s)]$ there are unique minimal and maximal mild solutions $\underline{u}(\cdot;s,x)$ and $\overline{u}(\cdot;s,x)$ of (6) such that $\overline{v} \leqq \underline{u} \leqq \overline{u} \leqq \underline{w}$ on [s,T]. Consequently, the family of operators $\underline{U} = \{\underline{U}(t,s)\colon 0 \leqq s \leqq t \leqq T\}$ in X_0 defined by

$$\underline{U}(t,s)x = \underline{u}(t;s,x) \quad \text{for } x \in [\overline{v}(s),\underline{w}(s)] \text{ and } 0 \leqq s \leqq t \leqq T$$

and that of operators $\overline{U} = \{\overline{U}(t,s)\colon 0 \leqq s \leqq t \leqq T\}$ in X_0 defined by

$$\overline{U}(t,s) = \overline{u}(t;s,x) \quad \text{for } x \in [\overline{v}(s),\underline{w}(s)] \text{ and } 0 \leqq s \leqq t \leqq T$$

form order-preserving nonlinear evolution operators constrained in the domain $\mathcal{D} = \cup_t(\{t\} \times \mathcal{D}(t))$ in the sense of Definition 7 below and

$$\mathcal{D}(t) = [\overline{v}(t),\underline{w}(t)] \quad \text{for } t \in [0,T].$$

Definition 7. A two-parameter family $\mathcal{U} = \{U(t,s)\colon 0 \leqq s \leqq t \leqq T\}$ of operators in X_0 is called an order-preserving evolution operator constrained in $\mathcal{D}$, if it has the three properties below:

(E1) $U(t,s)$ maps $\mathcal{D}(s)$ into $\mathcal{D}(t)$, $U(s,s)$ is the identity operator on $\mathcal{D}(s)$ and $U(t,r) = U(t,s)U(s,r)$ on $\mathcal{D}(r)$ for $0 \leq r \leq s \leq t \leq T$.

(E2) $U(\cdot,s)x \in \mathbf{C}([s,T];X)$ for $x \in \mathcal{D}(s)$ and $s \in [0,T)$.

(E3) $U(t,s)x_1 \leq U(t,s)x_2$ for $0 \leq s \leq t \leq T$ and x_1, $x_2 \in \mathcal{D}(s)$ with $x_1 \leq x_2$.

6. Successive Approximation under Lipschitz type conditions.

We here discuss the construction of mild solutions of the problem (IVP) by applying the method of successive approximation. In addition to the basic hypotheses (F1), (F2), (L) and (N), we assume the following conditions:

(F4) $\tau\text{-}\lim x_n = 0$ whenever $\lim q(x_n) = 0$.

(N1) For each $\alpha > 0$ there is $L_\alpha > 0$ such that

$$q(B(t)x_1 - B(t)x_2) \leq L_\alpha q(x_1 - x_2) \tag{11}$$

for $t \in [0,T]$ and x_1, $x_2 \in D(B(t)) \cap X_\alpha$ with $x_1 \geq x_2$.

(S) There exist a q-bounded subsolution v and a q-bounded supersolution w of (6) such that $v \leq w$ on $[s,T]$.

A typical example of mixed topological lattice $(X_0,\{p_i\},q)$ satisfying (F4) is the vector sublattice $X_0 = BC(\mathbb{R}^n)$ of the Fréchet lattice $X = C(\mathbb{R}^n)$ equipped with the *l.s.c.* norm q which assigns to each $x \in X$ the supremum of $|x(\cdot)|$ over $\mathbb{R}^n$. Condition (F4) follows from (F5) below which is more convenient for applications:

(F5) For each i there is $C_i > 0$ such that $p_i(x) \leq C_i q(x)$ for $x \in X_0$.

In view of condition (F2), the order-Lipschitz condition (N2) below implies condition (N1):

(N2) For each $\alpha > 0$ there is $L_\alpha > 0$ such that

$$B(t)x_1 - B(t)x_2 \leq L_\alpha(x_1 - x_2)$$

whenever $t \in [0,T]$, x_1, $x_2 \in D(B(t)) \cap X_\alpha$ and $x_1 \geq x_2$.

Also, to get the continuous dependence of mild solutions on initial data, we often impose condition (N3) below.

(N3) For each $\alpha > 0$, there is $L_\alpha > 0$ such that (11) holds for $t \in [0,T]$ and x_1, $x_2 \in D(B(t)) \cap X_\alpha$.

Under conditions (F1), (F2), (F4), (L) and (N) a sequence of approximate solutions (u_k) can be constructed by (8) and (9). As mentioned in Section 5, conditions (L) and (N) together imply that $v \leq u_0 \leq \cdots \leq u_k \leq \cdots \leq w$ on $[s,T]$, and so $u_k(t) \in X_\alpha$ for $t \in [s,T]$ and some $\alpha > 0$. Using (9) and condition (N1), we obtain the estimates

$$q(u_k(t) - u_{k-1}(t)) \leq (k!)^{-1}(ML_\alpha)^k t^k \cdot 2\alpha$$

for $t \in [s,T]$ and $k = 1,2,\cdots$, where M is a constant as mentioned in (S3). But this means that the sequence $(u_k(t))$ is Cauchy with respect to the norm q for $t \in [s,T]$. From this, (F2) and (F4) it follows that there is an X_0-valued measurable function u on $[s,T]$ such that $\tau\text{-lim}\, u_k(t) = u(t)$ and $v \leq u \leq w$ on $[s,T]$. Thus we see in the same way as in Section 5 that the limit function u gives a mild solution of (IVP).

If in particular (F5) holds, then the convergence of the sequence (u_k) is uniform on $[s,T]$ and the above conclusion is directly obtained.

Moreover, let $v(s) \leq x_1 \leq x_2 \leq w(s)$ and let u_1, u_2 be the corresponding mild solutions of (IVP) such that $v \leq u_1 \leq u_2 \leq w$ on $[s,T]$. Then by (F4) and the standard estimation of Gronwall type we obtain

$$q(u_1(t) - u_2(t)) \leq M\exp(ML_\alpha(t-s))q(x_1 - x_2) \text{ for } t \in [s,T]. \qquad (12)$$

This shows that mild solutions constrained in the subdomain $\cup_{s\leq t\leq T}(\{t\} \times [v(t),w(t)])$ are uniquely determined by their initial data since the maximal and minimal solutions of (6) with the same initial value x coincide by (12).

If (N3) holds instead of (N1), we obtain the Lipschitz continuous dependence (with respect to q) of the mild solutions on initial data. Namely, if x_1, $x_2 \in [v(s),w(s)]$ and if u_1, u_2 are the corresponding mild solutions satisfying $v \leq u_1 \leq u_2 \leq w$ on $[s,T]$, then (12) holds for $t \in [s,T]$.

Combining the above-mentioned, we obtain the following result.

Theorem 11. Suppose that (F1), (F2), (F4), (L) and (N) hold.
(a) If (N1) and (S) holds, then for every $x \in [v(s),w(s)]$ there is a unique mild solution $u(\cdot;s,x)$ of (IVP) such that $v \leq u \leq w$ on $[s,T]$.
(b) Suppose in (a) that (S) holds with $s = 0$, and let $\overline{v}$ and $\underline{w}$ be defined by (10). Then there is an order-preserving evolution oeprator $\mathcal{U} = \{U(t,s)\}$ constrained in $\mathcal{D}$ such that $U(t,s)x = u(t;s,x)$ for $x \in \mathcal{D}(s)$ and $s \in [0,T)$.
(c) If in particular (N3) holds in (b), then for each $s \in [0,T)$ the family $\{U(t,s): t \geq s\}$ is equicontinuous on $\mathcal{D}(s)$ with respect to q. If in addition (F5) holds, then $\{U(t,s): t \geq s\}$ is equicontinuous from $(\mathcal{D}(s),q)$ into (X,τ).

7. Compact Integral Operators

In this section we discuss the existence of mild solutions of (IVP) in the case where the integral operators $\Gamma_{s,x}$ defined by (7) become compact. Assume that the basic conditions (F1), (F2), (L) and (N) hold. We put condition (S) as before and, this time, we impose the following condition on the semigroup $\mathfrak{T}$.

(L1) The semigroup $\mathfrak{T} \equiv \{T(t)\}$ on X_0 is compact.

(S) For each $s \in [0,T)$ there exist a q-bounded subsolution v and a q-bounded supersolution w of (6) such that $v \leq w$ on $[s,T]$.

For $s \in [0,T)$ and $x \in [v(s),w(s)]$, let $\mathfrak{W}_{s,x}$ be the family of all functions u in $\mathfrak{C}([s,T];X)$ such that $u(s) = x$ and $v \leq u \leq w$ on $[s,T]$, where v, w are the sub- and supersolutions of (6) as mentioned in (S), respectively. On the nonlinear operators $B(t)$ we put condition (N4) below.

(N4) If $s \in [0,T)$, $v(s) \leq x \leq w(s)$, $\mathfrak{W}' \subset \mathfrak{W}_{s,x}$ and $\mathfrak{W}'$ is τ-equicontinuous on $[s,T]$, then $\{B(\cdot)u(\cdot): u \in \mathfrak{W}'\}$ is τ-equicontinuous on $[s,T]$.

Let $s \in [0,T)$ and suppose there is $\alpha > 0$ such that $v(t), w(t) \in X_\alpha$ for $t \in [s,T]$, the family of operators $\{B(t): t \in [s,T]\}$ is uniformly τ-equicontinuous from X_α into X, and such that the family of functions $\{B(\cdot)z: z \in X_\alpha\}$ is equicontinuous from $[s,T]$ into (X,τ). Then (N4) is satisfied. There are many nonlinear operators which have the above properties and arise from applications to pde.

To formulate Lemma 12 below we note that the space $\mathfrak{C}([s,T];X)$ becomes a Fréchet space with respect to the seminorms $\mathfrak{p}_i$ defined by $\mathfrak{p}_i(u) = \sup\{p_i(u(t)): t \in [s,T]\}$, $u \in \mathfrak{C}([s,T];X)$, $i = 1,2,\cdots$.

<u>Lemma 12.</u> Let $s \in [0,T)$, $v(s) \leq x \leq w(s)$ and let $\mathfrak{W}_{s,x}$ be as above.
(a) The family $\mathfrak{F} \equiv \{\Gamma_{s,x}u: u \in \mathfrak{W}_{s,x}\}$ is τ-equicontinuous over $[s,T]$.
(b) $\Gamma_{s,x}$ maps the convex closure $\mathfrak{K}$ (in the Fréchet space $\mathfrak{C}([s,T];X)$) of $\mathfrak{F}$ into itself and, for each $t \in [s,T]$, the set of values $\{[\Gamma_{s,x}u](t): u \in \mathfrak{K}\}$ is sequentially compact in (X,τ).

By condition (N) there is $\beta > 0$ such that $Z \equiv \{B(t)u(t): u \in \mathfrak{W}_{s,x}, t \in [s,T]\} \subset X_\beta$. Hence $K \equiv T(\varepsilon)[Z]$ is sequentially τ-compact in $X_{M\beta}$ by (L1), where M is a constant as mentioned in (S3); and so the family $\{T(\cdot)z: z \in K\}$ is τ-equicontinuous on $[s,T]$. Using these facts we obtain (a). Since $\mathfrak{W}_{s,x}$ is closed and convex in $\mathfrak{C}([s,T];X)$, $\mathfrak{K} \subset \mathfrak{W}_{s,x}$ and $\Gamma_{s,x}$ maps $\mathfrak{K}$ into itself. Moreover, it is seen from (a) that the family of functions $\mathfrak{K}$ is τ-equicontinuous over $[s,T]$. Using these facts and applying the compactness condition (L1) we see that for any sequence (u_n) in $\mathfrak{K}$ there is a subsequence $(u_{n(k)})$ such that $([\Gamma_{s,x}u_{n(k)}](t))_{k=1}^{\infty}$ is convergent in (X,τ) for each $t \in [s,T]$. Assertion (b) is thereby completed.

In view of Lemma 12 we can apply the Ascoli-Arzelà theorem to conclude that $\Gamma_{s,x}$ maps the closed convex set $\mathbb{K}$ in $\mathbb{C}([s,T];X)$ into a sequentially compact subset of $\mathbb{K}$. Therefore the Schauder-Tychonoff fixed point theorem implies the following existence theorem of mild solutions.

Theorem 13. Suppose that (F1), (F2), (L), (L1), (N), (N4) and (S) hold. Then for each $x \in [v(s), w(s)]$ there exist at least one mild solution u of (IVP) such that $v \leq u \leq w$ on $[s,T]$.

Suppose that there exists a pair of q-bounded sub- and supersolutions v, w of (6) such that $v \leq w$ on $[0,T]$, and let $\overline{v}$, $\underline{w}$ be defined by (10). If in Theorem 13 the uniqueness of solutions of the integral equation (6) is guaranteed for $s \in [0,T)$ and $x \in \mathcal{D}(s)$, then one can construct in the same way as in Section 6 an evolution operator $\mathfrak{U} \equiv \{U(t,s)\}$ constrained in $\mathcal{D}$. Here the order-preserving property of $\mathfrak{U}$ follows from the existence and uniqueness of solutions of (6). In fact, let $s \in [0,T)$ and $\overline{v}(s) \leq x_1 \leq x_2 \leq \underline{w}(s)$. Then $u_1(t) \equiv U(t,s)x_1$ gives a q-bounded subsolution of (6) such that $u_1 \leq \underline{w}$ on $[s,T]$ and so by Theorem 13 there is a solution u_2 of (6) such that $u_1 \leq u_2 \leq \underline{w}$ on $[s,T]$ and $u_2(s) = x_2$. But $u_2(t) \equiv U(t,s)x_2$ on $[s,T]$ by the uniqueness of $U(\cdot,s)x_2$. This means that $U(t,s)$ is order-preserving on $\mathcal{D}(s)$.

8. Remarks.

Here we make some remarks on the existence results for mild solutions stated in the last three sections. More general and precise studies will be made in the forthcoming paper [6] (in which the detailed proofs of Theorems 10, 11 and 13 will also be given).

(1) Consider the initial-value problem in X_0 of the form

$$(\mathrm{IVP})_0 \quad \begin{cases} u'(t) = (\Lambda + \Phi(t))u(t), \quad s < t < T, \\ u(s) = x \in X_0 \end{cases}$$

where Λ is a linear operator in X_0 and $\Phi(t)$, $t \in [0,T]$, are nonlinear operators with domains $D(\Phi(t))$ and ranges $R(\Phi(t))$ in X_0. In general Λ does not satisfy condition (L) and $\{\Phi(t)\}$ does not satisfy (N), although the family of semilinear operators $\{\Lambda + \Phi(t)\}$ is of class $\mathfrak{S}(X_0)$ if there is a τ-continuous linear operator S from X_0 into itself such that $q(Sx) \leq \omega q(x)$ for $x \in X_0$ and some $\omega > 0$, $A \equiv \Lambda - S$ satisfies (L), and such that $B(t) \equiv \Phi(t) + S$, $t \in [0,T]$, satisfy condition (N). In this case $(\mathrm{IVP})_0$ is reduced to the problem of type (IVP). We give some simple examples of this case in the next section.

(2) Let u be a q-bounded mild solution of (IVP) obtained by one of

Theorems 10, 11 and 13. Suppose that for some $C > 0$ and $\theta \in (0,1]$,

$$q(B(t)u(t) - B(t')u(t')) \leqq C|t - t'|^{\theta} \quad \text{for } t, t' \in [s,T]. \qquad (13)$$

If the semigroup $\mathfrak{T} \equiv \{T(t)\}$ on X_0 is analytic, then u becomes a C^1-solution on (s,T) of the probelm (IVP) by Proposition 7.

(3) Let $s \in [0,T)$, $x \in D(B(s))$ and consider the integral equation (6). Suppose that there is $v \in D(A)$ such that $v \leqq x$, $v \in \cap_{s \leqq t \leqq T} D(B(t))$ and $0 \leqq Av + B(t)v$ for $t \in [s,T]$. Then the constant fucntion $v(t) \equiv v$ gives a q-bounded subsolution of (6). Likewise, if there is $w \in D(A)$ such that $w \geqq x$, $w \in \cap_{s \leqq t \leqq T} D(B(t))$ and $0 \geqq Aw + B(t)w$ for $t \in [s,T]$, then the constant function $w(t) \equiv w$ is a q-bounded supersolution of (6). There is a direct way for finding such elements v and w: Let $s \in [0,T)$ and assume that there are two operators $\underline{B}$ and $\overline{B}$ such that $D(\underline{B}) = D(\overline{B}) \subset \cap_{s \leqq t \leqq T} D(B(t))$ and $\underline{B}z \leqq B(t)z \leqq \overline{B}z$ for $z \in D(\underline{B})$. If the equations $0 = Av + \underline{B}v = Aw + \overline{B}w$ have solutions v, w with $v \leqq x \leqq w$, then the constant functions $v(t) \equiv v$ and $w(t) \equiv w$ give a pair of sub- and supersolutions of (6) on $[s,T]$ such that $v \leqq w$ on $[s,T]$. In view of these facts, we obtain the following.

Corollary 14. Let $s \in [0,T)$ and suppose in each of Theorems 10, 11 and 13 that the following condition holds instead of (S):

(S') There exist $v, w \in D(A)$ such that $v, w \in \cap_{s \leqq t \leqq T} D(B(t))$, $v \leqq w$, and $Av + B(t)v \geqq 0 \geqq Aw + B(t)w$ for $t \in [s,T]$.

Then for each $x \in [v,w]$ there is a mild solution of (IVP).

9. Applications to Semilinear Parabolic Systems.

In this section we consider some semilinear parabolic equations and make brief mention of the application of our results to such equations. For the detailed argument we refer to the work of Kusano and the author [9].

9.1. We first consider the parabolic initial-value problem

$$\begin{aligned} u_t &= \Delta u + f(t,x,u), \quad (t,x) \in (0,T) \times \mathbb{R}^n, \\ u(0,x) &= u_0(x), \quad x \in \mathbb{R}^n, \end{aligned} \qquad (14)$$

where $n \geqq 3$, Δ denotes the n-dimensional Laplace operator and $f(t,x,\eta)$ is defined on $[0,T] \times \mathbb{R}^n \times (0,\infty)$. On the nonlinear function f we put the following conditions.

(C1) f is locally Hölder continuous in (t,x,η); and for $0 < \alpha < \beta$ there exist $C > 0$ and $\theta \in (0,1)$ such that $|f(t,x,\eta) - f(\hat{t},x,\hat{\eta})| \leqq C[|t - \hat{t}|^{\theta} + |\eta - \hat{\eta}|]$ for (t,x,η) and $(\hat{t},x,\hat{\eta}) \in [0,T] \times \mathbb{R}^n \times [\alpha,\beta]$.

(C2) For $\alpha, \beta \in \mathbb{R}$ with $0 < \alpha < \beta$ there is $L_{\alpha,\beta} \geqq 0$ such that

$$f(t,x,\xi) - f(t,x,\eta) \geq -L_{\alpha,\beta}(\xi - \eta)$$

for $\alpha \leq \eta \leq \xi \leq \beta$, $t \in [0,T]$ and $x \in \mathbb{R}^n$.

(C3) There exist a locally Hölder continuous function $\phi : [0,\infty) \to [0,\infty)$ and a locally Lipschitz continuous function $F : (0,\infty) \to (0,\infty)$ such that $\int_0^\infty r\phi(r)dr < \infty$ and $|f(t,x,\eta)| \leq \phi(|x|)F(\eta)$ on $[0,T] \times \mathbb{R}^n \times (0,\infty)$.

The above problem (14) can be formulated as an initial-value problem of the form (IVP) in a mixed topological lattice as follows: Let X be the Fréchet lattice $L^1_{\ell oc}(\mathbb{R}^n)$ with the natural ordering "$\leq$" and the seminorms $p_i(z) = \int_{|x|\leq i}|z(x)|ds$, $i = 1,2,\cdots$. Let $q(z) = \|z\|_\infty$ for $z \in X$, where $\|z\|_\infty$ is understood to be ∞ unless $x \in L^\infty(\mathbb{R}^n)$. Then $X_0 = L^\infty(\mathbb{R}^n)$ and $(X_0,\{p_i\},q)$ forms a mixed topological vector lattice with the properties (F1) and (F2). Also, it is well-known that the Fréchet lattice $(X,\leq)$ has the property (F3).

Let $0 < \alpha < \beta$ and let $L_{\alpha,\beta}$ be the constant as mentioned in (C2). In what follows, α and β are regarded as constant functions defined on $\mathbb{R}^n$ and $[\alpha,\beta]$ stands for the order interval with respect to the lattice ordering "$\leq$" in X. First we define for each $t > 0$ a linear operator $T(t)$ from X_0 into itself by

$$[T(t)z](x) = \exp(-L_{\alpha,\beta}t)(2\pi t)^{-n/2}\int \exp(-(2t)^{-1}\xi\cdot\xi)z(x+\xi)d\xi$$

for $z \in X_0$, where the integral is taken over $\mathbb{R}^n$. Defining $T(0) = I$ (the identity operator on X_0), we see that the one-parameter family $\mathfrak{T} \equiv \{T(t)\}$ forms a q-bounded continuous semigroup on X_0 such that (S3) holds with $M = 1$. The semigroup $\mathfrak{T}$ is order-preserving and analytic in such a way that (S7) and (S8) hold for $\omega = 0$. Further, if A is the *i.g.* of $\mathfrak{T}$, then $Az = \Delta z - L_{\alpha,\beta}z$ for $z \in D(A)$.

Next, to treat the nonlinear term $f(t,x,u)$ in an operator theoretic fashion, we define the operators $B(t): [\alpha,\beta] \to X_0$, $t \in [0,T]$, by

$$[B(t)z](x) = f(t,x,z(x)) + L_{\alpha,\beta}z(x) \qquad \text{for } x \in \mathbb{R}^n.$$

Under conditions (C1) - (C3), the operators $B(t)$ satisfy condition (N) and are q-Lipschitz continuous over the set $[\alpha,\beta]$ in the sense that (13) holds for some $\theta \in (0,1)$. Hence the family $\mathfrak{G} = \{A + B(t)\}$ is of class $\mathfrak{F}(X_0)$ and $[Az + B(t)z](x) = \Delta z(x) + f(t,x,z(x))$ for $z \in D(A) \cap [\alpha,\beta]$. Also, any function $v(t,x)$ on $[0,T] \times \mathbb{R}^n$ arising from this discussion can be interpreted as an X_0-valued function v via the relation $[v(t)](x) = v(t,x)$. This means that the problem (14) can be reduced to the problem (IVP) (with $s = 0$) in the space X_0.

Thus, Theorem 10 and Remark (2) in Section 8 can be applied to find C^1-solutions of (IVP), if a pair of sub- and supersolutions v, w

of the associated integral equation (6) exists in such a way that $\alpha \leq v(0) \leq u_0 \leq w(0) \leq \beta$ on $\mathbb{R}^n$ and $\alpha \leq v(t) \leq w(t) \leq \beta$ for $t \in [0,T]$. With regard this problem, there is a powerful method for finding such v, w which is based on Corollary 14. Consider the differential systems:

$$y'' + (n-1)r^{-1}y' - \phi(r)F(y) = 0, \quad y(0) = \alpha, \; y'(0) = 0; \tag{15}$$

$$z'' + (n-1)r^{-1}z' + \phi(r)F(z) = 0, \quad z(0) = \beta, \; z'(0) = 0. \tag{16}$$

If $y(r)$ (resp. $z(r)$) is a solution of (15) (resp. (16)) on $[0,\infty)$, then we see from (C3) that condition (S') in Corollary 14 holds for v, w defined respectively by $v(x) = y(|x|)$ and $w(x) = z(|x|)$, $x \in \mathbb{R}^n$. We now apply the recent results obtained in [9]. Namely, in addition to (C1) - (C3), we consider the following four conditions on the dominant F:

(C4) F is nondecreasing on some $(0,\eta_1)$ and $\eta^{-1}F(\eta) \to 0$ as $\eta \to 0+$.
(C5) F is nondecreasing on some (η_2,∞) and $\eta^{-1}F(\eta) \to 0$ as $\eta \to +\infty$.
(C6) F is nonincreasing on some (η_3,∞).
(C7) F is nondecreasing and there exist $\alpha, \beta > 0$ such that $4\alpha \leq \beta$, $2(n-2)^{-1}\beta^{-1}F(\beta)\int_0^\infty r\phi(r)dr \leq 1$, and the problem

$$y' = (n-2)^{-1}r\phi(r)F(y), \quad r > 0; \quad y(0) = \alpha$$

has a solution $\tilde{y}$ satisfying $\alpha \leq \tilde{y}(r) \leq 2\alpha$ for $r \geq 0$. In case (C4) holds, there exist a pair of (small) numbers α, β, a solution y of (15) and a solution z of (16) such that

$$\alpha \leq y(r) \leq 2\alpha \leq \beta/2 \leq z(r) \leq \beta \quad \text{for } r \geq 0. \tag{17}$$

Hence Theorem 10 and Remarks in Section 8 together imply that for α, β specified above and u_0 in the set $[\alpha,\beta] \cap D(A)$ there is a C^1-solution u on $(0,T)$ of (IVP) (with $s = 0$) such that $u(t) \in [v,w]$ for $t \in [0,T]$. From this it follows that the original problem (14) admits a solution u associated with the initial-value u_0. If either of (C5) and (C6) is satisfied, then there exist a pair of (large) numbers α, β, a solution y of (15) and a solution z of (16) satisfying the relation (17). Hence we obtain the same conclusion as above. Finally, if (C7) holds, then for α, β stated in (C7) there exist a solution y of (15) and a solution z of (16) satisfying $\alpha \leq y(r) \leq \tilde{y}(r) \leq 2\alpha \leq \beta/2 \leq z(r) \leq \beta$ for $r > 0$, so that we again obtain the same conclusion as above.

<u>Example 1.</u> Consider the parabolic equation

$$u_t = \Delta u + \psi(t,x)u^\gamma$$

where $\gamma \neq 1$ is a nonzero constant, ψ is locally Hölder continuous on $[0,T] \times \mathbb{R}^n$. Suppose that $|\psi(t,x)| \leq \phi(|x|)$ for $(t,x) \in [0,T] \times \mathbb{R}^n$ and some locally Hölder continuous function ϕ on $[0,\infty)$ with $\int_0^\infty r\phi(r)dr < \infty$. Then, defining $F(\eta) = \eta^\gamma$ for $\eta > 0$, we see that conditions (C1) - (C3) are satisfied. If $\gamma > 1$, then (C4) holds; if $0 < \gamma < 1$, then (C5) holds, while if $\gamma < 0$, then F satisfies (C6).

Example 2. Next consider the equation

$$u_t = \Delta u + \varepsilon\psi(t,x)e^u,$$

where $\varepsilon \neq 0$ and ψ is a function on $[0,T] \times \mathbb{R}^n$ satisfying all of the conditions imposed in Example 1. Then, putting $F(\eta) = e^\eta$ for $\eta > 0$, we see that (C1) - (C3) are satisfied. In this case, it is shown (see [9], Example 3) that condition (C7) holds for $|\varepsilon|$ sufficiently small.

9.2. We here consider the weakly coupled parabolic system

$$u_t = \Delta u + f(x,u,v), \quad v_t = \Delta v + g(x,u,v) \tag{18}$$

in $(0,\infty) \times \mathbb{R}^n$, $n \geq 3$, where Δ is the n-dimensional Laplace operator and f, g are functions from $\mathbb{R}^n \times (0,\infty)^2$ into $[0,\infty)$. On the functions $f(x,\xi,\eta)$, $g(x,\xi,\eta)$ we put the following conditions.

(H1) f, g are locally Hölder continuous in x and their partial derivatives f_ξ, f_η, g_ξ, g_η are continuous in (x,ξ,η).

(H2) f_ξ and g_η are bounded on $\mathbb{R}^n \times [\alpha,\beta]^2$ for $0 < \alpha < \beta$; and $f_\eta \geq 0$ and $g_\xi \geq 0$ on $\mathbb{R}^n \times (0,\infty)^2$.

(H3) There exist locally Hölder continuous functions ϕ, $\psi : [0,\infty) \to [0,\infty)$ and C^1-functions $F, G : (0,\infty)^2 \to [0,\infty)$ such that $\int_0^\infty r\phi(r)dr < \infty$, $\int_0^\infty r\psi(r)dr < \infty$, F_ξ, F_η, G_ξ, $G_\eta \geq 0$ on $(0,\infty)^2$ and

$$f(x,\xi,\eta) \leq \phi(|x|)F(\xi,\eta), \; g(x,\xi,\eta) \leq \psi(|x|)G(\xi,\eta)$$

for $(x,\xi,\eta) \in \mathbb{R}^n \times (0,\infty)^2$.

Under conditions (H1) - (H3) the initial-value problem for (18) can also be reduced to a problem of the form (IVP) in a mixed topological vector lattice in the following way: Let X be the Fréchet lattice $C(\mathbb{R}^n)^2$ with the natural ordering "$\leq$" and the seminorms $p_i((u,v)) = \sup\{\max\{|u(x)|,|v(x)|\} : |x| \leq i\}$, $i = 1,2,\cdots$. Let $q((u,v)) = \max\{\|u\|,\|v\|\}$ for $(u,v) \in X$, where $\|u\|$ denotes the supremum of $|u(x)|$ over $\mathbb{R}^n$. Then $X_0 = BC(\mathbb{R}^n)^2$ and $(X_0,\{p_i\},q)$ forms a mixed topological vector lattice for which conditions (F1), (F2) and (F5) are satisfied.

Let $\alpha > 0$ and let $L_{\alpha,\beta}$ be the supremum of $|f_\xi(x,\xi,\eta)| + |g_\eta(x,\xi,\eta)|$ over $\mathbb{R}^n \times [\alpha,\beta]^2$. We use the same symbol α to denote the constant function on $\mathbb{R}^n$ whose value is α and $[\alpha,\beta]$ stands for the order interval in the Fréchet lattice $C(\mathbb{R}^n)$. Now, in the same manner as in Section 9.1, we define for each t a linear operator T(t) from X_0 into itself by

$$T(t)[u,v] = \exp(-L_{\alpha,2\alpha}t)[e^{t\Delta}u, \; e^{t\Delta}v]$$

and

$$[e^{t\Delta}u](x) = (2\pi t)^{-n/2}\int \exp(-(2t)^{-1}\xi\cdot\xi)u(x+\xi)d\xi$$

for $[u,v] \in X_0$. Putting $T(0) = I$, we see that the family $\mathfrak{T} = \{T(t)\}$ forms a q-bounded continuous semigroup on X_0 which is both order-preserving and analytic. Also, if A is the *i.g.* of $\mathfrak{T}$, then $A(u,v) = (\Delta u, \Delta v) - L_{\alpha,2\alpha}(u,v)$ for $(u,v) \in D(A)$.

To treat the nonlinear terms $f(x,u,v)$ and $g(x,u,v)$ in the space X_0, we define an operator $B : [\alpha,2\alpha] \to X_0$ by

$$B(u,v) = (f(\cdot,u,v),\ g(\cdot,u,v)) + L_{\alpha,2\alpha}(u,v)$$

for $u,\ v \in [\alpha,2\alpha]$. Then (H1) - (H3) together imply that $B(t) \equiv B$ satisfies conditions (N) and (N3). Hence $A + B$ is of class $\mathfrak{G}(X_0)$ and $[A + B](u,v) = (\Delta u + f(\cdot,u,v),\ \Delta v + g(\cdot,u,v))$ for $(u,v) \in D(A) \cap [\alpha,2\alpha]$. On the other hand, any vector-valued function $(u(t,x),\ v(t,x))$ on $[0,\infty) \times \mathbb{R}^n$ treated in this problem can be interpreted as an X_0-valued function $(u(t),v(t))$ in a natural way. Accordingly, the problem (18) can be reduced to the problem (IVP) (with $s = 0$) in X_0.

Thus, Theorem 11 and Remark (2) in Section 8 can be applied to find C^1-solutions of (IVP) (which are uniquely determined by the initial data in $[\alpha,2\alpha]$), if $(u_0,v_0) \in D(A)$ and there is a pair of sub- and supersolutions $(\underline{u},\underline{v})$, $(\overline{u},\overline{v})$ of the associated integral equation (6) with x and u replaced respectively by (u_0,v_0) and (u,v), and if $(\alpha,\alpha) \leqq (\underline{u}(0), \underline{v}(0)) \leqq (u_0,v_0) \leqq (\overline{v}(0),\overline{u}(0)) \leqq (2\alpha,2\alpha)$ and $(\alpha,\alpha) \leqq (\underline{v}(t),\underline{u}(t)) \leqq (\overline{u}(t), \overline{v}(t)) \leqq (2\alpha,2\alpha)$ for $t \in [0,T]$.

To find such sub- and supersolutions we apply the recent result of Kawano and Kusano [7]. Consider the ordinary differential system

$$\begin{aligned} y'' + (n-1)r^{-1}y' + \phi(r)F(y,z) &= 0 \\ z'' + (n-1)r^{-1}z' + \psi(r)G(y,z) &= 0 \end{aligned} \tag{19}$$

under the initial conditions $y(0) = z(0) = 2\alpha$ and $y'(0) = z'(0) = 0$. If (y,z) is a solution of the problem for (19) such that $\alpha \leqq y(r)$, $z(r) \leqq 2\alpha$ for $r \geqq 0$, then $\underline{u}$, $\underline{v}$, $\overline{u}$, $\overline{v}$ are obtained by defining $\underline{u}(t) \equiv \lim_{r\to\infty} y(r)$, $\underline{v}(t) \equiv \lim_{r\to\infty} z(r)$ (viewed as constant functions on $\mathbb{R}^n$), $\overline{u}(t) \equiv y(|\cdot|)$ and $\overline{u}(t) \equiv z(|\cdot|)$ on $\mathbb{R}^n$.

We then consider the following conditions on F and G:

(H4) $\lim_{\xi\to 0} \xi^{-1}F(\xi,\xi) = \lim_{\xi\to 0} \xi^{-1}G(\xi,\xi) = 0.$

(H5) $\lim_{\xi\to\infty} \xi^{-1}F(\xi,\xi) = \lim_{\xi\to\infty} \xi^{-1}G(\xi,\xi) = 0.$

In case (H4) holds, there exist a (small) number $\alpha > 0$ and a solution (y,z) of the problem for (19) as mentioned above, while if (H5) holds, then there exist a (large) number $\alpha > 0$ and a solution (y,z) of the problem for (19) possessing the desired properties.

Example 3. Consider the parabolic system

$$u_t = \Delta u + p(x)u^a v^b, \quad \Delta v_t = v + q(x)u^c u^d,$$

where a, b, c, d ≥ 0 and p, q are nonnegative and locally Hölder continuous on $\mathbb{R}^n$. Suppose that $p(x) \leq \phi(|x|)$ and $q(x) \leq \psi(|x|)$ on $\mathbb{R}^n$ for some locally Hölder continuous functions ϕ, ψ on $[0,\infty)$ satisfying $\int_0^\infty r\phi(r)dr < \infty$ and $\int_0^\infty r\psi(r)dr < \infty$. Defining $F(\xi,\eta) = \xi^a\eta^b$ and $G(\xi,\eta) = \xi^c\eta^d$, we see that (H1) - (H3) are satisfied. If $a + b > 1$ and $c + d > 1$, then (H4) holds; and if $a + b < 1$ and $c + d < 1$, then (H5) holds.

References

[1] J. B. Cooper, Saks Spaces and Applications to Functional Analysis, Mathematics Studies 28, North-Holland, 1978.

[2] M. Crandall, A. Pazy and L. Tartar, Remarks on generators of analytic semigroups, Israel J. Math., 32 (4) (1979), 365-374.

[3] D. H. Fremlin, Topological Riesz Spaces and Measure Theory, Cambridge University Press, 1974.

[4] H. G. Garnir, M. De Wilde and J. Schmets, Analyse Fonctionnelle, T. II, Measure et intégration dans l'espace euclidien, Birkhäuser Verlag, Basel, 1972.

[5] K. Hashimoto and S. Oharu, Integration in Fréchet lattices with applications to operator semigroups, to appear.

[6] K. Hashimoto, T. Kusano and S. Oharu, Semilinear evolution equations in Fréchet lattices, to appear.

[7] N. Kawano and T. Kusano, On positive entire solutions of a class of second order semilinear elliptic systems, Math. Z., 186 (1984), 287-297.

[8] T. Kusano and S. Oharu, Semilinear evolution equations in Fréchet lattices with applications to parabolic systems, to appear.

[9] T. Kusano and S. Oharu, Bounded entire solutions of second order semilinear elliptic equations with application to a parabolic initial value problem, Indiana Univ. Math. J., 34 (1985), 85-95.

[10] A. Pazy, Semigroups of Linear Operators and Applications to Partial Differential Equations, Applied Mathematical Sciences 44, Springer-Verlag, 1983.

[11] T. Takahashi and S. Oharu, Approximation of operator semigroups in a Banach space, Tôhoku Math. J., 24 (4) (1972), 505-528.

[12] K. Yosida, Functional Analysis (6th edition), Springer-Verlag, New York, 1980.

ON SOME SINGULAR HYPERBOLIC EVOLUTION EQUATIONS IN HILBERT SPACES

M. Povoas
C.M.A.F.
Av. Gama Pinto, 2
1699 Lisboa Codex

0. Introduction. Let H be a Hilbert space, A and C operators in H. We consider evolution equations of the form

(0.1) $$\frac{d}{dt} Cy + Ay = f\ , \quad f \in L^2(0,T;H),\ T < +\infty,$$

which are hyperbolic in the sense of Kato, that is, $-A$ is the infinitesimal generator of a strongly continuous semigroup of contractions in H (which is not analytical). The operator C is self-adjoint and nonnegative, but is not invertible (and in some cases not necessarily bounded). Equations of the form (0.1) with the initial data $Cy(0)=0$ have been studied by the author in [4]. We consider now two cases of (0.1) where the operator C depends on t. We shall study the Cauchy problems

(0.2) $$\begin{cases} \frac{d}{dt} \varphi(t)Cy(t) + Ay(t) = f(t) \\ Cy(0) = 0 \end{cases}$$

where $C \in \mathscr{L}(H)$ and φ is a real regular function of t, and

(0.3) $$\begin{cases} \frac{d}{dt} (C_0+C(t))y(t) + Ay(t) = f(t) \\ C_0 y(0) = 0 \end{cases}$$

where C_0 is a self-adjoint, nonnegative unbounded operator in H and $C(t) \in \mathscr{L}(H)$ is a family of bounded operators in H depending smoothly on t, with application to symmetric systems.

1. The problem (0.2). We assume that

(H1) $A \in g(1,0;H)$,

(H2) $C \in \mathscr{L}(H)$, $C^* = C$, $(Cy,y) \geq 0$, $\forall y \in H$,

(H3) $\varphi \in C^2(0,T)$, $\varphi(t) = \psi(t)^2$, with $0 < m \leq \psi(t) \leq M$, $\forall t \in [0,T]$.

We denote by $\mathscr{A}, \mathscr{C}$ the operators in $\mathscr{H}$ defined respectively by

$$D(\mathscr{A}) = L^2(0,T;D(A)),\ (\mathscr{A}y)(t) = Ay(t) \quad \text{a.e.},$$

where $D(A)$ is endowed with the graph norm,

$$(\mathscr{C}y)(t) = Cy(t) \quad \text{a.e.}$$

Then $\mathscr{C}$ is also self-adjoint and nonnegative. Let

$$\vartheta = \begin{cases} 0\ , & \text{if } \varphi' \text{ is nonnegative} \\ \frac{1}{2} \|\varphi'\mathscr{C}\|_{\mathscr{L}(\mathscr{H})}\ , & \text{otherwise} \end{cases}$$

and define the operator Λ in $\mathscr{H}$ by

$$D(\Lambda) = \{ y \in \mathscr{H} : \mathscr{C}y \in H^1(0,T;H),\ Cy(0) = 0 \},$$

$$\Lambda y = \psi \frac{d}{dt}(\psi \mathscr{C} y) + \frac{1}{2}\varphi' \mathscr{C} y + \vartheta y.$$

Lemma 1.1. Under the hypotheses (H2),(H3) one has $\Lambda \in g(1,0;\mathscr{H})$.□

Proof. $\mathscr{D}(0,T;H) \subset D(\Lambda)$, so that $D(\Lambda)$ is dense in $\mathscr{H}$, and it is easily seen that Λ is closed. Also, by considering the semigroup $G(s)$ $(s \geq 0)$ of the right-translactions (generated by the operator $\Phi = \frac{d}{dt}$, $D(\Phi) = \{y \in H^1(0,T;H) : y(0) = 0\}$), it is easily shown, as in [4; Lemma 1.1], that both Λ and its adjoint Λ^*, defined by

$$D(\Lambda^*) = \{ z \in \mathscr{H} : \mathscr{C} z \in H^1(0,T;H),\ Cz(T) = 0\}$$

$$(\Lambda^* z)(t) = -\psi(t) \frac{d}{dt} \psi(t) Cz(t) + \frac{1}{2}\varphi'(t) Cz(t) + \vartheta z(t) \quad \text{a.e.},$$

are nonnegative operators, and so $\Lambda \in g(1,0;\mathscr{H})$.□

Lemma 1.2. Let (H2),(H3) be satisfied. Then there exists $\alpha > 0$ such that $Y = D(\Phi)$ is stable for $(\Lambda+\xi)^{-1}$ (that is $y \in D(\Phi) \Rightarrow (\Lambda+\xi)^{-1} y \in D(\Phi)$) and $u \in (\Lambda+\xi)^{-1}(D(\Phi)) \Rightarrow \frac{d}{dt}(\Lambda+\xi)u = (\Lambda+\xi)u' + \varphi'' \mathscr{C} u + \varphi' \mathscr{C} u'$, for every $\xi > \alpha$. Furthermore

$$\|(\Lambda+\xi)^{-1}\|_{\mathscr{L}(Y)} \leq \frac{1}{\xi-\alpha}, \ \forall\, \xi > \alpha. \square$$

Proof. Let $y \in D(\Phi)$, $u = (\Lambda+\xi)^{-1} y$. Then, by (H3), $\psi y \in D(\Phi)$. For every $\varepsilon > 0$ let $C_\varepsilon = C + \varepsilon I$, $\mathscr{C}_\varepsilon = \mathscr{C} + \varepsilon \mathscr{I}$, where I and $\mathscr{I}$ denote the identities of H and $\mathscr{H}$ respectively, and define the operator Λ_ε by

$$D(\Lambda_\varepsilon) = \{ y \in \mathscr{H} : \mathscr{C}_\varepsilon\, y \in D(\Phi)\}$$

$$\Lambda_\varepsilon y = \psi \Phi(\psi \mathscr{C}_\varepsilon\, y) + \frac{1}{2}\, \varphi' \mathscr{C}_\varepsilon\, y + \vartheta y.$$

We first observe that $D(\Lambda_\varepsilon) = D(\Phi)$. One has obviously $D(\Phi) \subset D(\Lambda_\varepsilon)$ and conversely if $y \in D(\Lambda_\varepsilon)$ one concludes, since C_ε commutes with $G(s)$, $\forall\, s > 0$, $\psi \mathscr{C}_\varepsilon \frac{I-G(s)}{s}(\psi y) + \frac{1}{2}\, \varphi' \mathscr{C}_\varepsilon\, y + \vartheta y = \psi \frac{I-G(s)}{s} \mathscr{C}_\varepsilon \psi y + \frac{1}{2}\, \varphi' \mathscr{C}_\varepsilon\, y + \vartheta y \xrightarrow[s \to 0]{\mathscr{H}} \Lambda_\varepsilon y$. Since $0 \in \varrho(\mathscr{C}_\varepsilon)$ one has, in view of (H3),

$$\frac{I-G(s)}{s}(\psi y) \overset{\mathscr{H}}{\to} \mathscr{C}_\varepsilon^{-1} \frac{1}{\psi}(\Lambda_\varepsilon y - \frac{1}{2}\mathscr{C}_\varepsilon y - \vartheta y) \quad \text{as} \quad s \to 0,$$

whence $\psi y \in D(\Phi)$ and $\Lambda_\varepsilon y = \psi \mathscr{C}_\varepsilon \Phi(\psi y) + \frac{1}{2}\, \mathscr{C}_\varepsilon\, y + \vartheta y$, $\forall \varepsilon > 0$.

Now, since $\mathscr{C}_\varepsilon$ is also self-adjoint and nonnegative, one has by Lemma1.1 $\Lambda_\varepsilon \in g(1,0;\mathscr{H})$ and so for every $y \in \mathscr{H}$, $\varepsilon, \xi > 0$, there exists a unique solution $u_\varepsilon \in D(\Lambda_\varepsilon)$ of the equation $(\Lambda_\varepsilon + \xi) u_\varepsilon = y$, that is

$$(1.1) \quad \psi \Phi \circ \mathscr{C}_\varepsilon \psi u_\varepsilon + \frac{1}{2}\, \varphi' \mathscr{C}_\varepsilon\, u_\varepsilon + (\vartheta+\xi) u_\varepsilon = \psi \mathscr{C}_\varepsilon \circ \Phi(\psi u_\varepsilon) + \frac{1}{2}\, \varphi' \mathscr{C}_\varepsilon\, u_\varepsilon + (\vartheta+\xi) u_\varepsilon = y.$$

By multiplying by u_ε and integrating over $[0,T]$ we conclude that

$$(1.2) \qquad \|u_\varepsilon\|_{\mathscr{H}} \leq \frac{1}{\xi} \|y\|_{\mathscr{H}}, \quad \forall\, \varepsilon > 0,$$

and so there exists a sequence (still denoted by ε) such that

$$u_\varepsilon \to u \quad \text{weakly in } \mathscr{H} \text{ as } \varepsilon \to 0.$$

Then $\frac{d}{dt}(\mathscr{C}(\psi u_\varepsilon)) \to \frac{d}{dt}\mathscr{C}(\psi u)$ weakly in $H^{-1}(0,T;H)$ as $\varepsilon \to 0$. Also $(\psi u_\varepsilon)' \to (\psi u)'$ weakly in $H^{-1}(0,T;H)$, so that $\varepsilon(\varphi u_\varepsilon)' \to 0$ strongly in $H^{-1}(0,T;H)$ as $\varepsilon \to 0$. Since $\varphi' \mathscr{C}_\varepsilon u_\varepsilon \to \varphi' \mathscr{C} u$ weakly in $\mathscr{H}$, we obtain, by taking (1.1) to the limit weakly in $H^{-1}(0,T;H)$ as $\varepsilon \to 0$,

$$\psi \frac{d}{dt}\mathscr{C}\psi u_\varepsilon + \varepsilon\varphi u'_\varepsilon + \frac{1}{2}\varphi' \mathscr{C} u_\varepsilon + \varepsilon\varphi' u_\varepsilon + (\vartheta+\xi)u_\varepsilon \to \psi\frac{d}{dt}\mathscr{C}\psi u + \frac{1}{2}\varphi'\mathscr{C}u + (\vartheta+\xi)u = y$$

whence $\frac{d}{dt}\mathscr{C}\psi u \in \mathscr{H}$, so that $\mathscr{C}\psi u \in H^1(0,T;H)$ and

$$\psi\frac{d}{dt}\mathscr{C}\psi u + \frac{1}{2}\varphi'\mathscr{C}u + (\vartheta+\xi)u = y. \tag{1.3}$$

Also, in a way similar to that used in [4; Lemma 1.2], it is easily seen that $Cu(0) = 0$. Then $u \in D(\Lambda)$ and $(\Lambda+\xi)u = y$, $u = (\Lambda+\xi)^{-1}y$. We have thus shown that the solution u_ε of $(\Lambda_\varepsilon+\xi)u_\varepsilon = y$ converges weakly in $\mathscr{H}$ to the solution of $(\Lambda+\xi)u = y$ as $\varepsilon \to 0$.

We next prove that $y \in D(\Phi)$ implies $u = (\Lambda+\xi)^{-1}y \in D(\Phi)$. Indeed, by derivating (1.1) with respect to t we obtain

$$\frac{d}{dt}\psi\mathscr{C}_\varepsilon\frac{d}{dt}(\psi u_\varepsilon) + \frac{1}{2}\frac{d}{dt}(\varphi'\mathscr{C}_\varepsilon u_\varepsilon) + (\vartheta+\xi)u'_\varepsilon = y' \in \mathscr{H} \tag{1.4}$$

and since $u_\varepsilon \in D(\Phi)$ one has $\frac{d}{dt}\varphi'\mathscr{C}_\varepsilon u_\varepsilon = \varphi'\mathscr{C}_\varepsilon u'_\varepsilon + \varphi''\mathscr{C}_\varepsilon u_\varepsilon$, so that the second and third terms of the sum belong to $\mathscr{H}$ and so $\mathscr{C}_\varepsilon\frac{d}{dt}(\psi u_\varepsilon) \in H^1(0,T;H)$, whence $u'_\varepsilon \in D(\Lambda_\varepsilon) = D(\Phi)$. Also from (1.4) we conclude

$$\Lambda_\varepsilon u'_\varepsilon + (\varphi'\mathscr{I} + \varphi''\Phi^{-1})\mathscr{C}_\varepsilon u'_\varepsilon + \xi u'_\varepsilon = y', \quad \forall\varepsilon > 0 \tag{1.5}$$

and, by multiplying by u'_ε and integrating over $[0,T]$, we have

$$\xi\|u'_\varepsilon\|^2_{\mathscr{H}} \leqslant \mathrm{Re}(y', u'_\varepsilon)_{\mathscr{H}} + \|(\varphi''\Phi^{-1} + \varphi'\mathscr{I})\circ\mathscr{C}\|_{\mathscr{L}(\mathscr{H})}\|u'_\varepsilon\|^2_{\mathscr{H}} + \varepsilon\|\varphi''\Phi^{-1} + \varphi'\mathscr{I}\|_{\mathscr{L}(\mathscr{H})}\|u'_\varepsilon\|^2_{\mathscr{H}}.$$

By putting $\alpha = \|(\varphi''\Phi^{-1} + \varphi'\mathscr{I})\circ\mathscr{C}\|_{\mathscr{L}(\mathscr{H})}$, $k = \|\varphi''\Phi^{-1} + \varphi'\mathscr{I}\|_{\mathscr{L}(\mathscr{H})}$ we obtain

$$\|u'_\varepsilon\|_{\mathscr{H}} \leqslant \frac{1}{\xi-\alpha-\varepsilon k}\|y'\|_{\mathscr{H}} \leqslant \frac{1}{\xi-\alpha-\varepsilon_0 k}\|y'\|_{\mathscr{H}}, \quad \forall\varepsilon < \varepsilon_0$$

and thus there exists a sequence (still denoted by ε) such that u'_ε converges weakly in $\mathscr{H}$ as $\varepsilon \to 0$, so that $u' \in \mathscr{H}$, $u'_\varepsilon \to u'$ weakly in $\mathscr{H}$ and

$$\|u'\|_{\mathscr{H}} \leqslant \frac{1}{\xi-\alpha}\|y'\|_{\mathscr{H}}. \tag{1.6}$$

By taking (1.5) to the limit (in $\mathscr{H}$ endowed with the weak topology) as $\varepsilon \to 0$ we obtain

$$(\Lambda_\varepsilon + \xi)u'_\varepsilon = y' - (\varphi''\Phi^{-1} + \varphi'\mathscr{I})\mathscr{C}_\varepsilon u'_\varepsilon \overset{\mathscr{H}}{\to} y' - \varphi'\mathscr{C}u' - \varphi''\mathscr{C}u$$

whence

$$(\Lambda+\xi)u' = y' - \varphi'\mathscr{C}u' - \varphi''\mathscr{C}u.$$

Moreover from (1.2) and (1.6) it follows that

$$\|u\|_{D(\Phi)} \leq \frac{1}{\xi-\alpha} \|y\|_{D(\Phi)} , \quad \forall \xi > \alpha$$

and thus

$$\|(\Lambda+\xi)^{-1}\|_{\mathscr{L}(Y)} \leq \frac{1}{\xi-\alpha} , \quad \forall \xi > \alpha$$

as claimed.□

On the other hand one has by (H1) $\mathscr{A} \in g(1,0;\mathscr{H})$. Let us assume now that

(H4) $\exists\, \lambda,\chi \in \mathbb{R}$: $\chi > \vartheta$ and $\mathscr{A}+\lambda\varphi\mathscr{C} \geq \chi\mathscr{I}$, that is

$$\int_0^T (A+\lambda\varphi(t)Cy(t),\, y(t))_H \geq \chi\|y\|^2_{\mathscr{H}} , \quad \forall\, y \in D(\mathscr{A}).$$

Then one has

<u>Lemma 1.3</u>. Under the hypotheses (H1),...,(H4), there exists $\beta > 0$ such that $(\mathscr{A}+\lambda\varphi\mathscr{C}+\xi)^{-1} \in \mathscr{L}(Y)$ and

$$\|(\mathscr{A}+\lambda\varphi\mathscr{C}+\xi)^{-1}\|_{\mathscr{L}(Y)} \leq \frac{1}{\xi+\chi-\beta} , \ \forall \xi > \beta-\chi .□$$

<u>Proof</u>. Let $y \in Y$ and $u = (\mathscr{A}+\lambda\varphi\mathscr{C}+\xi)^{-1}y$, that is

(1.7) $$(\mathscr{A}+\lambda\varphi\mathscr{C}+\xi)u = y \in D(\Phi).$$

Then $$\frac{I-G(s)}{s}\, u \overset{\mathscr{H}}{\to} (\mathscr{A}+\lambda\varphi\mathscr{C}+\xi)^{-1}(y'-\lambda\varphi'\mathscr{C}u)$$

and so $u' \in \mathscr{H}$ and

(1.8) $$(\mathscr{A}+\lambda\varphi\mathscr{C}+\xi)u' = y' - \lambda\varphi'\mathscr{C}u .$$

By multiplying (1.7) and (1.8) respectively by u and u' and integrating over $[0,T]$, one concludes

$$\|u\|_{\mathscr{H}} \leq \frac{1}{\xi+\chi} \|y\|_{\mathscr{H}}$$

and

$$\|u'\|_{\mathscr{H}} \leq \frac{1}{\xi+\chi-\beta} \|y'\|_{\mathscr{H}} ,$$

where $\beta = |\lambda|\, c\, \|\mathscr{C}\circ\Phi^{-1}\|_{\mathscr{L}(\mathscr{H})}$, $c \geq |\varphi'(t)|$, $\forall t \in [0,T]$.
Then

$$\|u\|_Y \leq \frac{1}{\xi+\chi-\beta} \|y\|_Y , \ \forall \xi > \beta-\chi ,$$

as asserted.□

<u>Theorem 1.1</u>. Let the hypotheses (H1),...,(H4) be satisfied. Then for every $f \in \mathscr{H}$ there exists a unique strong solution $y \in \mathscr{H}$ of the problem

$$\begin{cases} \frac{d}{dt}\varphi(t)Cy(t) + Ay(t) = f(t) \quad \text{a.e.} \\ Cy(0) = 0 \end{cases}$$

that is, there exists a sequence $(y_n)_{n\in\mathbb{N}}$ such that

$y_n \in L^2(0,T;D(A))$, $\mathscr{C}y_n \in H^1(0,T;H)$, $Cy_n(0) = 0$, $\forall n \in \mathbb{N}$,

$y_n \to y$, $\frac{d}{dt}\varphi\mathscr{C}y_n + \mathscr{A}y_n \to f$ in $\mathscr{H}$ as $n \to \infty$.

Moreover if $f \in H^1(0,T;H)$, $f(0) = 0$ and $\chi - \vartheta > \omega = \alpha+\beta$ then $y \in H^1(0,T;H) \cap L^2(0,T;D(A))$, $y(0) = 0$, so that y is a strict solution.□

Proof. Let $\tilde{\mathscr{A}}$ be the operator defined by

$$D(\tilde{\mathscr{A}}) = D(\mathscr{A}); \quad (\tilde{\mathscr{A}}y)(t) = Ay(t) + \lambda\varphi(t)Cy(t) - \chi y(t), \ \forall y \in D(\mathscr{A}).$$

Then $\Lambda, \tilde{\mathscr{A}} \in g(1,0;\mathscr{H})$ and by Lemmas 1.2 and 1.3 one has

$$\|(\Lambda+\xi)^{-1}\|_{\mathscr{L}(Y)} \leqslant \frac{1}{\xi-\alpha}, \ \forall \xi > \alpha, \tag{1.9}$$

$$\|(\tilde{\mathscr{A}}+\xi)^{-1}\|_{\mathscr{L}(Y)} \leqslant \frac{1}{\xi-\beta}, \ \forall \xi > \beta. \tag{1.10}$$

Let Λ_n be the Yosida approximation $\Lambda_n = n\Lambda(\Lambda+n)^{-1}$ and let $D(L_n) = D(\tilde{\mathscr{A}})$, $L_n y = \Lambda_n y + \tilde{\mathscr{A}}y$. Since $-\Lambda$, $-\tilde{\mathscr{A}}$ are generators of semigroups of contractions in $\mathscr{H}$, the series $\sum_{j\geqslant 0} n^{2j}(\Lambda+n)^{-j}(\tilde{\mathscr{A}}+\xi+n)^{-j}$ converges in $\mathscr{L}(\mathscr{H})$ and

$$(L_n+\xi)^{-1} = (\tilde{\mathscr{A}}+\xi+n)^{-1}\left(I-n^2(\Lambda+n)^{-1}(\tilde{\mathscr{A}}+\xi+n)^{-1}\right)^{-1}.$$

On the other hand from (1.9) and (1.10) it follows that the series also converges in $\mathscr{L}(Y)$ for every $\xi > \omega$, $n > n_0 = \frac{\alpha(\xi-\beta)}{\xi-\alpha-\beta}$, so that

$$\|(L_n+\xi)^{-1}\|_{\mathscr{L}(Y)} \leqslant \frac{n-\alpha}{n(\xi-\omega)-\alpha(\xi-\beta)} \underset{n\to\infty}{\to} \frac{1}{\xi-\omega}.$$

Then there exists $K > 0$ such that

$$\|(L_n+\xi)^{-1}\|_{\mathscr{L}(Y)} \leqslant K, \quad \forall n \in \mathbb{N},$$

and thus the hypothesis (i) of [1; Theorem 5.6] is satisfied, by taking $Y = D(\Phi)$, $B = \Lambda$. It follows that the operator $\tilde{L}$ defined by

$$D(\tilde{L}) = D(\Lambda) \cap D(\tilde{\mathscr{A}}) = D(L), \quad \tilde{L}y = \Lambda y + \tilde{\mathscr{A}}y$$

satisfies $\varrho(-\bar{\tilde{L}}) \supset]0,+\infty[$, $\|(\bar{\tilde{L}}+\xi)^{-1}\|_{\mathscr{L}(\mathscr{H})} \leqslant \frac{1}{\xi}$, $\forall \xi > 0$ and $(L_n+\xi)^{-1} \to (\bar{\tilde{L}}+\xi)^{-1}$ in $\mathscr{L}(\mathscr{H})$ as $n \to \infty$, $\forall \xi > 0$. In particular we conclude that $z = (\bar{\tilde{L}}+\chi-\vartheta)^{-1}(\tilde{f})$ is the unique strong solution of the problem

$$\begin{cases} \psi(t)\dfrac{d}{dt}\psi(t)\,Cz(t) + \dfrac{1}{2}\varphi'(t)\,Cz(t) + Az(t) + \lambda\varphi(t)\,Cz(t) = e^{-\lambda t}f(t) \\ \qquad\qquad = \tilde{f}(t) \text{ a.e.} \\ Cz(0) = 0 \end{cases}$$

that is

$$\begin{cases} \dfrac{d}{dt}\varphi(t)\,Cz(t) + (A+\lambda\varphi(t)C)z(t) = e^{-\lambda t}f(t) \\ Cz(0) = 0 \end{cases}$$

and so $y(t) = e^{\lambda t}z(t)$ is the unique strong solution of $Lu=f$. Furthermore

$$\| z \|_{\mathscr{H}} = \| e^{-\lambda t} y \|_{\mathscr{H}} \leq \frac{1}{\chi - \vartheta} \| \tilde{f} \|_{\mathscr{H}}.$$

Finally, if $f \in D(\Phi) = Y$ and $\chi - \vartheta > \omega$, it follows from [1; Prop. 5.4] that y is a strict solution and $y \in D(\Phi)$.□

2. The problem (0.3). We suppose that $A \in g(1,0:H)$, C_0 is a self-adjoint and nonnegative (unbounded) operator in H, C(t) is a family of bounded operators satisfying $C \in W^{2,\infty}(0,T;\mathscr{L}(H))$, C(t) is self-adjoint and $C_0+C(t)$ is nonnegative, $\forall t \in [0,T]$. Moreover we assume that

$$\exists\ \lambda, \chi\ :\ \chi > \frac{1}{2} \| C' \| \quad \text{and} \quad \mathscr{A} + \lambda (\mathscr{C}_0 + \mathscr{C}) - \chi \in g(1,0;\mathscr{H}),$$

where $\mathscr{A}, \mathscr{C}$ are defined as before. We define $\tilde{\Lambda}, \tilde{\mathscr{A}}$ in $\mathscr{H}$ by $D(\tilde{\Lambda}) = \{y \in L^2(0,T;W): (\mathscr{C}_0+\mathscr{C})y \in H^1(0,T;H),\ C_0 y(0) = 0\}$, $\tilde{\Lambda} y = \frac{d}{dt} (\mathscr{C}_0+\mathscr{C})y + \vartheta y$, $D(\tilde{\mathscr{A}}) = L^2(0,T;D(A)) \cap L^2(0,T;W)$, $\tilde{\mathscr{A}} = \mathscr{A} + \lambda(\mathscr{C}_0+\mathscr{C}) - \chi$, where W denotes the domain of C_0, endowed with the graph norm, and ϑ is a suitable constant. The existence and uniqueness of the solution of (0.3) is then proved by applying [1; Theorem 5.6] to the sum of operators $L = \tilde{\Lambda} + \tilde{\mathscr{A}}$, under some further assumptions concerning in particular the commutators $[C_0+C(t), C'(t)]$ and $[C_0+C(t), C''(t)]$. For the proofs and more details we refer to [5].

3. Symmetric systems. We consider now the case where the operator A is of the form

(3.1) $$A = A_0(x) + \sum_{i=1}^{N} A_i(x) \frac{\partial}{\partial x_i}$$

where $A_i(x)$ $(i=0,\dots,N)$ are m×m matrices whose elements are defined in a bounded set $\Omega \subset \mathbb{R}^N$, with boundary $\Gamma = \partial\Omega$ of class C^2, Ω lying locally on one side of Γ. Let $Q = \Omega \times]0,T[$, $H = (L^2(\Omega))^m$, $\mathscr{H} = (L^2(Q))^m$ and let us assume that

The matrices $A_i(x)$ $(i=1,\dots,N)$ are hermitian and their elements are of class C^1 in $\bar{\Omega}$.

The elements of $A_0(x)$ belong to $L^\infty(\Omega)$.

Suppose further that $-A$ is formally dissipative, that is

$$(Ky,y)_H \geq 0, \quad \forall y \in H,$$

where $K \in \mathscr{L}(L^2(\Omega))^m$ is defined by

$$(Ky)(x) = (A_0(x) + A_0^*(x) - \sum_{i=1}^{N} \frac{\partial A_i}{\partial x_i}(x))y(x) \quad \text{a.e.}$$

and that the characteristic matrix

$$A_\nu(x) = \sum_{i=1}^{N} \nu_i(x) A_i(x)$$

($\nu(x) = (\nu_1(x),\dots,\nu_N(x))$ is the outward normal to the boundary Γ in $x\in\Gamma$) has constant rank q near the boundary. We consider a subspace $\mathscr{B}(x)$ of $\mathbb{C}^m$ maximal nonnegative with respect to $A_\nu(x)$ and varying smoothly near Γ. Then the operator A defined in H by

$$D(A) = \{ y\in H:\ \sum_{i=1}^{N} A_i(x)\frac{\partial y}{\partial x_i}\in H,\ y(x)\in\mathscr{B}(x)\quad\text{on }\Gamma\}$$

and (3.1) is maximal nonnegative, so that (H1) is satisfied.

Now let C,φ satisfy (H2),(H3) and assume that

(H4') $\exists\,\lambda,\chi:\chi>\vartheta$ and $\int_0^T ((\frac{1}{2}K+\lambda\varphi(t)C)y(t),y(t))_H dt\geq\chi\|y\|^2_{\mathscr{H}},\quad\forall y\in\mathscr{H}.$

Then (H4) is satisfied and from Theorem 1.1 it follows, for every $f\in\mathscr{H}$:

Corollary 3.1. Under the preceding hypotheses, there exists a unique strong solution $y\in\mathscr{H}$ of the initial-boundary value problem

$$\begin{cases}\frac{d}{dt}\varphi(t)Cy(x,t)+\sum_{i=0}^{N}A_i(x)\frac{\partial y}{\partial x_i}(x,t)=f(x,t) & \text{in } Q,\text{ a.e.},\\ C\,y(x,0)=0 & \text{in }\Omega,\text{ a.e.},\\ y(x,t)\in\mathscr{B}(x) & \text{on }\Sigma=\Gamma\times[0,T]\text{ a.e..}\end{cases}$$

If furthermore $f\in H^1(0,T;H)$, $f(0)=0$ and $\chi-\vartheta>\omega$, then y is a strict solution, that is

$$(3.2)\qquad \sum_{i=1}^{N}A_i(x)\frac{\partial y}{\partial x_i}\in L^2(Q)^m,\ y\in H^1(0,T;L^2(\Omega)^m)\quad\text{and}\quad y(x,0)=0\text{ a.e..}$$

□

In particular let $C(x)$ be a positive semidefinite hermitian matrix whose elements belong to $L^\infty(\Omega)$. Then the operator C defined in H by $(Cy)(x)=C(x)y(x)$ a.e. $x\in\Omega$, satisfies the preceding hypotheses and, by assuming the compatibility condition (H4'), we conclude from Corollary 3.1

Corollary 3.2. Let the hypotheses above be satisfied. Then $\forall f\in L^2(Q)^m$ there exists a unique strong solution $y\in L^2(Q)^m$ of the initial-boundary value problem

$$\begin{cases}\frac{d}{dt}\varphi(t)C(x)y(x,t)+\sum_{i=0}^{N}A_i(x)\frac{\partial y}{\partial x_i}(x,t)=f(x,t) & \text{in } Q,\text{ a.e.},\\ C(x)y(x,0)=0 & \text{in }\Omega,\text{ a.e.},\\ y(x,t)\in\mathscr{B}(x) & \text{on }\Sigma,\text{ a.e..}\end{cases}$$

If $f\in H^1(0,T;L^2(\Omega)^m)$, $f(0)=0$ and $\chi-\vartheta>\omega$, then (3.2) holds.□

Let us consider now Maxwell's equations

$$(3.3)\qquad\begin{cases}\operatorname{curl} H=-S+\sigma E+\frac{dD}{dt}\\ \operatorname{curl} E=-\frac{dB}{dt}\end{cases}$$

in an open bounded set $\Omega\subset\mathbb{R}^3$, with the constitutive laws

(3.4) $$D = \varphi(t)\,\varepsilon E\ ,\quad B = \varphi(t)\,\mu H$$

where E and H denote the electric and magnetic fields, D is the electric displacement, B is the magnetic induction, ε,μ are bounded operators in $L^2(\Omega)^3$ (for instance

(3.5) $$(\varepsilon E)(x,t) = \varepsilon(x)E(x,t),\ (\mu H)(x,t) = \mu(x)H(x,t) \text{ a.e.},$$

$\varepsilon(x),\mu(x)$ being positive semidefinite hermitian 3×3 matrices with elements in $L^\infty(\Omega)$), σ is the electrical conductivity and S is a given current density. By putting

$$y = \begin{bmatrix} E \\ H \end{bmatrix},\ A_0 = \begin{bmatrix} \sigma & 0 \\ 0 & 0 \end{bmatrix},\quad f = \begin{bmatrix} S \\ 0 \end{bmatrix} \quad\text{and}\quad C = \begin{bmatrix} \varepsilon & 0 \\ 0 & \mu \end{bmatrix}$$

the equations take the form

$$\frac{d}{dt}\,\varphi(t)Cy(x,t) + \sum_{i=0}^{3} A_i \frac{\partial y}{\partial x_i}(x,t) = f(x,t)$$

where A_i (i=1,2,3) are symmetric constant matrices. From Corollary 3.1 (or Corollary 3.2, in the case (3.5)) it then follows that

Corollary 3.3. Under the preceding hypotheses $\forall S\in L^2(Q)^3$ there exists a unique strong solution $(E,H)\in L^2(Q)^6$ of Maxwell's equations (3.3) with the constitutive laws (3.4), satisfying the initial data $D(x,0) = 0$, $B(x,0) = 0$ a.e. in Ω, and maximal nonnegative boundary conditions $(E(x,t),H(x,t))\in\mathscr{B}(x)$ a.e. on Σ.

If $S\in H^1(0,T;L^2(\Omega)^3)$, $S(0) = 0$ and $\chi-\vartheta>\omega$, then $E,H\in H^1(0,T;L^2(\Omega)^3)$, $E(0) = 0$, $H(0) = 0$ and (E,H) is a strict solution.□

We have analogous results concerning problem (0.3) in the case where A is a symmetric system, applying in particular to Maxwell's equations in a moving dielectric (see [5]).

REFERENCES

[1] G. Da Prato and P. Grisvard, Sommes d'opérateurs linéaires et équations différentielles opérationnelles (J. Math. Pures et Appliquées, vol. 54, 1975,pp. 305-387).

[2] T. Kato, Perturbation Theory for Linear Operators, Grundlehren, B. 132, Springer, Berlin, 1966.

[3] M. Povoas, Some remarks on Maxwell's equations (J. Math. Anal. Appl., Vol. 68, 1979, pp 35-50).

[4] M. Povoas, On some singular hyperbolic evolution equations (J. Math. Pures et Appliquées, vol. 60, 1981,pp. 133-192).

[5] M. Povoas, On some hyperbolic evolution equations with time-dependent singularity (J. Differential Equations, vol. 59, 1985).

PERIODIC SOLUTIONS OF THE THERMOSTAT PROBLEM

Jan Prüss

Fachbereich 17 der Universität - Gesamthochschule Paderborn, Warburger Str. 100, D 4790 Paderborn, FRG.

1. Statement of the problem

Many heat conductions processes are regulated by a thermostat, i.e. a temperature - dependent switch which activates the burner of a heater. A very simple model for this was proposed by Glashoff and Sprekels [2]. Consider a rod of length 2l, insulated from the sides, and heated at its ends and let the thermostat respond to the temperature at the mid point of the rod. If this temperature falls below some value u_o, the thermostat switches to "ON", and if it increases beyond some u_1 the thermostat switches to "OFF". Assuming a linear heating law at the ends of the rod, by symmetry we obtain the following equations for the temperature u(t,x) of the rod

$$\begin{aligned} u_t(t,x) &= au_{xx}(t,x) \quad , \qquad t \in \mathbb{R}, x \in (o,1) \\ u_x(t,o) &= o \qquad\qquad\qquad\qquad\qquad\qquad (1) \\ \alpha u_x(t,1) &= y(t) - u(t,1). \end{aligned}$$

Here y(t) denotes the temperature of a heater which in turn is determined by the mode of the switcher according to

$$\beta \dot{y}(t) = mv(t) - \gamma y(t) \qquad (2)$$

where m corresponds to the mass converted by the burner in one time unit, γ is related to largest possible temperature of the heater and β measures how fast this temperature is reached. v(t) denotes the switching mode of the thermostat at time t, i.e. v(t) = o means "OFF" and v(t) = 1 corresponds to "ON". The action of the thermostat is described by the <u>switching</u>

<u>conditions</u>

$$\begin{cases} u(t,o) > u_1 \Rightarrow & v(t) = o \\ u(t,o) < u_0 \Rightarrow & v(t) = 1 \;, \end{cases} \tag{3}$$

where $o < u_0 < u_1$ are prescribed temperature values, and switching only occurs for $u(t,o) = u_0, u_1$. Introducing dimensionless variables we may assume $l = a = m = \gamma = 1$ and are then left with two parameters $\alpha, \beta \in [o,\infty)$ only.

Of course, equations (1), (2), (3) are still far away from reality; in general (1) and (2) have to be replaced by a second order parabolic equation on a three dimensional domain and (2) by a possibly higher order ordinary differential equation, however, the important feature are the switching conditions (3) which will not change then. For this reason we may consider (1), (2), (3) as the prototype of the thermostat problem and confine our attention to these equations.

Everybody knows that thermostats work and that the temperatures in question exhibit a periodic behavior. Since numerical computations for (1), (2), (3) carried out by Glashoff and Sprekels also support this, the following questions are natural.

(i) Is problem (1),(2),(3) wellposed ?

(ii) Does there exist a periodic solution ?

(iii) Is this periodic solution globally attracting ?

Introducing a somewhat different switching law, Glashoff and Sprekles [2], [3] reduced the thermostat problem to a set - valued integral equation and applied a certain fixed point theorem for set - valued maps to obtain the wellposedness of the problem. However, no results on periodic solutions have been obtained this way. Also in their approach switching is allowed also in case $u_0 < u(t,o) < u_1$ but $u_t(t,o)$ changes sign, a kind of switching which is not observed in reality.

On the other hand, Seidman [4] has developped a general theory of switching systems including wellposedness, existence of periodic solutions and asymptotic behavior. His results are based on certain transversality assumptions on the flows involved, but these are not fulfilled for problem (1),(2),(3).

In this note we restrict our attention to (ii), the existence of periodic solutions, although by means of similar methods also (i) and (ii) could be answered partly.

2. Reformulation of the problem

We want to reformulate equations (1) and (2) as an integral equation involving the switching function v(t) and the temperature u(t) = u(t,o) at the thermostat. Integrating (2) we obtain in case $\beta > o$

$$y(t) = \beta^{-1} \int_0^\infty e^{-\tau/\beta} v(t-\tau)d\tau$$

and (4)

$$y(t) = v(t) \quad \text{for} \quad \beta = o.$$

To obtain a representation of u(t,x) we use eigenfunction expansion. Let $\lambda_j, j = o,1,2,...$ denote the positive roots of the equation $tg\lambda = (\alpha\lambda)^{-1}$; then $-\lambda_j^2$ are the eigenvalues of (1) with corresponding eigenfunctions $\varphi_j(x) = \cos \lambda_j x$. We have $\lambda_j \in [j\pi,(2j+1)\pi/2]$ for all j and $\lambda_j = (2j+1)\pi/2$ in case $\alpha = o$. Defining

$$g(t,x) = \sum_{n=0}^\infty c_n e^{-\lambda_n^2 t} \varphi_n(x)$$

with

$$c_n = 2(-1)^n \lambda_n(1+\alpha^2\lambda_n^2)^{1/2} \cdot (1+\alpha+\alpha^2\lambda_n^2)^{-1} ,$$

the solution of (1) on the line is given by

$$u(t,x) = \int_0^\infty g(\tau,x)y(t-\tau)d\tau. \tag{5}$$

Combining (4) and (5) we obtain

$$u(t) = \int_0^\infty k(\tau)\, v(t-\tau)d\tau , \tag{6}$$

where

$$k(t) = \beta^{-1} \int_0^t e^{-\tau/\beta} g(t-\tau,o)d\tau. \tag{7}$$

This derivation shows that any periodic solution of (1),(2),(3) by restriction yields a periodic solution of (6),(3), and conversely a periodic solution of (6),(3) extends by (4) and (5) to a periodic solution of (1), (2),(3) with the same period. Therefore, (1),(2),(3) and (6),(3) are equivalent in this respect.

For the sake of future reference, we now collect some properties of the kernels $g(t) = g(t,o)$ and $k(t)$ which can be derived easily.

$$k \in C_0^1(\mathbb{R}_+) \cap L^1(\mathbb{R}_+), \quad \int_0^\infty k(t)dt = 1, \quad k(o) = o,$$
$$o < k(t) \leqq k_0 e^{-\kappa_0 t} \quad \text{for} \quad t > o \text{ and some } k_o, \kappa_o > o. \tag{8}$$

Typically, as a heat kernel $g(t)$ is increasing for $o < t < t_0$, reaches its maximum $g_0 = g(t,o)$ at $t = t_0$, and is decreasing to zero for $t > t_0$. Convolution with $\beta^{-1}e^{-t/\beta}$ does not change the shape of g but flattens it and rises the value of t_0. In the special case $\alpha = \beta = o$ more specific information can be derived from the approximation

$$k(t) = g(t) \sim \begin{cases} (\pi t^3)^{-1/2} e^{-1/4t} & \text{for} \quad t < 1/\pi \\ \pi e^{-\pi^2 t/4} & \text{for} \quad t > 1/\pi \end{cases}$$

which yields values of relative error less than o,5%; from this one obtains $t_0 \approx 1/6$ and $g_0 \approx 1.85$. Note that the case $\alpha = \beta = o$ corresponds to that of fastest response since there is no resistance neither in the heater nor at the wall of the rod.

The properties (8) yield already a condition which is necessary for existence of nonconstant periodic solutions of (6),(3). In fact, if $u \in C(\mathbb{R},\mathbb{R}_+)$, $v \in L^\infty(\mathbb{R})$, $v(\mathbb{R}) \subset \{o,1\}$ is a periodic solution, which is nonconstant, then

$$u = k * v < k * 1 = \int_0^\infty k(t)dt = 1.$$

Hence in case $u_1 \geqq 1$ there is no switching of v from 1 to o at all, and by periodicity we obtain $u \equiv v \equiv 1$, a contradiction.

Proposition 1: *A necessary condition for existence of nonconstant periodic solutions of* (6),(3) *is* $u_1 < 1$. *In case* $u_1 \geq 1$ *holds,* $u \equiv v \equiv 1$ *is an equilibrium solution of* (6),(3).

3. Periodic solutions

To overcome the discontinuities of v(t) induced by the switching law (3) which acts instanteneously, we replace (3) by an ordinary differential equation for v which corresponds to a finite speed of switching. This equation should be very fast and be such that for fixed u the modes v = o for $u > u_1$ and $v = 1$ for $u < u_o$ are the only stable equilibria, globally attracting, while for $u \in [u_o, u_1]$ both of these modes should be stable equilibria. This leads to a reasonable approximation of switching law (3).

The simplest right hand side f(u,v) which meets these requirements is

$$f(u,v) = \begin{cases} -v & \text{for } v \leq z \\ v - 2z & \text{for } z \leq v \leq z + 1/2 \\ 1 - v & \text{for } z + 1/2 \leq v \end{cases} ,$$

where

$$z = \frac{1}{2} \; \frac{u - u_o}{u_1 - u_o} \quad .$$

We therefore study the approximating problem

$$\begin{cases} \varepsilon \dot{v} = f(u,v) \\ u = k * v \end{cases} \tag{9}$$

first, derive conditions ensuring existence of periodic solutions for (9), and then let $\varepsilon \to o$ to obtain a periodic solution of the original problem (6),(3). This approach is completely different from those used in [2] ,[3] [4] .

To understand the qualitative behavior of (9) we let $\eta > o$ be small and consider the set

$$\Omega_\eta = \{(u,v) \in [o,1]^2 : \; |f(u,v)| \geq \eta\}.$$

If $\varepsilon > 0$ is much smaller than η, any solution $(u(t),v(t))$ of (9) with $(u(o),v(o)) \in \Omega_\eta$ will move very rapidly to either $\{o \leq v \leq \eta\}$ or to $\{1-\eta \leq v \leq 1\}$ and then remain there, until eventually the line $v = 2z$ is reached. In case this line is crossed with speed bounded from below, then the solution hits Ω_η again and another rapid change of v follows. Thus the solution oscillates as indicated in Fig. 1 . Using this property of (9), by means of a fixed point argument for the Poincare' map of (9) existence of periodic solutions will be established, once conditions are known which ensure strict crossing of the line $v = 2z$.

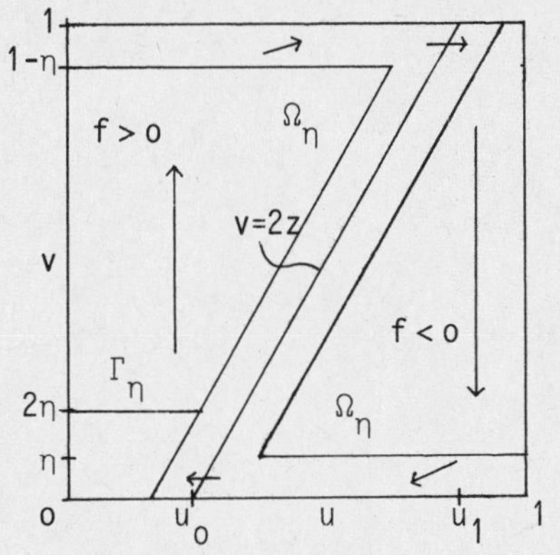

Figure 1

Theorem 1: *Let* $u_1 < 1$ *and let* g *and* k *be as above. Suppose the following conditions are satisfied.*

(C1) $\quad 1 - u_1 > \int_{t_1}^{\infty} g(\tau)d\tau \quad , \quad t_1 = \beta \log \frac{1-u_0}{1-u_1}$

(C2) $\quad u_0 > \int_{t_2}^{\infty} g(\tau)d\tau \quad , \quad t_2 = \beta \log \frac{u_1}{u_0}$.

Then the thermostat problem (6),(3) *has a* τ*-periodic solution* $(u,v) \in C(\mathbb{R}, \mathbb{R}_+) \times L^\infty(\mathbb{R})$ *of minimal type, i.e. which satisfies* $v(t) \equiv 1$ *on* (o,τ_0), $v(t) \equiv o$ *on* (t_0, τ) *for some* $\tau_0 \in (o,\tau)$.

Note that conditions (C1) and (C2) are fulfilled whenever β is large enough, whatever the shape of g may be. On the other hand, if we take the special shape of g into account a sharper result is obtained.

Theorem 2 : *Let* $u_1 < 1$ *and* g *and* k *be as above. Suppose the following conditions are satisfied.*

(C3) *There is* $t_0 > 0$ *such that* $g(t)$ *is increasing on* $(0,t_0)$ *and decreasing on* (t_0,∞).

(C4) *The solution* $\rho_0 > 0$ *of* $\rho - 1 + e^{-\rho} = (u_1 - u_0)/g_0\beta$ *satisfies* $t_0 < \beta\rho_0$, *where* $g_0 = g(t_0)$; $\beta\rho_0 = (u_1 - u_0)/g_0$ *for* $\beta = 0$.

Then the thermostat problem admits a periodic solution of minimal type.

The proof of Theorems 1 and 2 differ only in the argument used to obtain strict crossing at the line $v = 2z$ and are given in the next section. It should be clear that this approach can also be used to obtain existence of solutions of the initial value problem for (1),(2),(3). However, no uniqueness can be expected for general history functions since strict crossing requires some restrictions on the history values of a solution, even in case the assumptions of Theorems 1 and 2 are satisfied.

4. Proofs of Theorems 1 and 2

The proof consists of three steps. In the first step we show by means of the Tychonov fixed piont theorem (cp. e.g. Deimling [1]) that for small but fixed $\varepsilon > 0$ problem (9) admits a nonconstant periodic solution. This requires the Poincare' map to be welldefined and for this in turn strict crossing at the line $v = 2z$ is needed. The verification of the latter by hypotheses (C1), (C2) or (C3),(C4) is carried out in step 2. Finally, in step 3 we let $\varepsilon \to 0+$ to obtain the periodic solution of minimal type of (6), (3).

(i) Let $X = C(\mathbb{R}_-)$, the Frechet space of all realvalued continous functions defined on $\mathbb{R}_- = (-\infty,0]$ equipped with the topology of uniform convergence on compact subsets and let

$$D = \{\varphi \in X : 0 \leq \varphi \leq 1,\ |\varphi| = \sup_{t \neq \bar{t}} |t - \bar{t}|^{-1} \cdot |\varphi(t) - \varphi(\bar{t})| \leq M,$$
$$(\int_0^\infty k(\rho)\varphi(-\rho)d\rho\ ,\ \varphi(0)) \in \Gamma_\eta\},$$

where

$$\Gamma_\eta = \{(u,v) \in [0,1]^2 :\ u \leq u_0 + (v - \eta)(u_1 - u_0),\ v = 2\eta\}$$

and $M = \varepsilon^{-1}(1 + (u_1 - u_0)^{-1})$; obviously $D \subset X$ is compact and convex. For $\varphi \in D$, we let $(u(\cdot,\varphi), v(\cdot,\varphi))$ denote the solution of (9) on $\mathbb{R}_+$ such that $v(t,\varphi) = \varphi(t)$ for $t \leq 0$. Since f is Lipschitz this solution exists, is unique, and depends continuously on φ; moreover it is easy to see that $0 \leq v(\cdot,\varphi) \leq 1$ on $\mathbb{R}$ as well as $|v(\cdot,\varphi)|_L \leq M$ hold again.

Now, if there were a continuous function $\tau : D \to (0,\infty)$ such that $(u(\tau(\varphi),\varphi), v(\tau(\varphi),\varphi)) \in \Gamma_\eta$ for each $\varphi \in D$ then the Poincaré map $T : D \to X$ defined by

$$(T\varphi)(t) = v(\tau(\varphi) + t,\varphi) \quad , \quad t \leq 0$$

would be continuous and satisfy $TD \subset D$; hence by Tychonoff's fixed point theorem there would exist a fixed point φ_0 of T, i.e. a periodic solution of (9) in D.

(ii) Suppose (C1) and (C2) holds, in particular $\beta > 0$. To obtain the function $\tau(\varphi)$, we show that for any $\varphi \in D$ there is a time of first return $\tau(\varphi) > 0$ such that $(u(\tau(\varphi),\varphi), v(\tau(\varphi),\varphi)) \in \Gamma_\eta$. If $\varepsilon > 0$ is sufficiently small, say $\varepsilon < \eta^2 \beta(u_1 - u_0)$, any solution of (9) starting in Γ_η will leave Ω_η through the set $v = 1 - \eta, u \leq u_0 + 2\eta(u_1 - u_0)$, this will not happen before $t_\varepsilon = \varepsilon(1 - 3\eta)/\eta$ since $|\dot{v}| \geq \eta/\varepsilon$ on Ω_η holds. The solution then stays in the strip $1-\eta \leq v \leq 1$ until it reaches the line $v = 2z$. Since we have

$$\beta\dot{u} = g * v - u \leq 1 - u$$

this will take at least

$$t_1^\eta = \beta \cdot \log \frac{1 - u_0 - 2\eta(u_1 - u_0)}{1 - u_1 + \eta(u_1 - u_0)}$$

time units, and in virtue of $u_1 < 1$ and $\int_0^\infty g(s)ds = 1$, the solution must reach this line. If $t^* \geq t_\varepsilon + t_1^\eta$ denotes the first time when $v = 2z$ holds, we have

$$\beta u(t^*) \geq (1-\eta)\int_{t_\varepsilon}^{t^*} g(t^*-s)ds - u_1 \geq (1-\eta)(1 - \int_{t_1^\eta}^{\infty} g(s)ds) - u_1.$$

Hence, choosing $\eta > 0$ small enough, t_1^η is close to t_1 and by (C1) we obtain $\beta u(t^*) \geq \sigma > 0$ for some $\sigma > 0$, i.e. the solution crosses the line $v = 2z$ strictly. Possible after decreasing $\eta > 0$ once more, the solution then must reach $\partial\Omega_\eta$, and repeating these arguments with (C2) instead it becomes apparent that the time of first return $\tau(\varphi)$ exist. The continuity of $\tau(\varphi)$ follows from the continuous dependence of the solutions on the history values.

(iii) Suppose (C3) and (C4) hold; then $g \in L^1(\mathbb{R}_+)$ and $|g|_{L^1} = 2g_0$ where $g_0 = g(t_0)$ denotes the maximum value of g. The relations $u = k * v$ and $k = \beta^{-1} g * e^{-s/\beta}$ therefore imply $|u| \leq g_0$, hence as before any solution of (9) starting in Γ_η will leave Ω_η through $v = 1 - \eta$, $u \leq u_0 + 2\eta(u_1 - u_0)$ before $t_\varepsilon = \varepsilon(1-3\eta)/\eta$ provided $\varepsilon < \eta^2(u_1-u_0)/g_0$ is satisfied. Again the solution then stays in the strip $1 - \eta \leq v \leq 1$ until it reaches the line $v = 2z$, and by means of $\dot{u} \leq g_0$ it is easily seen that this requires at least $t_3^\eta = (1-3\eta)(u_1-u_0)/g_0$ time units. For $\beta > 0$ this estimate can be improved by

$$\beta\ddot{u} = \dot{g} * v - \dot{u} \leq g_0 - \dot{u}$$

in case we also have $\dot{u}(0) \leq 0$; in fact integrating this equation we obtain the lower bound $t_3^\eta = \beta\rho_0^\eta$, where ρ_0^η denotes the positive solution of

$$\rho - 1 + e^{-\rho} = (1-3\eta)^2(u_1-u_0)/g_0 \quad ;$$

obviously $\rho_0^\eta \to \rho_0$ as $\eta \to 0$ with ρ_0 as defined in (C4), and $\beta\rho_0 \geq (u_1-u_0)/g_0$ for all $\beta > 0$.

Now, suppose $\varphi \in D$ is a history function such that $u(0,\varphi) \leq 0$ and $\varphi(t) \leq 2\eta$ for $t \in [-t_0, 0]$, and let t^* denote the first time at which the solution reaches the line $v = 2z$; we then have $t^* \geq t_3^\eta + t_\varepsilon > t_0 + t_\varepsilon$ by (C4) if $\eta > 0$ is chosen sufficiently small. To obtain strict crossing we estimate $g * v$ from below according to

$$(\dot{g} * v)(t^*) = [\int_{t_\varepsilon}^{t^*} + \int_{0}^{t_\varepsilon} + \int_{-t_0}^{0} + \int_{-\infty}^{-t_0}] \dot{g}(t^*-s)v(s)ds$$

$$\geq g(t^*)v(t_\varepsilon) + \int_{0}^{t_\varepsilon} \dot{g}(t^*-s)v(s)ds + 2\eta(g(t^*+t_0)-g(t^*)) - g(t^*+t_0)$$

$$\geq (1-2\eta)(g(t^*)-g(t^*+t_0)) - \eta g_0 - \int_{t^*-t_\varepsilon}^{t^*} |\dot{g}(s)|ds$$

since $\dot{v}(t) \geq o$ in $[o,t^*]$, $v(t_\varepsilon) \geq 1-\eta$ and $g(o) = o$. This inequality shows that $(\dot{g} * v)(t^*) \geq \sigma > o$ holds for some σ as $\eta \to o$, independently of $t^* \in [t_0,T]$ since by (C3) g is decreasing for $t > t_0$. Therefore, in case $\beta = o$ we obtain $\dot{u}(t^*) = (\dot{g} * v)(t^*) \geq \sigma > o$ and in case $\beta > o$ the relation $\beta\ddot{u}(t^*) + \dot{u}(t^*) \geq \sigma > o$, hence in both cases we have strict crossing and $\partial\Omega_\eta$ is reached if $\eta > o$ is chosen sufficiently small. Repeating this argument we see that the time of first return $\tau(\varphi)$ exists for all $\varphi \in D_1$ where $D_1 = \{\varphi \in D : \dot{u}(o,\varphi) \leq o$ and $\varphi(t) \leq 2\eta$ on $[-t_0,o]\}$.

(iv) Let $\eta > o$ be sufficiently small and $\varepsilon < \eta^2(u_1-u_0)/g_0$; by (i),(ii),(iii) above (use D_1 instead of D in the situation of Theorem 2) there are τ_ε-periodic solutions $(u_\varepsilon,v_\varepsilon)$ such that $1-2\eta \leq v_\varepsilon(t) \leq 1$ in $[t_\varepsilon,\tau_\varepsilon^0 + t_\varepsilon]$, $o \leq v_\varepsilon(t) \leq 2\eta$ in $[\tau_\varepsilon^0+2t_\varepsilon, \tau_\varepsilon]$, $o \leq u_\varepsilon \leq 1$ and $|\dot{u}_\varepsilon| \leq C$ for some constant C independent of ε and η. Since $\{\tau_\varepsilon^0\}$ and $\{\tau_\varepsilon\}$ are bounded and bounded away from zero, $\{u_\varepsilon\}$ is bounded and equicontinuous and $\{v_\varepsilon\} \subset L^\infty(\mathbb{R})$ is bounded, there are $\tau^0, \tau \in (o,\infty)$, $u \in C(\mathbb{R};\mathbb{R}_+)$ and $v \in L^\infty(\mathbb{R})$ τ-periodic, and subsequences $\eta_n \to o$, $\varepsilon_n \to o$ such that $\tau_{\varepsilon_n}^0 \to \tau^0$, $\tau_{\varepsilon_n} \to \tau$, $u_{\varepsilon_n} \to u$ uniformly w.r. to t, and $v_{\varepsilon_n} \xrightarrow{*} v$ weak-star in $L^\infty(\mathbb{R})$.

Hence the equations $u_\varepsilon = k * v_\varepsilon$ imply $u = k * v$ on $\mathbb{R}$, and since $L^\infty(o,\tau) \subset L^1(o,\tau)$ we also have $v_{\varepsilon_n} \rightharpoonup v$ weakly in $L^1(o,\tau)$, in particular $v \in \bigcap_m \overline{co}\,\{v_{\varepsilon_n} : n \geq m\}$ where '$\overline{co}$' denotes 'closed convex hull'. But this yields $v(t) = 1$ a.e. in (o,τ^0) and $v(t) = o$ a.e. in (τ^0,τ), i.e. (u,v) is a τ - periodic solution of the thermostat problem (6),(3) of minimal type. ∎

References :

[1] Deimling,K. : Nonlinear Functional Analysis. Springer Verlag, Heidelberg 1985.

[2] Glashoff,K. and Sprekels,J. : The regulation of temperature by thermostats and set - valued integral equations. J. Integral Equ. 4,95-112 (1982).

[3] Glashoff,K. and Sprekels,J. : An application of Glicksberg's Theorem to set - valued integral equations arising in the theory of thermostats. SIAM J. Math. Anal. 12,477-486 (1981).

[4] Seidman,T.I. : Switching systems : thermostats and periodicity. Preprint.

SOME QUESTIONS ON THE INTEGRODIFFERENTIAL EQUATION $u' = AK*u + BM*u$

Andrea Pugliese

Dipartimento di Matematica
Universita' degli Studi di Trento
38050 POVO (TRENTO) - ITALIA

1. Introduction and notation.

We study the problem

$$(1) \qquad \begin{cases} u'(t) = AK*u(t) + BM*u(t) \\ u(0) = x \end{cases}$$

in a Banach space X, where A and B are closed linear operators with domains (D_A and D_B respectively) dense in X and $*$ denotes the convolution. K and M are scalar kernels, which are absolutely Laplace transformable in some half-plane.

As a motivation for this study, we remark that Da Prato and Iannelli [3] have considered equations of the form

$$u'(t) = AK*u(t)$$

with the same meanings of the symbols. This paper is therefore an extension of their work to equation (1).

On the other hand several authors(see [1], [5], [8]) have studied equations of the form

$$(2) \qquad u'(t) = Au(t) + B*u(t)$$

where $B(t)$ are bounded linear operators from D_A (with the graph norm) to X. Therefore if we set $A=0$, it must be $B(t) \in L(X)$.

Grimmer and Pritchard [7] have considered equations of the form (2) also with $A=0$ and $B(t)$ unbounded. However for our particular formulation ($B(t) = \overline{K(t)A + M(t)B}$) assumption (V1)-(V3) of [7] will not necessarily be satisfied.

We will construct a resolvent family of operators for this equation, through its formal Laplace transform. We will need an extension of the result of Da Prato and Grisvard [2] on the sum of closed operators, that will be discussed in the next Paragraph.

The definition of a resolvent family is analogous to that of several authors (see [3], [5], [7], [8]).

Definition 1. Let $\{U(t)\}_{t\geq 0}$ be a family of linear operators $U(t) \in L(X)$. It is said to be a (M,ω)-resolvent family for (1) if the following properties are verified :

(R1) $U(0) = I$

(R2) $\forall x \in X$, $u(t) = U(t)x \in C([0,+\infty);X)$

(R3) $\forall x \in D_A \cap D_B$, $u(t) = U(t)x \in C^1([0,+\infty);X) \cap C([0,+\infty);D_A \cap D_B)$ and

(R3a) $U'(t)x = A\int_0^t K(t-s)U(s)x ds + B\int_0^t M(t-s)U(s)x ds$

(R3b) $U'(t)x = \int_0^t K(t-s)U(s)Ax\, ds + \int_0^t M(t-s)U(s)Bx\, ds$

(R4) $\|U(t)\| \leq Me^{\omega t}$ $\forall t \geq 0$

A resolvent family will be constructed when A and B are generator of analytic semigroups as precised later. Here only the case where A and B commute is considered.

For later use we define the sectors

$$\Sigma_{\vartheta;\psi} = \{ \lambda \in C, \lambda \neq 0, -\vartheta < \arg \lambda < \psi\} \qquad (\vartheta,\psi > 0)$$

and the paths

$$\Gamma_{\varepsilon,\vartheta,\psi} = \{\rho e^{-i\vartheta}, \rho \geq \varepsilon\} \cup \{\varepsilon e^{i\varphi}, -\vartheta \leq \varphi \leq \psi\} \cup \{\rho e^{i\psi}, \rho \geq \varepsilon\}$$

taken counterclockwise. When we use the closure of a sector, it is to be intended as the closure in the set $\{\lambda \neq 0\}$. Moreover by Σ_ϑ we mean $\Sigma_{\vartheta,\vartheta}$ and by $\Gamma_{\varepsilon;\vartheta}$, $\Gamma_{\varepsilon;\vartheta;\vartheta}$

2. Sum of linear operators.

The results of Da Prato-Grisvard [2] for the parabolic case are here extended to

Theorem 1. Let A and B be linear, closed operators from $D_A \subset X$ (respectively D_B) into X, Banach space. Let $\rho(A) \supseteq \Sigma_{\vartheta_{1,A},\vartheta_{2,A}}$ and $\rho(B) \supseteq \Sigma_{\vartheta_{1,B},\vartheta_{2,B}}$ with the following assumptions

(H1) $\forall\ \vartheta, \psi$ with $-\vartheta_{1,A} < -\vartheta < \psi < \vartheta_{2,A}$ there exists $M_A(\vartheta,\psi)$ such that $\forall\ \lambda \in \Sigma_{\vartheta;\psi}$ it is $\|R(\lambda,A)\| \leq M_A(\vartheta,\psi)/|\lambda|$ and analogously for B.

(H2) $\vartheta_{1,A} + \vartheta_{2,B} > \pi$ and $\vartheta_{1,B} + \vartheta_{2,A} > \pi$

(H3) $[R(\lambda,A);\ R(\mu,B)] = 0 \quad \forall(\lambda,\mu) \in \rho(A) \times \rho(B)$

(H4) D_A and D_B dense in X.

Then $A+B$ is closable, its resolvent set includes $\Sigma_{\vartheta_1,\vartheta_2}$ with $\vartheta_i = \min\{\vartheta_{i,A}, \vartheta_{i,B}\}$.

Its resolvent operator is

$$(3) \qquad S_\lambda = \frac{1}{2\pi i} \int_{\Gamma_{\varepsilon_0;\vartheta_{1,0};\vartheta_{2,0}}} R(\zeta,B)\, R(\lambda-\zeta,A)\,d\zeta$$

with ε_0, $\vartheta_{1,0}$ and $\vartheta_{2,0}$ defined in the course of the proof.

Moreover it satisfies the following estimates :

$\forall\ \vartheta,\psi$ with $-\vartheta_1 < -\vartheta < \psi < \vartheta_1$ there exist $c(\vartheta,\psi)$; $c_1(\vartheta,\psi)$; $c_2(\vartheta,\psi,\gamma)$; $c_3(\vartheta,\psi,\gamma)$ such that when $-\vartheta < \arg(\lambda) < \vartheta$

(4) $$\|S_\lambda x\| \leq \frac{c}{|\lambda|} \|x\| ;$$

if $x \in D_B(\gamma;\infty)$

(5) $$\|AS_\lambda x\| \leq c_1 \|x\| + c_2 \frac{|x|_{B,\gamma}}{|\lambda|^\gamma}$$

(6) $$\|BS_\lambda x\| \leq c_3 \frac{|x|_{B,\gamma}}{|\lambda|^\gamma}$$

$D_B(\gamma;\infty)$ is an interpolation space between D_B and X (see [2]) and can be characterized as $D_B(\gamma;\infty) = \{x \in X : |x|_{B,\gamma} < \infty\}$ where

$$|x|_{B,\gamma} = \sup_{t>1} \|t^\gamma BR(t,B)x\|$$

The proof is only sketched here, because it is similar to [2]

First one chooses $\vartheta_{1,0}$ and $\vartheta_{2,0}$ such that

(7) $$\pi - \vartheta_{1,A} < \vartheta_{2,0} < \vartheta_{2,B} \qquad \pi - \vartheta_{2,A} < \vartheta_{1,0} < \vartheta_{1,B}$$

and

(8) $$-\vartheta_{1,0} < -\vartheta < \psi < \vartheta_{2,0}$$

and ε_0 such that $\{\lambda - \varepsilon_0 e^{i\varphi}\}$ $(-\vartheta_{1,0} \leq \varphi \leq \vartheta_{2,0})$ is away from the half lines

$\{\rho\, e^{i\vartheta_{2,A}}\}$, $\{\rho\, e^{-i\vartheta_{1,A}}\}$ $(\rho \geq 0)$.

By Cauchy's theorem any such choice of $\vartheta_{1,0}$, $\vartheta_{2,0}$, and ε_0 will not change the value of S_λ. With an accurate choice one can easily prove (4).

To see that S_λ is indeed the resolvent of $\overline{A+B}$, we repeat the arguments of Da Prato and Grisvard [2] and obtain that, if $x \in D_B$,

$$\lambda S_\lambda x - AS_\lambda x - BS_\lambda x = x$$

As the roles of A and B are symmetric, we have that

if $x \in D_A \cap D_B$ $\qquad (\lambda-(A+B))S_\lambda x = S_\lambda(\lambda-(A+B))x = x$

if $x \in D_A + D_B$ $\qquad (\lambda-(A+B))S_\lambda x = x$

Therefore the assumptions of Theorem 2.9 of Da Prato and Grisvard [2] are fulfilled, so that A+B is closable and it is clear that

$$(\lambda-(\overline{A+B}))S_\lambda x = x \qquad \forall\, x \in X.$$

As for the estimates (5)-(6) they are found noting, as in [2], that for $x \in D_B(\gamma;\infty)$

$$BS_\lambda x = \frac{1}{2\pi i} \int_{\Gamma_{\varepsilon_0;\vartheta_{1,0};\vartheta_{2,0}}} R(\lambda-\zeta,A)\zeta^\gamma\, BR(\zeta,B)x \frac{d\zeta}{\zeta^\gamma} \tag{9}$$

and

$$AS_\lambda x = AR(\lambda,A)x + \frac{1}{2\pi i} \int_{\Gamma_{\varepsilon_0;\vartheta_{1,0};\vartheta_{2,0}}} AR(\lambda-\zeta,A)\zeta^\gamma\, BR(\zeta,B)x \frac{d\zeta}{\zeta^{1+\gamma}} \tag{10}$$

3. Construction of the resolvent operator

We now apply the results of Section 2 to the study of (1), with the following assumptions

(V1) A and B are closed linear operators with domain (D_A and D_B respectively) dense in X. There exist ϑ_A and ϑ_B $0 \le \vartheta_A, \vartheta_B < \pi$, continuous non-decreasing functions $M_A(\bullet)$ (and $M_B(\bullet)$) defined on $[0,\pi-\vartheta_A)$(and $[0,\pi-\vartheta_B)$ respectively) such that

(11) $\lambda \in \Sigma_{\pi-\vartheta_A} \Rightarrow \lambda \in \rho(A)$ and $\|R(\lambda,A)\| \leq \dfrac{M_A(\vartheta)}{|\lambda|}$

(12) $\lambda \in \Sigma_{\pi-\vartheta_B} \Rightarrow \lambda \in \rho(B)$ and $\|R(\lambda,B)\| \leq \dfrac{M_B(\vartheta)}{|\lambda|}$

where $\vartheta = |\arg(\lambda)|$

(V2) $[R(\lambda,A); R(\mu,B)] = 0 \quad \forall\ (\lambda,\mu) \in \rho(A)\times\rho(B)$

(V3) K and $M \in L^1_{loc}[0,\infty)$ and have Laplace transform $\hat{K}(\lambda)$ and $\hat{M}(\lambda)$

(V4) There exists $\vartheta > \pi/2$ such that

(13) $\hat{K}(\lambda)$ and $\hat{M}(\lambda)$ have an analytic extension to $\bar{\Sigma}_\vartheta$ with $\hat{K}(\lambda)\neq 0$ and $\hat{M}(\lambda)\neq 0$ in $\bar{\Sigma}_\vartheta$

$\forall\ \lambda \in \bar{\Sigma}_\vartheta$

(14) $|\arg \lambda - \arg \hat{K}(\lambda)| \leq \varphi_1 < \pi - \vartheta_A$

(15) $|\arg \lambda - \arg \hat{M}(\lambda)| \leq \varphi_2 < \pi - \vartheta_B$

(16) $|\arg \hat{M}(\lambda) - \arg \hat{K}(\lambda)| \leq \varphi_3 < \pi - \vartheta_A - \vartheta_B$.

Under (V1)-(V4) we can apply Theorem 1 to the two operators A and $\dfrac{\hat{M}(\lambda)}{\hat{K}(\lambda)} B$.

Condition (16) guarantees that, when $\lambda \in \bar{\Sigma}_\vartheta$, (H2) are fulfilled.

Therefore $A + \frac{\hat{M}(\lambda)}{\hat{K}(\lambda)}B$ is closable and its resolvent set includes $\Sigma_{\vartheta_1^\lambda, \vartheta_2^\lambda}$ where

$$\vartheta_1^\lambda = \min\{\pi - \vartheta_A, \pi - \vartheta_B - \arg(\frac{\hat{M}(\lambda)}{\hat{K}(\lambda)})\}\tag{17}$$

$$\vartheta_2^\lambda = \min\{\pi - \vartheta_A, \pi - \vartheta_B + \arg(\frac{\hat{M}(\lambda)}{\hat{K}(\lambda)})\}$$

Note that (16) implies that both ϑ_1^λ and ϑ_2^λ are positive. Because of (14) and (15), we have, for $\lambda \in \Sigma_\vartheta$,

$$-\vartheta_1^\lambda < \arg\lambda - \arg\hat{K}(\lambda) < \vartheta_2^\lambda$$

Therefore, when $\lambda \in \Sigma_\vartheta$, we can define

$$T_\lambda = \frac{1}{\hat{K}(\lambda)} \quad R(\lambda/\hat{K}(\lambda), \overline{A + \frac{\hat{M}(\lambda)}{\hat{K}(\lambda)}B}).\tag{18}$$

Proposition 1. T_λ is well defined for $\lambda \in \bar{\Sigma}_\vartheta$. There exist $N, M_1, M_2(\lambda), M_3(\lambda)$ such that for $\lambda \in \bar{\Sigma}_\vartheta$ it is

$$\|T_\lambda x\| \le \frac{N}{|\lambda|}\|x\| \quad . \qquad \text{If } x \in D_B(\gamma;\infty)\tag{19}$$

$$\|AT_\lambda x\| \le \frac{M_1}{|\hat{K}(\lambda)|}\|x\| + \frac{M_2(\gamma)\,|\hat{M}(\lambda)|^\gamma |x|_{B,\gamma}}{|\hat{K}(\lambda)|\;|\lambda|^\gamma}\tag{20}$$

$$\|BT_\lambda x\| \le \frac{M_3(\gamma)\,|x|_{B,\gamma}}{|\lambda|^\gamma |\hat{M}(\lambda)|^{1-\gamma}}\tag{21}$$

Proof. The well definition comes from the above reasoning.

As for the estimates (19)-(21) we can not apply directly Theorem 1, because of the intricate dependence of T_λ on λ. However, we can repeat the construction of Theorem 1; to get uniformity in λ, we set

$$(22)\qquad \eta = \min\{\pi - \vartheta_A - \varphi_1, \pi - \vartheta_B - \varphi_2, \pi - \vartheta_B - \vartheta_A - \varphi_3\}$$

and select

$$(23)\qquad \vartheta^\lambda_{1,0} = \pi - \vartheta_B - \arg\left(\frac{\hat{M}(\lambda)}{\hat{K}(\lambda)}\right) - \eta/2 \qquad \vartheta^\lambda_{2,0} = \pi - \vartheta_B + \arg\left(\frac{\hat{M}(\lambda)}{\hat{K}(\lambda)}\right) - \eta/2$$

$$\varepsilon^\lambda_0 = \frac{|\lambda|}{2|\hat{K}(\lambda)|}\,\mathrm{sen}\,(\eta/2)$$

From Theorem 1 we have

$$(24)\qquad T_\lambda = \frac{1}{2\pi i \hat{K}(\lambda)} \int_{\Gamma_{\varepsilon^\lambda_0;\,\vartheta^\lambda_{1,0};\,\vartheta^\lambda_{2,0}}} R\left(\zeta, \frac{\hat{M}(\lambda)}{\hat{K}(\lambda)} B\right) R\left(\frac{\lambda}{\hat{K}(\lambda)} - \zeta, A\right) d\zeta$$

With some computations one can check that, when $\zeta \in \Gamma_{\varepsilon^\lambda_0;\,\vartheta^\lambda_{1,0};\,\vartheta^\lambda_{2,0}}$,

$$-(\pi - \vartheta_B - \eta/2) \le \arg\left(\frac{\hat{K}(\lambda)}{\hat{M}(\lambda)}\zeta\right) \le \pi - \vartheta_B - \eta/2$$

and

$$-(\pi - \vartheta_A - \eta/2) \le \arg\left(\frac{\lambda}{\hat{K}(\lambda)} - \zeta\right) \le \pi - \vartheta_A - \eta/2 \ .$$

From (24) one has

$$\|T_\lambda\| \le \frac{M_B(\pi-\vartheta_B-\eta/2)}{2\pi|\hat{R}(\lambda)|} M_A(\pi-\vartheta_A-\eta/2) \int_{\Gamma^{\lambda}_{\varepsilon_0;\vartheta^{\lambda}_{1,0};\vartheta^{\lambda}_{2,0}}} \frac{|d\zeta|}{|\zeta|\,|\frac{\lambda}{\hat{R}(\lambda)}-\zeta|} \le \frac{c}{|\lambda|}$$

The estimates (20)-(21) are obtained analogously.

Using the construction of T_λ we can obtain

Theorem 2. Let (V1)-(V4) hold. Then (1) has one and only one resolvent family. Moreover

(25) $U(t)$ has an analytic extension to $\Sigma_{\vartheta-\pi/2}$

(26) $\forall\ \varphi < \vartheta - \pi/2$ there exists $M_k(\varphi)$ such that

$$\left\|\frac{d^k}{dt^k} U(t)\right\| \le \frac{M_k(\varphi)}{|t|^k} \qquad k \in N \quad t \in \Sigma_\varphi$$

Proof. Set

$$U(t) = \frac{1}{2\pi i}\int_{\Gamma_{\varepsilon,\vartheta}} e^{\lambda t}\, T_\lambda\, d\lambda \qquad t \in \Sigma_{\vartheta-\pi/2} \tag{27}$$

The proof of [3] that $U(t)$ is a resolvent family and satisfies (26) can be repeated word by word. All that is used are the estimates of Proposition 1 and the identity

$$T_\lambda x = \frac{x}{\lambda} + \frac{T_\lambda}{\lambda}(\hat{R}(\lambda)A+\hat{M}(\lambda)B)x \qquad \forall\ x \in D_A \cap D_B .$$

As for the uniqueness, one can repeat, with obvious modification, the proof of [4] or [8] that a resolvent family satisfying (R1)-(R3) is unique.

Remark 1. Proposition 1 and Theorem 2 can be extended to the sum of n operators.

Precisely let $A_1, \ldots, A_n$ be closed linear operators satisfying (V1) and

(V2') $\quad [R(\lambda, A_i)R(\mu, A_j)] = 0 \qquad \forall\, (\lambda,\mu) \in \rho(A_i) \times \rho(A_j)$

(V3') $\quad K_1, \ldots, K_n \in L^1_{loc}[0,\infty)$ and are L-transformable

(V4') $\quad$ There exists $\vartheta > \pi/2$ such that $\hat{K}_i(\lambda)$ have an analytic extension to $\bar{\Sigma}_\vartheta$ with $\hat{K}_i(\lambda) \neq 0$ and $\forall\, \lambda \in \bar{\Sigma}_\vartheta$

$$|\arg \lambda - \arg \hat{K}_i(\lambda)| \leq \varphi_i < \pi - \vartheta A_i \tag{28}$$

$$|\arg \hat{K}_i(\lambda) - \arg \hat{K}_j(\lambda)| \leq \varphi_{i,j} < \pi - \vartheta_{A_i} - \vartheta_{A_j} \tag{29}$$

Then we define $B_1(\lambda) = A$ and inductively, reasoning as in Proposition 1, we can prove that for $\lambda \in \bar{\Sigma}_\vartheta$,

$$B_{i-1}(\lambda) + \frac{\hat{K}_i(\lambda)}{\hat{K}_1(\lambda)} A_i$$ is closable and by defining

$B_i(\lambda)$ this closure, we have that $\lambda/\hat{K}_1(\lambda) \in \rho(B_i(\lambda))$ $i = 1 \ldots n$.

Therefore we can define $T_\lambda = R(\lambda/\hat{K}_1(\lambda), B_n(\lambda))$ and obtain Proposition 1 and Theorem 2.

Remark 2. Above we have obtained a resolvent family with $\omega=0$. If K and M are such that (14)-(16) hold for $\lambda \in \Sigma_{\vartheta}+\omega$, $\omega > 0$, one can repeat the same arguments obtaining $\|U(t)\| \le M\, e^{\omega t}$.

On the other hand, if A and B are such that (V1) holds for $\tilde{A}=A-\omega$, $\tilde{B}=B-\eta$, with $\omega,\eta > 0$, there is no straightforward extension.

It is possible to apply Theorem 1 to $\tilde{A}$ and $\tilde{B}(\hat{M}(\lambda)/\hat{K}(\lambda))$ to handle this case; however the conditions analogous to (14)-(16) that would result, seem difficult to check.

Otherwise one can use the resolvent family $\tilde{R}(t)$ relative to $\tilde{A}$ and $\tilde{B}$; the solution u(t) of (1) can be obtained as the solution of the integral equation

$$u(t) = \tilde{R}(t)u_o + \omega\tilde{R} * (K * u) + \eta\tilde{R} * (M * u)$$

that can be solved (at least locally) with a fixed point.

4. Some regularity results.

We have seen that, for $x \in D_A \cap D_B$, $U(t)x$ is a strict solution of (1). For a generic $x \in X$, all we can say is that $U(t)x \in C^1((0,+\infty); X)$ and that the operator P_t, defined by

$$D_{P_t} = \{x \in X : K*U(t)x \in D_A \text{ and } M*U(t)x \in D_B\} \text{ and } P_t x = A\, K*U(t)x + BM*U(t)x,$$

has a continuous extension P_t such that $\forall\, x \in X$ $U'(t)x=P_t x$.

The problem arises of finding when $K*U(t)x \in D_A$ and $M*U(t)x \in D_B$.

Proposition 2. Let $x \in D_B(\gamma,\infty)$.

Let for some ψ, $\frac{\pi}{2} < \psi \le \vartheta$

$$\text{(30)} \qquad \frac{|\hat{M}(\lambda)|}{|\lambda|^2} \le \varphi(|\lambda|) \qquad \text{when } \frac{\pi}{2} \le |\arg(\lambda)| \le \psi$$

with $\lim_{t\to\infty} \varphi(t) = 0$.

Then $M*U(t)x \in D_B$.

If moreover (30) holds with $\hat{K}(\lambda)$ in place of $\hat{M}(\lambda)$

$$K*U(t)x \in D_{\hat{A}} .$$

Proof. As $M(t)$ and $U(t)x$ are absolutely Laplace transformable, we have (see [9])

$$M*U(t)x = \frac{1}{2\pi i}\frac{d}{dt}\int_{\sigma-i\infty}^{\sigma+i\infty} e^{\lambda t}\hat{M}(\lambda)T_\lambda x\frac{d\lambda}{\lambda} \tag{31}$$

for any $\sigma > 0$. (30) guarantees (see [6]) that we can change the contour of integration in (31) to $\Gamma_{\varepsilon,\psi}$ and, differentiating under the integral, we obtain

$$M*U(t)x = \frac{1}{2\pi i}\int_{\Gamma_{\varepsilon,\psi}} e^{\lambda t}\hat{M}(\lambda)T_\lambda x\, d\lambda \tag{32}$$

If $x \in D_B(\gamma,\infty)$, from Proposition 1, we have that the integrand belongs to D_B and

$$\|B e^{\lambda t}\hat{M}(\lambda)T_\lambda x\| \le M_\varepsilon(\gamma)\,|e^{\lambda t}|\,\frac{|\hat{M}(\lambda)|^\gamma}{|\lambda|^\gamma}\,|x|_{B,\gamma}$$

Using (30) we see that this expression is convergent over $\Gamma_{\varepsilon,\varphi}$.

Therefore $M*U(t)x \in D_B$.

If $\hat{K}(\lambda)$ satisfies (48), repeating the argument, we have

$$K*U(t)x = \frac{1}{2\pi i}\int_{\Gamma_{\varepsilon,\psi}} e^{\lambda t}\hat{K}(\lambda)T_\lambda x\, d\lambda$$

From (20) we obtain

$$\|A e^{\lambda t} \hat{K}(\lambda)T_\lambda x\| \leq M_1 |e^{\lambda t}| \, \|x\| + M_2(\gamma) \, | e^{\lambda t} | \, \frac{|\hat{M}(\lambda)|^\gamma}{|\lambda|^\gamma} |x|_{B,\gamma}$$

Using (30), this expression is integrable over $\Gamma_{\varepsilon,\varphi}$, therefore the thesis.

Exchanging the role of A and B , and examining more closely the proof, one can get the somewhat stronger

Proposition 3. If $x \in D_A(\delta;\infty) + D_B(\gamma;\infty)$, (30) holds and $|e^{\lambda t}| \dfrac{|\hat{K}(\lambda)|^\delta}{|\lambda|^\delta}$ is integrable over $\Gamma_{\varepsilon;\psi}$, then $M{*}U(t)\,x \in D_B$. For the same x , if (30) holds with $\hat{K}(\lambda)$ in place of $\hat{M}(\lambda)$ and $|e^{\lambda t}| \dfrac{|\hat{M}(\lambda)|^\gamma}{|\lambda|^\gamma}$ is integrable over $\Gamma_{\varepsilon;\psi}$, $K{*}U(t)x \in D_A$.

We next consider the time regularity of U(t)x at t=0. With exactly the same proof as in [4] we obtain

Proposition 4. Let x be such that

(33) $$\sup_{\lambda\in\Sigma_\varphi,\ |\lambda|>1} |\lambda|^\alpha |\lambda T_\lambda x - x| < +\infty \quad \text{with } \frac{\pi}{2} < \varphi \leq \vartheta$$

Then $U(t)x \in C^\alpha([0;\infty); x)$.

To use Proposition 4, we note that, if $x \in D_A(\gamma;\infty) + D_B(\delta;\infty)$, then $T_\lambda x \in D_A \cap D_B$ and

(34) $$\lambda T_\lambda x - x = \hat{K}(\lambda)AT_\lambda x + \hat{M}(\lambda)BT_\lambda x \ .$$

Therefore to check whether (33) holds, we can use (20)-(21), obtaining

Corollary 1. Let $x \in D_A(\delta;\infty) \cap D_B(\gamma;\infty)$.
If $|\lambda|^{\alpha-\gamma}|\hat{M}(\lambda)|^{\gamma}$ and $|\lambda|^{\alpha-\delta}|\hat{K}(\lambda)|^{\gamma}$ are bounded in $\{\lambda \in \Sigma_{\varphi}\ ,\ |\lambda| > 1\}$, then Proposition 4 holds.

5. An example.

Let $K(t) = C_0 t^{-(1-\beta)}$, $M(t) = C_1 t^{-(1-\gamma)}$. Let A and B verify (V1)-(V2) with $\vartheta_A, \vartheta_B < \pi/2$. This means that A and B are commuting generators of analytic semigroups.

Suppose (35) $\vartheta_A < \frac{\pi}{2}(1-\beta)$ and $\vartheta_B < \frac{\pi}{2}(1-\gamma)$.

Fix ϑ with (36) $\pi/2 < \vartheta < \min\left\{\frac{\pi-\vartheta_A}{1-\beta}, \frac{\pi-\vartheta_B}{1+\gamma}\right\}$.

For $\lambda \in \Sigma_{\vartheta}$, (14) and (15) are satisfied with $\varphi_1 = \vartheta(1+\beta)$, $\varphi_2 = \vartheta(1+\gamma)$. As for (16) we have $\vartheta_A + \vartheta_B < \pi(1-\frac{(\beta+\gamma)}{2})$, (36) implies $\vartheta < \frac{(2\pi - (\vartheta_A + \vartheta_B))}{2+\beta+\gamma}$ and clearly

$$\max_{x\in[0,\pi(1-\frac{(\beta+\gamma)}{2})]} \frac{2\pi-x}{\pi-x} = \frac{2+\beta+\gamma}{\beta+\gamma} .$$

Therefore $(2\pi - \vartheta_A + \vartheta_B) \leq \frac{2+\beta+\gamma}{\beta+\gamma}(\pi-(\vartheta_A+\vartheta_B))$

and so $$\vartheta < \frac{(2\pi-(\vartheta_A+\vartheta_B))}{2+\beta+\gamma} \leq \frac{(\pi-(\vartheta_A+\vartheta_B))}{\beta+\gamma} < \frac{\pi-(\vartheta_A+\vartheta_B)}{|\beta-\gamma|}$$

Therefore $|\arg \hat{K}(\lambda) - \arg \hat{M}(\lambda)| \leq \vartheta\, |\beta - \gamma| < \pi - (\vartheta_A + \vartheta_B)$.

Note that (35) are, respectively, the conditions that $u' = AK*u$ and $v' = B\,M*v$ admit an analytic resolvent. For the case of these kernels, therefore, there is no added condition. As for the regularity, (30) holds with K and M ; thus, if $x \in D_A(\delta;\infty) + D_B(\alpha;\infty)$, $K*U(t)\,x \in D_A$ and $M*U(t)x \in D_B$, for all $t > 0$.

Finally we can use Corollary 1, to obtain that if

$$x \in D_A\left(\frac{\alpha}{1+\beta};\infty\right) \cap D_B\left(\frac{\alpha}{1+\gamma};\infty\right), \text{ then } U(t)x \in C^{\alpha}([0,+\infty); x).$$

As an example of A and B, we set $X = L^p(R^2)$, $D(A) = \{u : u_{xx} \in L^p\}$, $D(B) = \{u : u_{yy} \in L^p\}$ (where the derivatives are in the sense of distributions), $Au = u_{xx}$, $Bu = u_{yy}$. We are thus studying the equation

$$(37)\begin{cases} u_t(t,x,y) = C_0 \dfrac{\partial^2}{\partial x^2} \displaystyle\int_0^t (t-s)^{-(1-\beta)} u(s,x,y)ds + C_1 \dfrac{\partial^2}{\partial y^2} \displaystyle\int_0^t (t-s)^{-(1-\gamma)} u(s,x,y)ds \\ u(0,x,y) = u_0(x,y) \end{cases}$$

As (V1)-(V2) are fulfilled with $\vartheta_A = \vartheta_B = 0$, we can construct a resolvent family $U(t)$ for any $0 < \beta,\gamma < 1$; we have that $u(t,x,y) = U(t)u_0$ is a solution of (37) as long as

$u_0 = u_1 + u_2$ with $u_1 \in D_A(\delta;\infty), u_2 \in D_B(\alpha;\infty)$, for some $0 < \alpha,\delta < 1$.

(37) can be studied also with the methods of [7], setting $Y = W^{2,p}(R^2)$ and $A = \Delta$. Assumptions (V1)-(V3) of [7] can be verified, obtaining that $R(t)u_0$ is a solution of (37) when $u_0 \in D((-A)^{\alpha})$, $\alpha > 0$. Therefore in this work we have a slightly weaker condition. For a case where (V1) of [7] is not satisfied, while this work applies, one can modify the above example to $D(A) = \{u : \partial_x^4 u \in L^p\}$, $Au = -\partial_x^4 u$.

REFERENCES

[1] G. Chen and R. Grimmer: Semigroups and integral equations, J. Integral Equations 2(1980), 133-154.

[2] G. Da Prato and P. Grisvard: Sommes d'operateurs lineaires et equations differentialles operationalles, J. Math. pures et appl. 54(1975), 305-387.

[3] G. Da Prato and M. Iannelli: Linear integro-differential equations in Banach spaces, Rend. Sem. Mat. Univ. Padova 62(1980), 207-219.

[4] G. Da Prato, M. Iannelli and E. Sinestrari: Regularity of solutions of a class of linear integrodifferential equations in Banach spaces, J. Integral Equations 8(1985), 27-40.

[5] W. Desch and W. Schappacher: The semigroup approach to integrodifferential equations, to appear.

[6] G. Doetsch: Introduction to the theory and application of the Laplace transformation, Springer-Verlag, Berlin (1970).

[7] R. Grimmer and A. Pritchard: Analitic resolvent operators for integral equations in Banach space, J. Differential Equations 50(1983), 234-254.

[8] R. Grimmer and J. Pruss: On linear Volterra Equations in Banach spaces, Int. J. Comp. and Appl. Math. (to appear).

[9] E. Hille and R.S. Phillips: Functional analysis and semigroups, American Mathematical Society, Providence (1957).

ON FUCHSIAN HYPERBOLIC PARTIAL DIFFERENTIAL EQUATIONS

Hidetoshi TAHARA

Department of Mathematics
Sophia University
Kioicho, Chiyoda-ku, Tokyo 102, Japan

In this paper, I shall give some results concerning the following problem: construct a theory of " regular singularities " for partial differential equations.

In the case of ordinary differential equations, " regular singularities " have been completely investigated and the theory is one of the most beautiful theories in analysis (see Wasow [6]). I want to develop an analogous theory in the category of partial differential equations.

Notations to be remarked are as follows: $(t,x)=(t,x_1,\dots,x_n)\in \mathbb{R}\times\mathbb{R}^n$, $\mathbb{N}=\{1,2,3,\dots\}$, $\mathbb{Z}_+=\{0,1,2,\dots\}$ etc.

1. REGULAR SINGULARITIES IN C^∞ SPACES

First, we shall define regular singularities in the C^∞ sense. For an open subset U of $\mathbb{R}^n$, $u(t,x)\in C^\infty((0,T)\times U)$ and $\lambda(x)\in C^\infty(U)$, we define

$$u(t,x)=o(t^{\lambda(x)};\nabla^\infty) \quad \text{on } U \text{ (as } t\longrightarrow +0)$$

by the following: for any $j\in\mathbb{Z}_+$, $\alpha\in\mathbb{Z}_+^n$ and a compact subset K of U we have

$$(t\partial_t)^j\partial_x^\alpha(t^{-\lambda(x)}u(t,x))\big|_K \longrightarrow 0 \quad \text{uniformly on } K$$
$$(\text{as } t\longrightarrow +0).$$

Let $P=P(t,x,\partial_t,\partial_x)$ be a linear partial differential operator on $(0,T)\times\mathbb{R}^n$ with coefficients in $C^\infty((0,T)\times\mathbb{R}^n)$. Then, we define

Definition. We say that P has regular singularities on $\{t=0\}$, if P satisfies the following two conditions.

(1)(Tempered growth condition). If $u(t,x)\in C^\infty((0,T)\times\mathbb{R}^n)$ satisfies $Pu=0$, then we have $u(t,x)=o(t^{\lambda(x)};\nabla^\infty)$ on $\mathbb{R}^n$ (as $t\longrightarrow +0$) for some $\lambda(x)\in C^\infty(\mathbb{R}^n)$.

(2)(Non-flat condition). If $u(t,x)\in C^{\infty}((0,T)\times\mathbb{R}^n)$ satisfies $Pu=0$ and $u(t,x)=o(t^{\infty};\nabla^{\infty})$ on U (as $t\longrightarrow +0$) for some open subset U of $\mathbb{R}^n$, then we have $u(t,x)=0$ in a neighbourhood of $\{0\}\times U$ in $[0,T)\times\mathbb{R}^n$.

This is our definition of regular singularities in the C^{∞} sense. To make clear the meanings, we give a typical example.

Example. Let P be of the form

$$P = \partial_t + \alpha t^a \partial_x + \beta t^b,$$

where $(t,x)\in(0,T)\times\mathbb{R}$, $\alpha,\beta\in\mathbb{C}$, $\alpha\neq 0$, $\beta\neq 0$ and $a,b\in\mathbb{Z}$. Then, the following (i)$\sim$(iii) are equivalent to each other.

(i) P has regular singularities on $\{t=0\}$.

(ii) $\alpha\in\mathbb{R}$, $a\geqq 0$ and $b\geqq -1$.

(iii) tP is expressed in the form $tP=t\partial_t+\alpha t^k\partial_x+\beta t^h$ with $\alpha\in\mathbb{R}$, $k\geqq 1$ and $h\geqq 0$. (This means that tP is a Fuchsian hyperbolic operator in Section 2.)

Our interest lies in the following problem.

Problem 1. Characterize the class of operators with regular singularities on $\{t=0\}$.

In the general case, we don't know how to solve this problem. So, in the next section we will consider a typical model of operators with regular singularities, which is called a " Fuchsian hyperbolic operator in t ".

2. FUCHSIAN HYPERBOLIC OPERATORS

Let us consider

$$P = (t\partial_t)^m + \sum_{\substack{j+|\alpha|\leqq m\\ j<m}} a_{j,\alpha}(t,x)(t\partial_t)^j(t^k\partial_x)^{\alpha}, \tag{2.1}$$

where $(t,x)=(t,x_1,\ldots,x_n)\in[0,T]\times\mathbb{R}^n$, $m\in\mathbb{N}$, $\alpha\in\mathbb{Z}_+^n$, $a_{j,\alpha}(t,x)\in C^{\infty}([0,T]\times\mathbb{R}^n)$, $k=(k_1,\ldots,k_n)\in\mathbb{N}^n$, and

$$(t^k\partial_x)^{\alpha}=(t^{k_1}\partial_{x_1})^{\alpha_1}\cdots(t^{k_n}\partial_{x_n})^{\alpha_n}.$$

In addition, we impose the following condition:

Assumption (H). All the roots $\lambda_i(t,x,\xi)$ $(1\leqq i\leqq m)$ of

$$\lambda^m + \sum_{\substack{j+|\alpha|=m\\ j<m}} a_{j,\alpha}(t,x)\,\lambda^j\,\xi^{\alpha} = 0$$

are real, simple and bounded on $[0,T]\times\mathbb{R}^n\times\{\xi\in\mathbb{R}^n;\ |\xi|=1\}$.

This is the operator discussed from now on. We call this a " Fuchsian hyperbolic operator in t ". As to more general Fuchsian hyperbolic operators, see Tahara [2,4,5]. Since P is a Fuchsian type operator (in Baouendi-Goulaouic [1]), we can define the characteristic polynomial $C(\rho,x)$ by

$$C(\rho,x) = \rho^m + \sum_{j<m} a_{j,(0,\dots,0)}(0,x)\rho^j,$$

and the characteristic exponents $\rho_1(x),\dots,\rho_m(x)$ by the roots of $C(\rho,x)=0$.

3. BASIC RESULTS

The purpose of this section is to prove that our operator P in Section 2 has regular singularities on $\{t=0\}$ in our sense.

First, let us give a theorem of unique solvability of Pu=f. Let $\lambda(x),\ \mu(x)\in C^\infty(\mathbb{R}^n)$ such that

$$\max_{1\leq i\leq m} (\mathrm{Re}\ \rho_i(x)) < \mathrm{Re}\ \mu(x) < \mathrm{Re}\ \lambda(x) \quad \text{on } \mathbb{R}^n. \tag{3.1}$$

Then, we have

Theorem 1. Let P be as in Section 2, and let $\lambda(x),\ \mu(x)\in C^\infty(\mathbb{R}^n)$ be as in (3.1). Then, we have the following results.

(1) For any $f(t,x)\in C^\infty((0,T)\times\mathbb{R}^n)$ satisfying $f(t,x)=o(t^{\lambda(x)};\nabla^\infty)$ on $\mathbb{R}^n$ (as $t\longrightarrow +0$), there exists a unique solution $u(t,x)\in C^\infty((0,T)\times\mathbb{R}^n)$ of Pu=f such that $u(t,x)=o(t^{\mu(x)};\nabla^\infty)$ on $\mathbb{R}^n$ (as $t\longrightarrow +0$).

(2) The domain $D(t_0,x^0)$ defined by

$$D(t_0,x^0) = \left\{(t,x)\in(0,T)\times\mathbb{R}^n;\ |x^0-x| < \lambda_{max}T^{\kappa-1}(t_0-t)\right\}$$

(where $\lambda_{max}=\sup\{|\lambda_i(t,x,\xi)|;\ 1\leq i\leq m,\ (t,x)\in[0,T]\times\mathbb{R}^n,\ |\xi|=1\}$ and $\kappa=\max\{k_1,\dots,k_n\}$) is a dependence domain of $(t_0,x^0)\in(0,T)\times\mathbb{R}^n$. In other words, if $u(t,x)\in C^\infty((0,T)\times\mathbb{R}^n)$ satisfies Pu=0 in $D(t_0,x^0)$ and $u(t,x)=o(t^{\mu(x)};\nabla^\infty)$ on $U=\{x\in\mathbb{R}^n;(+0,x)\in D(t_0,x^0)\}$ (as $t\longrightarrow +0$), then $u(t,x)$ also satisfies $u(t,x)=0$ in $D(t_0,x^0)$.

Before the proof, we prepare some lemmas. For a compact subset K of $\mathbb{R}^n$, we write

$$C(K) = \left\{(t,x)\in(0,T)\times\mathbb{R}^n;\ \min_{y\in K}|x-y| \leq \lambda_{max}T^{\kappa-1}t\right\}.$$

Lemma 1. For any compact subset K of $\mathbb{R}^n$, there is an $A>0$ which satisfies the following condition. For any $f(t,x)\in C^\infty((0,T)\times\mathbb{R}^n)$ satisfying $\mathrm{supp}(f)\subset C(L)$ for some $L\subset K$ and $f(t,x)=o(t^a;\nabla^\infty)$ on $\mathbb{R}^n$ (as $t\longrightarrow +0$) for some $a>A$, there exists a unique solution $u(t,x)\in C^\infty((0,T)$

$\times\mathbb{R}^n$) *of* $Pu=f$ *such that* $\mathrm{supp}(u)\subset C(L)$ *and* $u(t,x)=o(t^b;\nabla^\infty)$ *on* $\mathbb{R}^n$ (*as* $t\longrightarrow +0$) *for any* $b<a$.

Proof. Assume that $A>0$ is sufficiently large. Let $f(t,x)$ be as in Lemma 1, and put $g(t,x)=t^{-a}f(t,x)$. Then, we have $g(t,x)\in C^\infty((0,T),H^\infty(\mathbb{R}^n))$, $(t\partial_t)^j g(t,x)\in C^0([0,T),H^\infty(\mathbb{R}^n))$ (for any $j\in\mathbb{Z}_+$) and $\mathrm{supp}(g)\subset C(L)$. Therefore, by applying Theorem 2.3 of Tahara [4] to

$$P(t^a v) = t^a g \tag{3.2}$$

we can find a solution $v(t,x)\in C^\infty((0,T),H^\infty(\mathbb{R}^n))$ of (3.2) such that $(t\partial_t)^j v(t,x)\in C^0([0,T),H^\infty(\mathbb{R}^n))$ (for any $j\in\mathbb{Z}_+$) and $\mathrm{supp}(v)\subset C(L)$. Thus, by putting $u(t,x)=t^a v(t,x)$ we obtain a desired solution $u(t,x)\in C^\infty((0,T)\times\mathbb{R}^n)$ of $Pu=f$. The uniqueness of solutions is proved in the same way from Theorem 2.3 of [4]. Q.E.D.

Lemma 2. *Let* $C(\rho,x)$ *be the characteristic polynomial of* P, *and let us consider*

$$C(t\partial_t,x)u = f. \tag{3.3}$$

Let $\lambda(x),\ \mu(x)\in C^\infty(\mathbb{R}^n)$ *be as in* (3.1). *Then, we have the following result. For any* $f(t,x)\in C^\infty((0,T)\times\mathbb{R}^n)$ *satisfying* $\mathrm{supp}(f)\subset C(K)$ *for some compact subset* K *of* $\mathbb{R}^n$ *and* $f(t,x)=o(t^{\lambda(x)};\nabla^\infty)$ *on* $\mathbb{R}^n$ (*as* $t\longrightarrow +0$), *there exists a unique solution* $u(t,x)\in C^\infty((0,T)\times\mathbb{R}^n)$ *of* (3.3) *such that* $\mathrm{supp}(u)\subset C(K)$ *and* $u(t,x)=o(t^{\mu(x)};\nabla^\infty)$ *on* $\mathbb{R}^n$ (*as* $t\longrightarrow +0$).

Proof. Since $C(t\partial_t,x)=(t\partial_t-\rho_1(x))\dots(t\partial_t-\rho_m(x))$ holds and since (3.3) is an ordinary differential equation, the proof of this lemma is easy. Note that the unique solution $u(t,x)$ is given by

$$u(t,x) = \int_0^1\cdots\int_0^1 E(s_1,\dots,s_m;x)f(s_1\dots s_m t,x)ds_1\dots ds_m,$$

where

$$E(s_1,\dots,s_m;x) = \frac{1}{m!}\sum_{\pi\in S_m} s_{\pi(1)}^{-\rho_1(x)-1}\cdots s_{\pi(m)}^{-\rho_m(x)-1}$$

and S_m is the permutation group of m-numbers. Q.E.D.

By combining Lemma 1 and Lemma 2, we have

Lemma 3. *Let* $\lambda(x),\ \mu(x)\in C^\infty(\mathbb{R}^n)$ *be as in* (3.1). *Then, for any* $f(t,x)\in C^\infty((0,T)\times\mathbb{R}^n)$ *satisfying* $\mathrm{supp}(f)\subset C(K)$ *for some compact subset* K *of* $\mathbb{R}^n$ *and* $f(t,x)=o(t^{\lambda(x)};\nabla^\infty)$ *on* $\mathbb{R}^n$ (*as* $t\longrightarrow +0$), *there exists a unique solution* $u(t,x)\in C^\infty((0,T)\times\mathbb{R}^n)$ *of* $Pu=f$ *such that* $\mathrm{supp}(u)\subset C(K)$ *and* $u(t,x)=o(t^{\mu(x)};\nabla^\infty)$ *on* $\mathbb{R}^n$ (*as* $t\longrightarrow +0$).

Proof. Note that P is decomposed into the form

$$P = C(t\partial_t,x)-tR(t,x,t\partial_t,\partial_x)$$

for some linear differential operator $R(t,x,\partial_t,\partial_x)$ with coefficients in $C^\infty([0,T]\times\mathbb{R}^n)$. For simplicity, we denote $R(t,x,t\partial_t,\partial_x)v$ by $R[v]$.

First, let us prove the existence part. Let $f(t,x)$ be as in Lemma 3, and let $A>0$ be the constant chosen in Lemma 1 corresponding to K. Take $N\in\mathbb{N}$ and $a>b>A$ so that

$$\min_{x\in K}(\mathrm{Re}\,\mu(x))+N > a > b > \max_{x\in K}(\mathrm{Re}\,\mu(x)). \tag{3.4}$$

Then, we can find a solution $u(t,x)$ of $Pu=f$ in the form

$$u = u_0+u_1+\dots+u_{N-1}+u_N \tag{3.5}$$

by solving

$$C(t\partial_t,x)u_0 = f, \qquad (3.6)_0$$

$$C(t\partial_t,x)u_1 = tR[u_0], \qquad (3.6)_1$$

....

....

....

$$C(t\partial_t,x)u_{N-1} = tR[u_{N-2}], \qquad (3.6)_{N-1}$$

$$Pu_N = tR[u_{N-1}] \qquad (3.6)_N$$

inductively under the following conditions:

$$\mathrm{supp}(u_i)\subset C(K) \qquad (3.7)_i$$

for $i=0,1,\dots,N$,

$$u_0(t,x)=o(t^{\mu(x)};\nabla^\infty) \text{ on } \mathbb{R}^n \text{ (as } t\longrightarrow+0) \qquad (3.8)_0$$

$$u_i(t,x)=o(t^{\mu(x)+i-\varepsilon};\nabla^\infty) \text{ on } \mathbb{R}^n \text{ (as } t\longrightarrow+0) \qquad (3.8)_i$$

for any $\varepsilon>0$ and $i=1,2,\dots,N-1$,

$$u_N(t,x)=o(t^b;\nabla^\infty) \text{ on } \mathbb{R}^n \text{ (as } t\longrightarrow+0). \qquad (3.8)_N$$

In fact, this is done as follows. Since $f(t,x)=o(t^{\lambda(x)};\nabla^\infty)$ on $\mathbb{R}^n$ (as $t\longrightarrow+0$), by Lemma 2 we have a solution $u_0(t,x)$ of $(3.6)_0$ such that $(3.7)_0$ and $(3.8)_0$ hold. Then, by $(3.8)_0$ we have $tR[u_0](t,x)=o(t^{\mu(x)+1-\varepsilon};\nabla^\infty)$ on $\mathbb{R}^n$ (as $t\longrightarrow+0$) for any $\varepsilon>0$. Therefore, by Lemma 2 we obtain a solution $u_1(t,x)$ of $(3.6)_1$ such that $(3.7)_1$ and $(3.8)_1$ hold. By repeating the same argument, we can obtain $u_2,\dots,u_{N-1}$ as above. Note that by $(3.8)_{N-1}$ we have $tR[u_{N-1}](t,x)=o(t^{\mu(x)+N-\varepsilon};\nabla^\infty)$ on $\mathbb{R}^n$ (as $t\longrightarrow+0$) for any $\varepsilon>0$ and therefore by (3.4) we have $tR[u_{N-1}](t,x)=o(t^a;\nabla^\infty)$ on $\mathbb{R}^n$ (as $t\longrightarrow+0$). Hence, by Lemma 1 we can obtain a solution $u_N(t,x)$ of $(3.6)_N$ such that $(3.7)_N$ and $(3.8)_N$ hold. Thus, we obtain a desired solution $u(t,x)$ of the form (3.5).

Next, let us prove the uniqueness part. Let $u(t,x)\in C^\infty((0,T)\times\mathbb{R}^n)$ such that $Pu=0$, $\mathrm{supp}(u)\subset C(K)$ and $u(t,x)=o(t^{\mu(x)};\nabla^\infty)$ on $\mathbb{R}^n$ (as $t\longrightarrow$

+0). Then, we have

$$C(t\partial_t,x)u = tR[u]$$

and $tR[u](t,x)=o(t^{\mu(x)+1-\varepsilon};\nabla^\infty)$ on $\mathbb{R}^n$ (as $t \longrightarrow +0$) for any $\varepsilon>0$. Therefore, by Lemma 2 we have $u(t,x)=o(t^{\mu(x)+1-\varepsilon};\nabla^\infty)$ on $\mathbb{R}^n$ (as $t \longrightarrow +0$) for any $\varepsilon>0$. This implies that $tR[u](t,x)=o(t^{\mu(x)+2-\varepsilon};\nabla^\infty)$ on $\mathbb{R}^n$ (as $t \longrightarrow +0$) for any $\varepsilon>0$. Hence, by Lemma 2 we have $u(t,x)=o(t^{\mu(x)+2-\varepsilon};\nabla^\infty)$ on $\mathbb{R}^n$ (as $t \longrightarrow +0$) for any $\varepsilon>0$. By repeating the same argument, we obtain

$$u(t,x)=o(t^{\mu(x)+N-\varepsilon};\nabla^\infty) \text{ on } \mathbb{R}^n \text{ (as } t \longrightarrow +0) \quad \text{for any } N \in \mathbb{N},$$

that is, $u(t,x)=o(t^\infty;\nabla^\infty)$ on $\mathbb{R}^n$ (as $t \longrightarrow +0$). Thus, by the uniqueness part of Lemma 1 we obtain $u(t,x)=0$. Q.E.D.

Proof of Theorem 1. Let $f(t,x) \in C^\infty((0,T)\times\mathbb{R}^n)$ such that $f(t,x)=o(t^{\lambda(x)};\nabla^\infty)$ on $\mathbb{R}^n$ (as $t \longrightarrow +0$). Let $\{\varphi_i(x)\}_{i=1}^\infty \subset C_0^\infty(\mathbb{R}^n)$ be a partition of unity on $\mathbb{R}^n$, and put $f_i(t,x)=\varphi_i(x)f(t,x)$. Then, by Lemma 3 we can find a solution $u_i(t,x) \in C^\infty((0,T)\times\mathbb{R}^n)$ of $Pu_i=f_i$ such that $\mathrm{supp}(u_i) \subset C(\mathrm{supp}(\varphi_i))$ and $u_i(t,x)=o(t^{\mu(x)};\nabla^\infty)$ on $\mathbb{R}^n$ (as $t \longrightarrow +0$). Hence, by putting $u(t,x)=\sum_{i=1}^\infty u_i(t,x)$ we obtain a desired solution of $Pu=f$. Thus, the existence part of (1) is proved.

To obtain the uniqueness, it is sufficient to prove (2). Let $u(t,x) \in C^\infty((0,T)\times\mathbb{R}^n)$ such that $Pu=0$ in $D(t_0,x^0)$ and $u(t,x)=o(t^{\mu(x)};\nabla^\infty)$ on $U=\{x \in \mathbb{R}^n;(+0,x) \in D(t_0,x^0)\}$ (as $t \longrightarrow +0$). Take any $s \in (0,t_0)$ and fix it. Put $U_s=\{x \in \mathbb{R}^n;(+0,x) \in D(s,x^0)\}$, choose $\nu(x)$, $\gamma(x) \in C^\infty(\mathbb{R}^n)$ so that

$$\max_{1\leq i\leq m} (\mathrm{Re}\, \rho_i(x)) < \mathrm{Re}\, \gamma(x) < \mathrm{Re}\, \nu(x) < \mathrm{Re}\, \mu(x) \quad \text{on } \mathbb{R}^n, \tag{3.9}$$

and let $\varphi(t,x) \in C^\infty([0,T)\times\mathbb{R}^n)$ such that $\varphi(t,x)=1$ in a neighbourhood of $\overline{D(s,x^0)}$ (in $[0,T)\times\mathbb{R}^n$) and that $\mathrm{supp}(\varphi) \subset D(t_0,x^0)\cup(\{0\}\times U)$. Put $g(t,x)=(P\varphi u)(t,x)$. Then, we have $\mathrm{supp}(g) \subset C(\overline{U}-U_s)$ and $g(t,x)=o(t^{\nu(x)};\nabla^\infty)$ on $\mathbb{R}^n$ (as $t \longrightarrow +0$). Therefore, by Lemma 3 we can find a solution $v(t,x) \in C^\infty((0,T)\times\mathbb{R}^n)$ of $Pv=g$ such that $\mathrm{supp}(v) \subset C(\overline{U}-U_s)$ and $v(t,x)=o(t^{\gamma(x)};\nabla^\infty)$ on $\mathbb{R}^n$ (as $t \longrightarrow +0$). Hence, by putting $w(t,x)=(\varphi u)(t,x)-v(t,x)$ we have $Pw=0$, $\mathrm{supp}(w) \subset C(\overline{U})$ and $w(t,x)=o(t^{\gamma(x)};\nabla^\infty)$ on $\mathbb{R}^n$ (as $t \longrightarrow +0$). Since $\gamma(x)$ satisfies (3.9), by the uniqueness part of Lemma 3 we obtain $w(t,x)=0$ on $(0,T)\times\mathbb{R}^n$. This implies that $u(t,x)=0$ in $D(s,x^0)$, because $w(t,x)=u(t,x)$ in $D(s,x^0)$. Since $s \in (0,t_0)$ is chosen arbitrarily, we can conclude that $u(t,x)=0$ in $D(t_0,x^0)$. Thus, (2) is proved. Q.E.D.

Denote by $C^{\infty}_{flat}((0,T)\times\mathbb{R}^n)$ the space of all functions $g(t,x)\in C^{\infty}((0,T)\times\mathbb{R}^n)$ satisfying $g(t,x)=o(t^{\infty};\nabla^{\infty})$ on $\mathbb{R}^n$ (as $t\longrightarrow+0$). Then, by putting $\lambda(x)=\infty$ and $\mu(x)=\infty$ in Theorem 1 we obtain

Corollary 1. *Pu=f is uniquely solvable in* $C^{\infty}_{flat}((0,T)\times\mathbb{R}^n)$.

Thus, by the uniqueness part we obtain

Corollary 2. *P satisfies " Non-flat condition ".*

Next, let us give a theorem on the tempered growth property of P. We say that $u(t,x)\in C^{\infty}((0,T)\times\mathbb{R}^n)$ is tempered (as $t\longrightarrow+0$), if $u(t,x)=o(t^{\lambda(x)};\nabla^{\infty})$ on $\mathbb{R}^n$ (as $t\longrightarrow+0$) for some $\lambda(x)\in C^{\infty}(\mathbb{R}^n)$. Then, we have

Theorem 2. *Let P be as in Section 2, and let* $u(t,x)\in C^{\infty}((0,T)\times\mathbb{R}^n)$. *Then, if* $(Pu)(t,x)$ *is tempered (as* $t\longrightarrow+0$), $u(t,x)$ *is also tempered (as* $t\longrightarrow+0$).

Note that this is a consequence of

Lemma 4. *Let* $u(t,x)\in C^{\infty}((0,T)\times\mathbb{R}^n)$ *such that* $\mathrm{supp}(u)\subset(0,T)\times K$ *for some compact subset K of* $\mathbb{R}^n$. *Then, if* $(Pu)(t,x)=o(t^{-a};\nabla^{\infty})$ *on* $\mathbb{R}^n$ *(as* $t\longrightarrow+0$) *for some* $a>0$, *we have* $u(t,x)=o(t^{-b};\nabla^{\infty})$ *on* $\mathbb{R}^n$ *(as* $t\longrightarrow+0$) *for some* $b>0$.

Proof of " Lemma 4 $\Longrightarrow$ Theorem 2 ". Let $u(t,x)\in C^{\infty}((0,T)\times\mathbb{R}^n)$ such that $(Pu)(t,x)$ is tempered (as $t\longrightarrow+0$). Our aim is to show the following: for any compact subset K of $\mathbb{R}^n$, there is a $b>0$ such that $u(t,x)=o(t^{-b};\nabla^{\infty})$ on K (as $t\longrightarrow+0$). For K, we choose compact subsets K_1, K_2 and L so that $K\subset K_1\subset K_2\subset L$, $((0,T)\times K)\cap C(K_2-\mathring{K}_1)=\phi$ and $C(K_2)\subset(0,T)\times L$. Let $\varphi(x)\in C^{\infty}_0(\mathbb{R}^n)$ such that $\varphi(x)=1$ in a neighbourhood of K_1 and $\mathrm{supp}(\varphi)\subset K_2$. Put $g(t,x)=[P,\varphi]u(t,x)$. Then, we have $\mathrm{supp}(g)\subset K_2-\mathring{K}_1$. Since P is a strictly hyperbolic operator in $(0,T)\times\mathbb{R}^n$, we can solve

$$\begin{cases} Pv = g(t,x), \\ \partial_t^i v\big|_{t=t_0} = 0, \quad i=0,1,\dots,m-1 \end{cases} \tag{3.10}$$

(where $0<t_0<T$) and obtain a solution $v(t,x)\in C^{\infty}((0,T)\times\mathbb{R}^n)$ of (3.10) such that $\mathrm{supp}(v)\subset(0,T)\times L$ and $\mathrm{supp}(v)\cap((0,T)\times K)=\phi$. Here, we put $w(t,x)=(\varphi u)(t,x)-v(t,x)$. Then,

$$\begin{aligned} Pw &= P\varphi u-Pv \\ &= \varphi Pu+[P,\varphi]u-g \\ &= \varphi Pu \end{aligned}$$

and $\mathrm{supp}(w)\subset(0,T)\times L$. Therefore, we have $(Pw)(t,x)=o(t^{-a};\nabla^{\infty})$ on $\mathbb{R}^n$ (as $t\longrightarrow+0$) for some $a>0$. Hence, by Lemma 4 we obtain $w(t,x)=o(t^{-b};\nabla^{\infty})$ on $\mathbb{R}^n$ (as $t\longrightarrow+0$) for some $b>0$. Thus, we can conclude that $u(t,x)=o(t^{-b};\nabla^{\infty})$ on K (as $t \longrightarrow +0$), because $w(t,x)=u(t,x)$ on $(0,T)\times K$. Q.E.D.

The proof of Lemma 4 is done by the L^2-argument, that is, Lemma 4 is a consequence of the following lemma.

Lemma 5. *Let* $u(t,x) \in C^{\infty}((0,T)\times\mathbb{R}^n)$ *such that* $\mathrm{supp}(u) \subset (0,T)\times K$ *for some compact subset* K *of* $\mathbb{R}^n$. *Then, if*

$$\|(t\partial_t)^j \partial_x^{\alpha}(Pu)(t)\| = o(t^{-a}) \ (\text{as } t \longrightarrow +0)$$

for any $j \in \mathbb{Z}_+$ *and* $\alpha \in \mathbb{Z}_+^n$

for some $a>0$, *we have*

$$\|(t\partial_t)^j \partial_x^{\alpha} u(t)\| = o(t^{-b}) \ (\text{as } t \longrightarrow +0)$$

for any $j \in \mathbb{Z}_+$ *and* $\alpha \in \mathbb{Z}_+^n$

for some $b>0$, *where* $\|\cdot\|$ *is the* L^2-*norm on* $\mathbb{R}^n$ *and* $\phi(t)=o(t^{-s})$ (*as* $t \longrightarrow +0$) *means that* $t^s\phi(t)$ *converges to zero* (*as* $t \longrightarrow +0$).

In the case of homogeneous solutions (that is, $Pu=0$), Lemma 5 is already proved in Proposition 5 of Tahara [5]. The inhomogeneous case can be proved in the same way as Proposition 5 of [5]. So, we omit the details.

As a corollary to Theorem 2, we have

Corollary 3. P *satisfies " Tempered growth condition "*.

Thus, by Corollaries 2 and 3 we have

Conclusion. *Our Fuchsian hyperbolic operator* P *has regular singularities on* $\{t=0\}$.

Remark. The proofs of Theorems 1 and 2 are based on the study of abstract singular differential equations of the form

$$t\frac{du}{dt}+tA(t)u+B(t)u = f(t), \quad 0<t<T,$$

where u, f(t), A(t) and B(t) satisfy the following conditions: (i) $u=u(t)$ and f(t) are functions on (0,T) with values in $L^2(\mathbb{R}^n)$, (ii) A(t) is a closed linear operator in $L^2(\mathbb{R}^n)$, (iii) $A(t)+A(t)^*$ is a bounded operator in $L^2(\mathbb{R}^n)$, and (iv) B(t) is a bounded linear operator in $L^2(\mathbb{R}^n)$. For details, see Tahara [3].

4. ASYMPTOTIC EXPANSIONS

In the case of ordinary differential equations, the singularity of solutions near a regular singularity is characterized explicitly by means of asymptotic expansions. Here, we want to show the similar characterization of the singularity of solutions for our Fuchsian hyperbolic operator P, by combining Theorem 1, Theorem 2 and a formal discussion (Lemma 6).

Theorem 3. Let P be as in Section 2, and assume that $\rho_i(x)-\rho_j(x)\notin \mathbb{Z}$ holds for any $x\in\mathbb{R}^n$ and $1\leq i\neq j\leq m$. Then, we have the following results.

(1) Any solution $u(t,x)\in C^\infty((0,T)\times\mathbb{R}^n)$ of $Pu=0$ can be expanded asymptotically into the form

$$u(t,x) \sim \sum_{i=1}^{m}\Big(\varphi_i(x)t^{\rho_i(x)}+\sum_{k=1}^{\infty}\sum_{h=0}^{k}\varphi_{i,k,h}(x)t^{\rho_i(x)+k}(\log t)^{k-h}\Big) \tag{4.1}$$

(as $t\longrightarrow +0$) for some unique $\varphi_i(x)$, $\varphi_{i,k,h}(x)\in C^\infty(\mathbb{R}^n)$.

(2) Conversely, for any $\varphi_1(x),\ldots,\varphi_m(x)\in C^\infty(\mathbb{R}^n)$ there exist a unique solution $u(t,x)\in C^\infty((0,T)\times\mathbb{R}^n)$ of $Pu=0$ and unique coefficients $\varphi_{i,k,h}(x)\in C^\infty(\mathbb{R}^n)$ such that the asymptotic relation in (1) holds.

Here, the meaning of the asymptotic relation (4.1) is as follows: for any $a>0$ and any compact subset K of $\mathbb{R}^n$, there is an $N_0\in\mathbb{N}$ such that

$$(t\partial_t)^j\partial_x^\alpha R_N(t,x)\big|_K \longrightarrow 0 \quad \text{uniformly on } K$$
$$(\text{as } t\longrightarrow +0)$$

for any $j\in\mathbb{Z}_+$, $\alpha\in\mathbb{Z}_+^n$ and $N\geqq N_0$, where $R_N(t,x)$ is defined by

$$R_N(t,x) = u(t,x)-\sum_{i=1}^{m}\Big(\varphi_i(x)t^{\rho_i(x)}+\sum_{k=1}^{N}\sum_{h=0}^{k}\varphi_{i,k,h}(x)t^{\rho_i(x)+k}(\log t)^{k-h}\Big).$$

Before the proof, we note the following lemma.

Lemma 6. Let P be as in (2.1) (here, the assumption (H) is not necessary), and assume that $\rho_i(x)-\rho_j(x)\notin\mathbb{Z}$ holds for any $x\in\mathbb{R}^n$ and $1\leqq i\neq j\leqq m$. Let $u(t,x)\in C^\infty((0,T)\times\mathbb{R}^n)$ such that $Pu=0$. Then, the following (i) and (ii) are equivalent to each other.

(i) $u(t,x)$ is tempered (as $t\longrightarrow +0$).

(ii) $u(t,x)$ is expanded asymptotically into the form (4.1).

Proof. (i) $\Longrightarrow$ (ii) is proved in the same way as Theorem 1 of Tahara [5]. (ii) $\Longrightarrow$ (i) is clear from the definition. Q.E.D.

Proof of Theorem 3. (1) is clear from Corollary 3 and Lemma 6. (2) is proved as follows. For any $\varphi_1(x),\ldots,\varphi_m(x)\in C^\infty(\mathbb{R}^n)$, we can construct a formal solution

$$\hat{u}(t,x) = \sum_{i=1}^{m}\left(\varphi_i(x)t^{\rho_i(x)} + \sum_{k=1}^{\infty}\sum_{h=0}^{k}\varphi_{i,k,h}(x)t^{\rho_i(x)+k}(\log t)^{k-h}\right)$$

and the coefficients $\varphi_{i,k,h}(x) \in C^{\infty}(\mathbb{R}^n)$ are uniquely determined. Moreover, we can construct a function $w(t,x) \in C^{\infty}((0,T)\times\mathbb{R}^n)$ such that $w(t,x) \sim \hat{u}(t,x)$ (as $t \longrightarrow +0$). Here, we consider

$$Pv = (Pw)(t,x), \quad v=\text{unknown}. \tag{4.2}$$

Since $(Pw)(t,x)=o(t^{\infty};\nabla^{\infty})$ on $\mathbb{R}^n$ (as $t \longrightarrow +0$), by Corollary 1 we can find a solution $v(t,x) \in C^{\infty}((0,T)\times\mathbb{R}^n)$ of (4.2) such that $v(t,x)=o(t^{\infty};\nabla^{\infty})$ on $\mathbb{R}^n$ (as $t \longrightarrow +0$). Thus, by putting $u(t,x)=w(t,x)-v(t,x)$ we obtain (2). See also Tahara [5]. Q.E.D.

5. CONCLUDING REMARKS

Finally, let us give some remarks. Put

$\mathcal{P}_R$ = the class of operators with regular singularities on $t=0$,

$\mathcal{P}_A$ = the class of operators for which Theorem 3 is valid.

Then, $\mathcal{P}_R \supset \mathcal{P}_A$ is clear. In Sections 2∼4, we have discussed a very simple example of Fuchsian hyperbolic operators. In general, we say that an operator P is a Fuchsian hyperbolic operator in t, if P satisfies the following two conditions: (i) P is a Fuchsian type operator in t (in the sense of Baouendi-Goulaouic [1]), and (ii) P satisfies some hyperbolicity condition (see Tahara [2,4,5]). Put

$\mathcal{P}_F$ = the class of Fuchsian hyperbolic operators in this generalized sense.

(The definition of $\mathcal{P}_F$ is not mathematical, because the naming of " Fuchsian hyperbolic operators " is vague and flexible. The author wishes to explain his thought only.) Then, by Tahara [5] we can say that $\mathcal{P}_A \supset \mathcal{P}_F$ holds at least in the spirit. Therefore, we have

$$\mathcal{P}_R \supset \mathcal{P}_A \supset \mathcal{P}_F.$$

Thus, the next problem to be attacked is summarized as follows.

Problem 2. Make clear the following parts.

(1) Whether $\mathcal{P}_R = \mathcal{P}_A$ or $\mathcal{P}_R \supsetneqq \mathcal{P}_A$?

(2) Whether $\mathcal{P}_A = \mathcal{P}_F$ or $\mathcal{P}_A \supsetneqq \mathcal{P}_F$?

REFERENCES

[1] M. S. Baouendi and C. Goulaouic : Cauchy problems with characteristic initial hypersurface, Comm. Pure Appl. Math., 26 (1973), 455-475.

[2] H. Tahara : Fuchsian type equations and Fuchsian hyperbolic equations, Japan. J. Math., New Ser. 5 (1979), 245-347.

[3] ———— : Singular hyperbolic systems, I. Existence, uniqueness and differentiability, J. Fac. Sci. Univ. Tokyo Sect. IA Math., 26 (1979), 213-238.

[4] ———— : Singular hyperbolic systems, III. On the Cauchy problem for Fuchsian hyperbolic partial differential equations, J. Fac. Sci. Univ. Tokyo Sect. IA Math., 27 (1980), 465-507.

[5] ———— : Singular hyperbolic systems, V. Asymptotic expansions for Fuchsian hyperbolic partial differential equations, J. Math. Soc. Japan, 36 (1984), 449-473.

[6] W. Wasow : Asymptotic expansions for ordinary differential equations, Interscience, 1965.

GLOBAL SOLUTIONS TO EVOLUTION EQUATIONS OF PARABOLIC TYPE

Wolf von Wahl
Mathematisches Institut
der Universität
Postfach 3008
8580 Bayreuth
Fed. Rep. of Germany

0. Introduction

We deal with the question of existence of global classical solutions to equations of type $u' + Au + M(u) = 0$, $u(0) = \varphi$, in a reflexive Banach space B. $-A$ generates an analytic semigroup e^{-tA}, $t \geq 0$, M is a nonlinear term which satisfies a suitable Lipschitz condition. Moreover A is supposed to be positive, i.e. the spectrum of A is contained in a set $\{\lambda | \mathrm{Re}\ \lambda \geq \delta\}$ for some $\delta > 0$. It is easy to show the existence of a local (in time) classical solution on a maximal interval of existence $[0, T(\varphi))$ with $0 < T(\varphi) \leq +\infty$, but it is difficult to give abstract conditions for the existence of global (in time) classical solutions which turn out to be sharp in the applications. The latter is our main concern here. We introduce a second Banach space V with $D(A) \subset V$ and essentially assume that $M(u)$ has the same order of magnitude with respect to V as Au; this means that $M(u)$ fulfills an estimate

$$(0.1) \quad \|M(u)\| \leq (\|Au\| + 1)\, g(\|u\|_V), \quad u \in D(A),$$

with a continuous function g. In the applications V plays the role of the space where a sufficiently nice solution can be estimated a-priori. Also a Lipschitz condition corresponding to (0.1) is supposed to hold. Our result then is as follows: If $T(\varphi)$ is finite then the mapping

$$u: [0, T(\varphi)) \to V$$

is not uniformly continuous. Thus the local solution may cease to exist by oscillating instead of tending to $+\infty$ in the V-norm as $T(\varphi)$ is approached. Let us remark that it is easy to show global existence if $M(u)$ is subordinate to Au, i.e. an estimate of the form

$$\|M(u)\| \leq (\|A^{1-\varepsilon} u\| + 1)\, g(\|u\|_V)$$

holds for some $\varepsilon > 0$. The applications show that our abstract access is sharp; it covers e.g. the case of second order parabolic equations $u' + Au + f(u,\nabla u) = 0$ (A elliptic) where f has quadratical growth with respect to ∇u.

I. The Abstract Theorem

Our first assumption is that M is a mapping from $D(A^{1-\rho})$ into B for some $\rho \in (0,1)$ which fulfills the Lipschitz condition

$$(I.1) \quad \|M(u)-M(v)\| \leq k(\|A^{1-\rho}u\| + \|A^{1-\rho}v\|)\cdot\|A^{1-\rho}(u-v)\|, \qquad u,v \in D(A^{1-\rho})$$

where k is continuous function from the nonnegative reals $\mathbb{R}^+$ into itself. Then the following well known theorem holds:

<u>Theorem I.1:</u> <u>Let</u> $\varphi \in D(A)$. <u>Then there is a number</u> $T(\varphi)$, $0 < T(\varphi) \leq +\infty$, <u>with the following properties</u>: <u>There is one and only one</u> $u \in C^1([0,T(\varphi)),B)$ <u>with</u>

$$u(t) \in D(A), \quad 0 \leq t < T(\varphi),$$

$$Au \in C^0([0,T(\varphi)),B),$$

$$u' + Au + M(u) = 0,$$

$$u(0) = \varphi.$$

<u>If</u> $T(\varphi) < +\infty$ <u>then</u>

$$\lim_{t\uparrow T(\varphi)} \|A^{1-\rho}u(t)\| = +\infty.$$

<u>Proof:</u> The proof rests on the consideration of the integral equation

$$u(t) = e^{-tA}\varphi - \int_0^t e^{-(t-s)A}M(u(s))\,ds$$

and is well known ([F, Part 2, 16], [W1, pp. 136-139]). □

Now we want to give an abstract criterion how to decide if $T(\varphi) = +\infty$. For this purpose we need some additional assumptions. As for the nonlinearity they are as follows: Let there be given a Banach space V with

$D(A) \subset V$ algebraically and topologically such that

(I.2) $\|M(u)\| \leq (\|Au\|+1)\cdot g_1(\|u\|_V)$, $u \in D(A)$,

with a continuous function $g_1 : \mathbb{R}^+ \to \mathbb{R}^+$. Moreover, let the following Lipschitz condition be fulfilled:

(I.3) $\|M(u)-M(v)\| \leq \|A(u-v)\| g_2(\|u-v\|_V, \|u\|_V + \|v\|_V) + R(u,v)$,

$$u,v \in D(A).$$

$g_2 : \mathbb{R}^+ \times \mathbb{R}^+ \to \mathbb{R}^+$ is continuous and has the property

$$g_2(r,s) \to 0, \quad r \to 0, \quad s \in \mathbb{R}^+.$$

$R(u,v)$ is a remainder of lower order having the form

$$R(u,v) = (\|A(u-v)\|^{1-\varepsilon}+1)\cdot g_3(\|Av\|, \|u\|_V + \|v\|_V)$$

for some $\varepsilon > 0$ and some continuous $g_3 : \mathbb{R}^+ \times \mathbb{R}^+ \to \mathbb{R}^+$. It is important that only one of the u,v enters in terms of the graph norm of A.

As for the linear part we assume that for any $u \in C^1([0,T],B)$ with $u(t) \in D(A)$, $0 \leq t \leq T$, $Au \in C^0([0,T],B)$ the estimate

(I.4) $$\int_0^T \|u'(t)\|^q \, dt + \int_0^T \|Au(t)\|^q \, dt \leq c(T,q)\left(\|Au(0)\|^q + \int_0^T \|f(t)\|^q \, dt\right)$$

holds for all $q \geq 2$; here we have set $f(t) = u' + Au(t)$. T is any positive number, $c(T,q)$ may be considered as bounded on bounded subsets of $T \geq 0$, $q \geq 2$.

Let us make a few remarks on our assumptions. In the applications, (I.1) is trivial, (I.3) is always an easy consequence of (I.2). (I.4) is known to be true for $q = 2$ if B is a Hilbert space; this follows by using the Fourier transformation for Hilbert space valued functions; for a different approach, see [W2, I, II]. It turns out that (I.4) then holds for any $q > 1$, see [W2, pp. 492-494]. If $B = L^p(\Omega)$ for some $p > 1$, if $\Omega \subset \mathbb{R}^n$ is a smooth bounded open set and if A is a positive elliptic operator then (I.4) is known to be true for $q = p$ (see [LUS, IV. and VII.]); it then follows that (I.4) is true for any $q > 1$, see again [W2, pp. 492-494].

We now can formulate very briefly our main abstract result.

Theorem I.2: *Let the assumptions* (I.1)-(I.4) *be fulfilled. Let* u *be the solution of* $u' + Au + M(u) = 0$, $u(0) = \varphi$, *on* $[0,T(\varphi))$ *which has been constructed in Theorem* I.1. *If* $T(\varphi) < +\infty$, *then*

$$u:[0,T(\varphi)) \to V$$

is not uniformly continuous.

Proof: Set $c(q) = c(2T(\varphi),q)$. We assume that $T(\varphi) < +\infty$ and that

$$u:[0,T(\varphi)) \to V$$

is uniformly continuous. We divide $[0,T(\varphi)]$ into possible small intervals of constant length:

$$[0,T(\varphi)] = \bigcup_{j=1}^{\hat{N}} [t_j,t_{j+1}],$$

$t_1 = 0$, $t_{\hat{N}+1} = T(\varphi)$. $|t_{j+1}-t_j| = \delta$ will be made small later on. For $\tilde{T} = \delta\cdot\tilde{N} = t_{\tilde{N}+1}$ we get

$$\int_0^{\tilde{T}} \|u'(t)\|^q\, dt + \int_0^{\tilde{T}} \|Au(t)\|^q\, dt =$$

$$= \sum_{j=1}^{\tilde{N}} \left(\int_{t_j}^{t_{j+1}} \|u'(t)\|^q\, dt + \int_{t_j}^{t_{j+1}} \|Au(t)\|^q\, dt\right),$$

$$\leq c(q) \sum_{j=1}^{\tilde{N}} \left(\int_{t_j}^{t_{j+1}} \|M(t,u(t))\|^q\, dt + \|A\varphi\|^q\right),$$

$$\leq c(q) \sum_{j=1}^{\tilde{N}} \int_{t_j}^{t_{j+1}} \|M(t,u(t))-M(t_j,u(t_j))\|^q\, dt +$$

$$+ c(q)\left(\sum_{j=1}^{\tilde{N}} \int_{t_j}^{t_{j+1}} \|M(t_j,u(t_j))\|^q\, dt + \|A\varphi\|^q\right),$$

$$\leq c(q)\left\{\sum_{j=1}^{\tilde{N}} \left(\int_{t_j}^{t_{j+1}} \|A(u(t)-u(t_j))\|^q g_2^q(\|u(t)-u(t_j)\|_V,\|u(t)\|_V+\|u(t_j)\|_V)\, dt\right.\right.$$

$$+ \int_{t_j}^{t_{j+1}} (\|A(u(t)-u(t_j))\|^{1-\varepsilon}+1)^q \cdot g_3^q(\|Au(t_j)\|, \|u(t)\|_V + \|u(t_j)\|_V)\ dt +$$

$$+ \int_{t_j}^{t_{j+1}} \|M(t_j,u(t_j))\|^q\ dt) + \|A\varphi\|^q)\Big\},$$

$$\leq c(q)\Big\{\sum_{j=1}^{\tilde{N}} (\int_{t_j}^{t_{j+1}} \|Au(t)\|^q g_2^q(\|u(t)-u(t_j)\|_V, \|u(t)\|_V + \|u(t_j)\|_V)\ dt +$$

$$+ \int_{t_j}^{t_{j+1}} \|Au(t_j)\|^q g_2^q(\|u(t)-u(t_j)\|_V, \|u(t)\|_V + \|u(t_j)\|_V)\ dt +$$

$$+ \varepsilon' \int_{t_j}^{t_{j+1}} (\|Au(t)\|+1)^q g_3^q(\|Au(t_j)\|, \|u(t)\|_V + \|u(t_j)\|_V)\ dt +$$

$$+ c(\varepsilon') \int_{t_j}^{t_{j+1}} (\|Au(t_j)\|+1)^q g_3^q(\|Au(t_j)\|, \|u(t)\|_V + \|u(t_j)\|_V)\ dt +$$

$$+ \int_{t_j}^{t_{j+1}} g_1^q(\|u(t_j)\|_V)(\|Au(t_j)\|+1)^q\ dt) + \|A\varphi\|^q)\Big\},\quad 0 < \varepsilon'.$$

Since $u:[0,T(\varphi)) \to V$ is uniformly continuous, u can be continued into $T(\varphi)$ in a unique way as a continuous function. Thus

$$\sup_{0 \leq t < T(\varphi)} \|u(t)\|_V \leq D.$$

Choosing δ so small that

$$\text{(I.5)}\qquad \sup_{\substack{|t-t_j| \leq \delta, \\ 1 \leq j \leq N, \\ 0 \leq t < T(\varphi), \\ 0 \leq s \leq 2D}} c(q) g_2^q(\|u(t)-u(t_j)\|_V, s) \leq \frac{1}{4}$$

and choosing

$$\varepsilon' = \frac{1}{4c(q)} \cdot \frac{1}{\sup\limits_{0 \leq s \leq 2D} g_3^q(\|Au(t_j)\|, s) + 1},$$

we see that

$$\int_{t_{\tilde{N}}}^{t_{\tilde{N}+1}} \|u'(t)\|^q \, dt + \int_{t_{\tilde{N}}}^{t_{\tilde{N}+1}} \|Au(t)\|^q \, dt, \ q \geq 2,$$

is estimated a priori in dependence of q and

$$D, \sup_{0 \leq t \leq t_{\tilde{N}}} \|Au(t)\|.$$

We want to emphasize that the smallness of δ needed in (I.5) does not depend on $t_{\tilde{N}}$. Now we use the integral representation for our local strong solution. First we get in the usual way that

$$\sup_{0 \leq t \leq t_{\tilde{N}}} \|Au(t)\| \leq c(\varphi),$$

secondly we have, using the equation,

$$\|A^{1-\rho}u(t)\| \leq \|A^{1-\rho}u(t_{\tilde{N}})\| + \int_{t_{\tilde{N}}}^{t} \frac{c}{(t-s)^{1-\rho}} \|M(u(s))\| \, ds,$$

$$\leq \|A^{1-\rho}u(t_{\tilde{N}})\| + c\Big(\int_{t_{\tilde{N}}}^{t} \|M(u(s))\|^q \, ds\Big)^{\frac{1}{q}},$$

$$\leq \|A^{1-\rho}u(t_{\tilde{N}})\| + c\Big(\int_{t_{\tilde{N}}}^{t_{\tilde{N}+1}} \|u'(\tilde{t})\|^q \, d\tilde{t} + \int_{t_{\tilde{N}}}^{t_{\tilde{N}+1}} \|Au(\tilde{t})\|^q \, d\tilde{t}\Big)^{\frac{1}{q}},$$

$t_{\tilde{N}} \leq t < t_{\tilde{N}+1}$, $q > \frac{1}{\rho}$. The last two integrals, however, have already been estimated. Thus proceeding stepwise we see that $\|A^{1-\rho}u(t)\|$ stays bounded if one approaches $T(\varphi)$ from below. Theorem I.1 shows that $T(\varphi)$ is not the maximal interval of existence. Theorem I.2 is proved. □

II. Examples

Our <u>first</u> example is a second order nonlinear parabolic equation: Let Ω be a smooth bounded open set of $\mathbb{R}^n$. Let there be given real functions $a_{ij} \in C^0(\overline{\Omega})$, $1 \leq i,j \leq n$, with

$$\sum_{1 \leq i,j \leq n} a_{ij}(x)\xi_i\xi_j \geq c|\xi|^2, \ x \in \overline{\Omega}, \ \xi \in \mathbb{R}^n,$$

for some $c>0$. Let $f:\mathbb{R}\times\mathbb{R}^n$ be a C^1-function satisfying the following growth condition:

$$|f(u,p)| \leq \mu(|u|)(|p|^2+1),\ u\in\mathbb{R},\ p\in\mathbb{R}^n,$$

and the sign condition

$$\frac{\partial f}{\partial u}(u,p) \geq -c,\ u\in\mathbb{R},\ p\in\mathbb{R}^n;$$

here μ is a monotonically increasing function from the nonnegative reals $\mathbb{R}^+$ into itself and c is again some positive constant. As Banach space B we choose $L^p(\Omega)$ with some $p>n+1$ and we consider the equation

$$u_t(t,x) - \sum_{1\leq i,j\leq n} a_{ij}(x)u_{x_ix_j}(t,x) + f(u(t,x),\nabla u(t,x)) = 0$$

$$u(0,x) = \varphi(x),\quad x\in\overline{\Omega},$$

$$u(t,x) = 0,\quad t\geq 0,\ x\in\partial\Omega,$$

as a nonlinear evolution equation

$$u' + Au + M(u) = 0,$$
$$u(0) = \varphi$$

in B where we have set

$$Au = -\sum_{1\leq i,j\leq n} a_{ij}(x)u_{x_ix_j},$$

$$u\in D(A) = H^{2,p}(\Omega)\cap \overset{\circ}{H}{}^{1,p}(\Omega).$$

Possibly after adding a term $\tilde{c}u$ ($\tilde{c}>0$) to Au we obtain an operator satisfying the assumptions in the beginning of 0. Since $\tilde{c}u$ may be subtracted from $f(u,\nabla u)$ without destroying the assumptions on the nonlinearity it is no loss of generality to assume that A is positive and that $-A$ generates an analytic semigroup. If $\rho\in(0,1)$ is sufficiently small then

$$D(A^{1-\rho}) \subset C^{1+\alpha}(\overline{\Omega})$$

for some $\alpha\in(0,1)$ (see [F, pp. 159, 177-178]). It is now almost trivial to show that

$$M(u) = f(u,\nabla u)$$

satisfies (I.1). The initial value φ is, for sake of simplicity, supposed to be in D(A). Then Theorem I.1 provides a local strong solution on a maximal interval of existence $[0,T(\varphi))$. We want to show now that $T(\varphi) = +\infty$. We have to check the assumptions (I.2), (I.3), (I.4). We choose $V = C^0(\overline{\Omega})$. Then, by the Gagliardo-Nirenberg inequality,

$$\|M(u)\| \leq \mu(\|u\|_{C^0(\overline{\Omega})})\cdot c(\|u\|_{H^{2,p}(\Omega)}\|u\|_{C^0(\overline{\Omega})}+1).$$

Since $c\|u\|_{H^{2,p}(\Omega)} \geq \|Au\| \geq \frac{1}{c}\|u\|_{H^{2,p}(\Omega)}$ for some $c > 0$, the assumption (I.2) is in fact fulfilled. (I.3) is a consequence of the triangle inequality:

$$\begin{aligned}
\|M(u)-M(v)\| &\leq \|M(u)\| + \|M(v)\|, \\
&\leq c\mu(\|u\|_{C^0(\overline{\Omega})})(\|\nabla u\|^2_{L^{2p}(\Omega)}+1) + c\mu(\|v\|_{C^0(\overline{\Omega})})(\|Av\|^2+1), \\
&\leq c\mu(\|u\|_{C^0(\overline{\Omega})})(\|\nabla(u-v)\|^2_{L^{2p}(\Omega)}+1) + c\mu(\|u\|_{C^0(\overline{\Omega})})\cdot \\
&\qquad \cdot\|\nabla v\|^2_{L^{2p}(\Omega)} + c\mu(\|v\|_{C^0(\overline{\Omega})})(\|Av\|^2+1), \\
&\leq c\mu(\|u\|_{C^0(\overline{\Omega})})\|A(u-v)\|\cdot\|u-v\|_{C^0(\overline{\Omega})} + (c\mu(\|u\|_{C^0(\overline{\Omega})})+ \\
&\qquad +c\mu(\|v\|_{C^0(\overline{\Omega})})+1)(\|Av\|^2+1),
\end{aligned}$$

where we have applied again the Gagliardo-Nirenberg inequality. Thus (I.3) is also fulfilled. (I.4) is well known for $q = p$ (see [LUS, IV]), but according to [W2] the estimate in question then follows for any $q > 1$. In view of Theorem I.2 we now can proceed as follows: Assume that $T(\varphi) < +\infty$. If we can show that $u:[0,T(\varphi)) \to C^0(\overline{\Omega})$ is uniformly continuous, then this is a contradiction to our assumption. Let h be a sufficiently small positive number. We consider the function

$$v(t) = u(t+h),\ 0 \leq t < T(\varphi)-h,$$

which fulfills

$$v' + Av + M(v) = 0,$$
$$v(0) = u(h)$$

on $[0,T(\varphi)-h)$. In view of $\frac{\partial f}{\partial u} \geq -c$ we can apply the classical maximum principle on u-v (cf. [Ki]) which gives

$$\|u(t)-u(t+h)\|_{C^0(\bar\Omega)} \le c\|\varphi-u(h)\|_{C^0(\bar\Omega)} ;$$

this is the desired uniform continuity, and thus we have shown that $T(\varphi) = +\infty$.

The <u>second</u> example stems from the field of parabolic equations with elliptic part of higher order. It has the form

$$u' + Au + M(u) = 0,$$
$$u(0) = \varphi, \ \varphi \text{ real},$$

where $Au = \sum_{\substack{|\alpha|\le m,\\ |\beta|\le m}} D^\alpha(A_{\alpha\beta}(x)D^\beta u)$ with

$$\sum_{\substack{|\alpha|=m,\\ |\beta|=m}} A_{\alpha\beta}(x)\xi^{\alpha+\beta} \ge c|\xi|^{2m}, \ x\in\bar\Omega, \ \xi\in\mathbb{R}^n$$

for some $c > 0$. Ω is as before. The $A_{\alpha\beta}\cdot(i)^{|\alpha|+|\beta|}$ are of class $C^{|\alpha|}(\bar\Omega)$ and real valued, A is assumed to be formally selfadjoint. Observe that $D^\alpha = \prod_{j=1}^{n} (\frac{1}{i}\frac{\partial}{\partial x_j})^{\alpha_j}$, where $\alpha = (\alpha_1,\dots,\alpha_n)$. The underlying Banach space $B = L^p(\Omega)$ for some $p > n+1$. We set

$$Au = \sum_{\substack{|\alpha|\le m,\\ |\beta|\le m}} D^\alpha(A_{\alpha\beta}D^\beta u),$$

$$u \in D(A) = H^{2m,p}(\Omega) \cap \overset{\circ}{H}{}^{m,p}(\Omega).$$

It is well known then that -A generates an analytic semigroup e^{-tA} in B. The nonlinearity M(u) is of the form f(u) with a real function f. f is supposed to be of class C^1 and to have a principal function $F(r) = \int_0^r f(s)\,ds$ with $F(r) \ge -cr^2$ for some $c > 0$. Moreover we assume that $n > 2m$ and that

$$f(u)\cdot u \le c(|u|^{\frac{n+2m}{n-2m}+1} + 1),$$

$$f(u)\cdot u \ge -c(|u|^{\frac{n+2m}{n-2m}+1-\varepsilon} + 1), \ |u| \ge 1,$$

for some $c>0$ and some $\varepsilon>0$. If we consider $(A+\tilde{c})u$ for a suitable $\tilde{c}>0$ we get that

$$\text{(II.1)} \quad ((A+\tilde{c})u,u) \geq c\|u\|^2_{H^{m,2}(\Omega)}$$

for some $c>0$; on the other hand $f(u)-\tilde{c}u$ satisfies the same assumptions as $f(u)$ does (with different constants). Thus it is no loss of generality to assume that (II.1) is fulfilled for A instead of $A+\tilde{c}$. As before it is easy to show that M fulfills (I.1). We now proceed as in the previous example, but now V is $L^{q^*}(\Omega)$ with $\frac{1}{q^*} = \frac{1}{2}-\frac{m}{n}$. It is also possible to set $V = H^{m,2}(\Omega)$. Assumption (I.2) is fulfilled since by the Gagliardo-Nirenberg inequality and our growth condition on f

$$\|M(u)\| \leq c(\|u\|_{H^{2m,p}(\Omega)} \|u\|^{4m/(n-2m)}_{L^{q^*}(\Omega)} + 1).$$

The necessary estimate for $\|M(u)-M(v)\|$ is derived as in the previous example by using the triangle inequality. Thus (I.3) also holds. (I.4) is known for $q=p$ ([LUS, VII, Theorem 10.4]) but from [W2] (I.4) follows for any $q>1$. Now we assume that $T(\varphi)<+\infty$. We have to prove the uniform continuity of $u:[0,T(\varphi)) \to V$. Scalar multiplication of $u'+Au+M(u) = 0$ by u' and the boundedness from below of F furnishes

$$\int_0^t \|u'(s)\|^2_{L^2(\Omega)} ds + \|A^{\frac{1}{2}}u(t)\|^2_{L^2(\Omega)}$$

$$\leq c\|u(t)\|^2_{L^2(\Omega)} + c\|A^{\frac{1}{2}}\varphi\|^2_{L^2(\Omega)} + \|F(\varphi)\|_{L^1(\Omega)}, \quad 0 \leq t < T(\varphi)$$

(Observe that A is selfadjoint in $L^2(\Omega)$). In particular we obtain

$$u' \in L^2((0,T(\varphi)),L^2(\Omega)).$$

Thus we have on $[0,T(\varphi))$

$$Au(t) + M(u(t)) = -u' \in L^2((0,T(\varphi)),L^2(\Omega)).$$

The calculations in [W3] now show that, by our growth conditions from above and from below on $f(u)\cdot u$,

$$\|Au(t)\|_{L^2(\Omega)} \leq c\|u(t)\|_{H^{2m,2}(\Omega)} \leq c\|u'(t)\|_{L^2(\Omega)}, \quad 0 \leq t < T(\varphi);$$

thus $Au \in L^2((0,T(\varphi)),L^2(\Omega))$. Since we already know that $u' \in L^2((0,T(\varphi)), L^2(\Omega))$ it follows that

$$A^{\frac{1}{2}}u \in C^{O}([O,T(\varphi)],L^{2}(\Omega))$$

(cf. [W2, p. 488]) and consequently

$$u \in C^{O}([O,T(\varphi)],H^{m,2}(\Omega));$$

by Sobolev

$$u \in C^{O}([O,T(\varphi)],L^{q^*}(\Omega))$$

which is a contradiction to the maximality of $T(\varphi)$.

Our <u>third</u> example deals with the equations of Navier-Stokes. In order not to overburden the presentation we make free use of the notion of a weak solution and some other results in connection with the Navier-Stokes system, which is as follows:

$$\text{(II.2)}\quad \begin{aligned} u' - \nu\Delta u + u\cdot\nabla u + \nabla\pi &= f, \\ \nabla\cdot u &= O. \end{aligned}$$

It describes the motion of a viscous incompressible fluid (with constant viscosity ν); u is the velocity, π the pressure and f an external force. The domain $\Omega \subset \mathbb{R}^n$ (this is the region being filled out by the fluid for $n = 3$) is assumed to be smooth and bounded. The boundary condition is $u|\partial\Omega = O$, and we also prescribe the initial velocity $u(O) = \varphi$. The Banach space $(L^p(\Omega))^n$ (in the sequel we omit the exponent n) is decomposed into the direct sum

$$L^p(\Omega) = H_p(\Omega) + \{\nabla g \mid g \in H^{1,p}(\Omega)\},$$

$p > 1$. $H_p(\Omega)$ is the closure of the divergence free $C_o^\infty(\Omega)$-vector fields in the $L^p(\Omega)$-norm. As Banach space B we choose $H_p(\Omega)$ for some $p > n+1$. If we apply the projection P_p of $L^p(\Omega)$ onto $H_p(\Omega)$ the equation (II.2) turns out to be a nonlinear evolution equation in B, namely

$$\text{(II.3)}\quad \begin{aligned} u' + Au + M(u) &= Pf, \\ u(O) &= \varphi. \end{aligned}$$

Here $Au = -P_p\Delta u$, $u \in H^{2,p}(\Omega) \cap \overset{o}{H}{}^{1,p}(\Omega) \cap H_p(\Omega) = D(A)$ is positive and generates an analytic semigroup in B. Since the $H^{2,p}(\Omega)$-norm is equivalent with $\|A.\|$ it can be shown in the usual way that

$$C^{1+\alpha}(\bar{\Omega}) \supset D(A^{1-\rho})$$

for some $\alpha, \rho \in (0,1)$. It is therefore easy to show that

$$M(u) = P_p(u\cdot\nabla u)$$

satisfies the Lipschitz condition (I.1). Theorem I.1 furnishes a solution of (II.3) on a maximal interval $[0,T(\varphi))$, where we have assumed that $\varphi \in D(A)$ and that f has an appropriate degree of regularity. Now consider a weak solution u of (II.2) over $(0,T)\times\Omega$ for some $T>0$. We assume that $\tilde{u} \in C^0([0,T],H_n(\Omega))$. Observe that this is just a marginal case which is not covered by Serrin's regularity and uniqueness criteria. By using our abstract theory we want to show that u is regular. Assume that $T(\varphi) \leqq T$. For V we choose the Banach space $H_n(\Omega)$. We have by the Gagliardo-Nirenberg inequality

$$\begin{aligned} \|M(u)\| &\leqq c(\|u\|^3_{L^{3p}(\Omega)} + \|\nabla u\|^{\frac{3}{2}}_{L^{3p/2}}), \\ &\leqq c(\|u\|_{H^{2,p}(\Omega)}\|u\|^2_{H_n(\Omega)} + \|u\|_{H^{2,p}(\Omega)}\|u\|^{\frac{1}{2}}_{H_n(\Omega)}), \\ &\leqq c\|Au\|(\|u\|^2_{H_n(\Omega)} + \|u\|^{\frac{1}{2}}_{H_n(\Omega)}). \end{aligned}$$

Thus (I.2) is fulfilled. (I.3) is easily derived as in the previous examples by using the triangle inequality. (I.4) was proved for $q=p$ and $n=3$ by Solonnikov [Sol] but is also true for $n \geqq 4$ ([W4, W5]). By [W2, pp. 492-494] (I.4) then follows for any $q>1$. It was proved in [SW, Theorem III.2] that

$$u(t) = \tilde{u}(t)$$

as long as u exists, i.e. on $[0,T(\varphi))$ (This is also a marginal case not being covered by Serrin's uniqueness theorem). This result provides the uniform continuity of $u:[0,T(\varphi)) \to H_n(\Omega)$. Thus $[0,T(\varphi))$ is not the maximal interval of existence and, in particular, $T(\varphi) > T$. Therefore $\tilde{u} \in C^0([0,T],H^{2,p}(\Omega))$.

One can cut down on the assumptions on φ. It is sufficient to assume that $\varphi \in H_n(\Omega)$. Using this result it is possible to show that any weak solution $\tilde{u} \in L^\infty((0,T),H_n(\Omega))$ has at most countably many singularities, i.e. points t where u(t) is not of class $C^{1+\alpha}(\overline{\Omega})$ (cf. [SW]).

References

[F] Friedman, A.: Partial Differential Equations. Holt, Rinehart and Winston: New York, Chicago, San Francisco (1969).

[Ki] Kilimann,N.: Ein Maximumprinzip für nichtlineare parabolische Systeme. Math. Z. 171, 227-230(1980).

[LUS] Ladyženskaja, O.A., Ural'ceva, N.N., and Solonnikov, V.A.: Linear and Quasilinear Equations of Parabolic Type. American Math. Soc.: Providence, R.I., Translations of Mathematical Monographs 23(1968).

[Sol] Solonnikov, V.A.: Estimates of the solutions of a nonstationary linearized system of Navier-Stokes equations. American Math. Soc. Transl. 75, 1-116(1968).

[SoW] Sohr, H., Wahl, W. von: On the Singular Set and the Uniqueness of Weak Solutions of the Navier-Stokes Equations. Manuscripta math. 49, 27-59(1984).

[W1] Wahl, W. von: Klassische Lösbarkeit im Großen für nichtlineare parabolische Systeme und das Verhalten der Lösungen für $t \to \infty$. Nachr. Ak. d. Wiss. Göttingen, II. Mathem.-Physik. Klasse, 131-177(1981).

[W2] Wahl, W. von: The equation $u' + A(t)u = f$ in a Hilbert space and and L^p-estimates for parabolic equations. J. London Math. Soc. (2)25, 483-497(1982).

[W3] Wahl, W. von: Regularity of Weak Solutions to Elliptic Equations of Arbitrary Order. J. Diff. Equations, 235-240(1978).

[W4] Wahl, W. von: Regularitätsfragen für die instationären Navier-Stokesschen Gleichungen in höheren Dimensionen. J. Math. Soc. Japan 32, 263-281(1980).

[W5] Wahl, W. von: Über das Verhalten für $t \to 0$ der Lösungen nichtlinearer parabolischer Gleichungen, insbesondere der Gleichungen von Navier-Stokes. Bayreuther Math. Schr. 16, 151-277(1984).

COMPACT PERTURBATIONS OF WEAKLY EQUICONTINUOUS SEMIGROUPS

I. I. Vrabie
Department of Mathematics
Polytechnic Institute of Iaşi
Iaşi 66oo, R. S. Romania

1. Introduction.

The main goal of the present paper is to prove a local existence result for a class of nonlinear evolution equations of the form

$$\begin{aligned} &\frac{du}{dt}(t) + Au(t) \ni B(t,u(t)), \quad o \leq t \leq T \\ &u(o) = u_o. \end{aligned} \tag{1}$$

Here A is an m-accretive operator acting on a real Hilbert space H, while B is a continuous mapping from $[o,T] \times H$ into H which carries bounded subsets in $[o,T] \times H$ into relatively compact subsets in H. Problems of this kind have been intensively studied over the past decade and the most important existence results obtained are those in [3-7].

Our approach based on a compactness argument involving the space of all weakly continuous functions from $[o,T]$ into H enables us to handle (in the Hilbert space frame only) all the results quoted above. In addition the class of all m-accretive operators A in (1) for which our result holds true is strictly broader than the corresponding classes considered in [4-7]. See Remark 1 below and the example in the last section. We begin with some notations and definitions we shall use later. Let H be a real Hilbert space with the inner product $(\cdot,\cdot)$ and norm $\|\cdot\|$. As usual, C(o,T;H) represents the real Banach space of all continuous functions from $[o,T]$ into H endowed with the sup norm. We denote by $C(o,T;H_w)$ the space of all continuous functions from $[o,T]$ into H, where H is endowed with its weak topology. Let F be the class of all <u>finite</u> subsets in H and K the class of all <u>compact</u> subsets in H. If G is an element either in F, or in K we define p_G: $C(o,T;H_w) \longrightarrow R_+$ by

$$p_G(f) := \sup\{\sup\{|(f(t),x)|;\ x \in G\};\ t \in [o,T]\}$$

for $f \in C(o,T;H_w)$. Clearly, if G is as above p_G is a semi-norm on the space $C(o,T;H_w)$. Moreover, $C(o,T;H_w)$ endowed with the family of seminorms $\{p_G;\ G\in F\}$ is a separated locally convex space. The same holds true if we replace $\{p_G;\ G\in F\}$ by $\{p_G;\ G\in K\}$. In all that follows we consider $C(o,T;H_w)$ equipped with the topology defined by $\{p_G;\ G\in F\}$. The next simple lemma proves useful later.

<u>Lemma 1</u>. <u>Both</u> $\{p_G;\ G\in F\}$ <u>and</u> $\{p_G;\ G\in K\}$ <u>define one and the same topology on</u> $C(o,T;H_w)$.

<u>Proof.</u> Let T_F and T_K be the topologies defined by $\{p_G;\ G\in F\}$ and by $\{p_G;\ G\in K\}$ respectively. Obviously $T_F\subset T_K$. To prove the converse inclusion we have merely to show that each neighborhood of the origin in T_K includes a neighborhood of the origin in T_F. To this aim it suffices to observe that a closed subset G in H is compact iff given $\varepsilon>o$ there exists $G_\varepsilon\in F$ such that for each $x\in G$ there is an $x_\varepsilon\in G_\varepsilon$ such that $\|x - x_\varepsilon\|<\varepsilon$. Q.E.D.

Let $A : D(A)\subset H \longrightarrow 2^H$ be an m-accretive operator and let $\{S(t)\ ;\ t\geq o\}$ be the C_o semigroup of nonexpansive operators generated by A on $\overline{D(A)}$ via Crandall-Liggett's Exponential Formula [1, p. 1o4] . Let us consider the Cauchy problem

$$\begin{aligned} &\frac{du}{dt}(t) + Au(t)\ni f(t),\quad o\leq t\leq T \\ &u(o) = u_o, \end{aligned} \tag{2}$$

where $f\in L^1(o,T;H)$ and $u_o\in \overline{D(A)}$. We recall that an <u>integral solution</u> of (2) on $[o,T]$ is a continuous function $u : [o,T] \longrightarrow \overline{D(A)}$ with $u(o) = u_o$ and satisfying

$$\|u(t) - x\|^2\leq \|u(s) - x\|^2 + 2\int_s^t (u(z) - x, f(z) - y)dz$$

for each $x\in D(A)$, $y\in Ax$ and $o\leq s\leq t\leq T$. See [1, p. 124] . It is well-known that if A is m-accretive then for each $f\in L^1(o,T;H)$ and $u_o\in \overline{D(A)}$ there exists a unique integral solution of (2) defined on $[o,T]$. We denote this solution by

$$u := I(f,u_o)$$

in order to exhibit its dependence on f and on u_o.
At this point it is quite transparent what an integral solution of (1) ought to be. Indeed, we define this as a continuous function u from $[o,T]$ into $\overline{D(A)}$ such that $B(\cdot,u(\cdot))\in L^1(o,T;H)$ and $u = I(B(\cdot,u(\cdot)),u_o)$.

The next two inequalities we shall very frequently make use later describe very precisely the continuous (in fact Lipschitz continuous) dependence of u on f and on u_0. They may be as well regarded as uniqueness results. Namely, we have

$$\|u(t) - v(t)\|^2 \leqslant \|u(s) - v(s)\|^2 + 2\int_s^t (u(z) - v(z), f(z) - g(z))dz \quad (3)$$

and

$$\|u(t) - v(t)\| \leqslant \|u(s) - v(s)\| + \int_s^t \|f(z) - g(z)\|dz \quad (4)$$

for every $f,g \in L^1(o,T;H)$, $u_0, v_0 \in \overline{D(A)}$ and $o \leqslant s \leqslant t \leqslant T$, where $u = I(f,u_0)$ and $v = I(g,v_0)$. See also [1, p. 124].

2. Statement of the main result.

We introduce first a special class of semigroups we shall need in the statement of our main existence theorem.

Definition 1. The semigroup $\{S(t) \; ; \; S(t) : \overline{D(A)} \longrightarrow \overline{D(A)}, \; t \geqslant o\}$ is called locally weakly equicontinuous - briefly LWE - if given u_0 in $\overline{D(A)}$ there is an $r > o$ such that the family $\{S(\cdot)u \; ; \; u \in \overline{D(A)}, \; \|u-u_0\| \leqslant r\}$ is equicontinuous at each $t \in]o,T]$ in the space $C(o,T;H_w)$, where $T > o$ is arbitrary.

Remark 1. The class of all m-accretive operators generating LWE semigroups is very large. We indicate below several subclasses occuring quite often in applications.

(i) The class of all m-accretive operators of the form $A_1 + A_2$ with A_1 linear and m-accretive and A_2 continuous, everywhere defined and accretive. This case has been considered in [5, 6] in a general Banach space frame.

(ii) The class of all m-accretive operators defined on a finite dimensional Hilbert space H.

(iii) The class of all subdifferentials of l.s.c. proper and convex functions from H into $\overline{R}$, with H arbitrary.

(iv) The class of all m-accretive operators which are restrictions to H of hemicontinuous, coercive and monotone operators from V into V'. Here V is a real reflexive Banach space densely and continuously embedded in H, while V' is the dual of V.

(v) The class of all (possible nonlinear) m-accretive operators A acting on H (with H arbitrary) for which there exists a linear operator

$A^*: D(A^*) \subset H \longrightarrow H$ with $D(A^*)$ dense in H and such that

$$(Au,u^*) = (u,A^*u^*)$$

for each $u \in D(A)$ and $u^* \in D(A^*)$. Roughly speaking this class includes the set of all (possible nonlinear) m-accretive operators whose restrictions to some dense subspace of H are linear.
The specific cases (ii) and (iii) have been considered in [7].

Now we are able to state our main result.

Theorem 1. Let $A : D(A) \subset H \longrightarrow 2^H$ be an m-accretive operator generating a LWE semigroup, let U be a nonempty, open subset in H and B a compact operator from $[o,T] \times U$ into H (i.e. B is continuous and carries every bounded subset in $[o,T] \times U$ into a relatively compact subset in H). Then for each $u_0 \in \overline{D(A)} \cap U$ there exists $T_0 \in]o,T]$ such that the problem (1) has at least one integral solution defined on $[o,T_0]$.

3. Proof of the main result.

In order to prove Theorem 1 we shall use a fixed point device based mainly on Schauder's Fixed Point Theorem. To this aim we need a compactness argument which is important by itself. We begin by recalling that a subset P in $L^1(o,T;H)$ is called uniformly integrable if given $\varepsilon > o$ there is a $\delta(\varepsilon) > o$ such that

$$\int_E \|f(t)\| dt < \varepsilon$$

for every measurable subset E in $[o,T]$ whose Lebesgue measure m(E) is less than $\delta(\varepsilon)$ and uniformly for $f \in P$. It is obvious that each uniformly integrable subset in $L^1(o,T;H)$ is bounded in this space. We note also that each bounded subset in $L^p(o,T;H)$ is uniformly integrable provided $1 < p \leq \infty$.
The main ingredient in the proof of Theorem 1 is the compactness result below which may be regarded as the "weak version" of Theorem 2.3 in [4] and in some sense of Theorem 1 in [8]. It should also be mentioned that, excepting some conceptual differences, the proof of this compactness result follows the same lines as those in the proof of Theorem 2.3 in [4].

Theorem 2. Let $A : D(A) \subset H \longrightarrow 2^H$ be an m-accretive operator generating a LWE semigroup. Then for each $u_0 \in \overline{D(A)}$ and each uniformly integrable *subset* P *in* $L^1(o,T;H)$ *there exists* $T_o \in]o,T]$ *depending on* u_o *and on* P *such that the restriction of the family* $\{I(f,u_o)\ ;\ f\in P\}$ *to* $[o,T_o]$ *is relatively compact in* $C(o,T_o;H_w)$.

Proof. Since P is a fortiori bounded in $L^1(o,T;H)$, in view of (4) the family $\{I(f,u_o)\ ;\ f\in P\}$ is bounded in $C(o,T;H)$. An appeal to Definition 1 shows that for u_o in $\overline{D(A)}$ there exists $r>o$ such that $\{S(\cdot)u\ ;\ u$ in $\overline{D(A)}, \|u-u_o\|\leqslant r\}$ is equicontinuous at each $t\in]o,T]$ in $C(o,T;H_w)$. Choose T_o in $]o,T]$ such that

$$\|S(t)u_o - u_o\| + \int_o^{T_o}\|f(s)\|ds \leqslant r$$

for every $f\in P$ and $t\in[o,T_o]$. This is always possible because $S(\cdot)u_o$ is strongly continuous at $t = o$ while P is uniformly integrable. Since $S(t)u_o = I(o,u_o)(t)$ for $t\in[o,T_o]$, by (4) and the above inequality, we get

$$\|I(f,u_o)(t) - u_o\| \leqslant \|I(f,u_o)(t) - S(t)u_o\| + \|S(t)u_o - u_o\|$$
$$\leqslant \|S(t)u_o - u_o\| + \int_o^{T_o}\|f(s)\|ds \leqslant r$$

for every $f\in P$ and $t\in[o,T_o]$. This inequality shows that the family $\{S(\cdot)I(f,u_o)(t)\ ;\ f\in P,\ t\in[o,T_o]\}\subset\{S(\cdot)u\ ;\ u$ in $\overline{D(A)}, \|u-u_o\|\leqslant r\}$ and therefore it is equicontinuous at each $h>o$ in $C(o,T;H_w)$. In addition, for each t in $[o,T_o]$ the set $\{I(f,u_o)(t)\ ;\ f\in P\}$ is weakly relatively compact in H because it is bounded in the norm of H. Hence, by Ascoli's Theorem [2, p. 34], to conclude the proof we have merely to show that $\{I(f,u_o)\ ;\ f\in P\}$ is equicontinuous at each $t\in[o,T_o]$ in the space $C(o,T_o;H_w)$. Inasmuch as the equicontinuity at $t = o$ follows from the fact that the above family is equicontinuous at $t = o$ in $C(o,T_o;H)$ (repeat the same routine as in the construction of T_o starting with an arbitrary $r>o$ to get this conclusion), we confine ourselves only to the case $t\in]o,T_o]$. Thus let $t\in]o,T_o]$, let $\varepsilon>o$ be arbitrary and choose $h>o$ such that $t-2h\in[o,T_o]$ and

$$\int_E\|f(t)\|dt < \varepsilon$$

for every measurable subset E in $[o,T_o]$ whose Lebesgue measure $m(E) < 2h$ and uniformly for $f\in P$.

Let G be an arbitrary finite subset in H and let us denote by $\|\cdot\|_G$ the corresponding semi-norm induced by G on H, i.e. $\|u\|_G := \max_{x\in G}|(u,x)|$ for $u\in H$. Since in all that follows u_o will be kept frozen in $\overline{D(A)}$, there is no danger of confusion if we simplify the notation by putting

$$u_f := I(f,u_o)$$

for each $f\in P$. Now, let us observe that to get the conclusion it suffices to show that for $t\in]o,T_o]$ and $\varepsilon>o$ as above there exists $\delta>o$ depending only on t, ε and G such that

$$\|u_f(t+s) - u_f(t)\|_G \leqslant k\varepsilon$$

for every $s\in R$, $|s|<\delta$, $t+s\in[o,T_o]$ and uniformly for $f\in P$, where $k>o$ depends only on G. First of all let us remark that

$$\|u_f(t+s) - u_f(t)\|_G \leqslant \|u_f(t+s) - S(h+s)u_f(t-h)\|_G$$

$$+ \|S(h+s)u_f(t-h) - S(h)u_f(t-h)\|_G + \|S(h)u_f(t-h) - u_f(t)\|_G$$

for every $f\in P$ and $s\in R$, $|s|<h$, $t+s\in[o,T_o]$. Taking into account that $\|u\|_G\leqslant m_G\|u\|$ for every $u\in H$, where $m_G := \max\{\|x\| ; x\in G\}$ we easily conclude that

$$\|u_f(t+s) - u_f(t)\|_G \leqslant m_G\|u_f(t+s) - S(h+s)u_f(t-h)\|$$

$$+ \|S(h+s)u_f(t-h) - S(h)u_f(t-h)\|_G + m_G\|S(h)u_f(t-h) - u_f(t)\| \qquad (5)$$

for every $f\in P$ and $s\in R$, $|s|<h$, $t+s\in[o,T_o]$.
Since the function $z\longmapsto S(h+z)u_f(t-h)$ is the unique integral solution of

$$\frac{dv}{dz}(z) + Av(z)\ni o, \quad -h\leqslant z\leqslant s$$
$$v(-h) = u_f(t-h),$$

a simple argument involving (4) implies both

$$\|u_f(t+s) - S(h+s)u_f(t-h)\| \leqslant \int_{t-h}^{t+s}\|f(z)\|dz$$

and

$$\|S(h)u_f(t-h) - u_f(t)\| \leqslant \int_{t-h}^{t}\|f(z)\|dz$$

for every $f\in P$ and $s\in R$, $|s|<h$, $t+s\in[o,T_o]$. Recalling the definition of $h>o$, the above inequalities lead to

$$\max\{\|u_f(t+s) - S(h+s)u_f(t-h)\|, \|S(h)u_f(t-h) - u_f(t)\|\} < \varepsilon \qquad (6)$$

for every $f \in P$ and $s \in R$, $|s| < h$, $t+s \in [o,T_o]$.
Now using the fact that $\{S(\cdot)u_f(t-h)\ ;\ f \in P\}$ is equicontinuous at $h > o$ in $C(o,T;H_w)$ we may infer that there exists $d(h,\varepsilon,G) > o$ such that

$$\|S(h+s)u_f(t-h) - S(h)u_f(t-h)\|_G < \varepsilon \qquad (7)$$

for every $s \in R$, $|s| < \min\{d(h,\varepsilon,G),h\}$, and uniformly for $f \in P$. Inasmuch as $h > o$ depends only on t and on ε , we are allowed to define $\delta(t,\varepsilon,G)$ as $\min\{d(h,\varepsilon,G),h\}$. Combining (5), (6) and (7) we conclude

$$\|u_f(t+s) - u_f(t)\|_G \leqslant (2m_G + 1)\varepsilon$$

for every $s \in R$, $|s| < \delta(t,\varepsilon,G)$, $t+s \in [o,T_o]$ and uniformly for $f \in P$, thereby completing the proof of Theorem 2. Q.E.D.

We are now able to proceed to the proof of Theorem 1. As we already mentioned, we shall use a fixed point device as follows. We define a nonempty, closed, convex and bounded subset C in $C(o,T_o;H)$ with T_o in $]o,T]$ and small enough such that the mapping $Q : C \longrightarrow C(o,T_o;H)$ defined by

$$(Qf)(t) := B(t,u_f(t))$$

for $f \in C$ and $t \in [o,T_o]$ maps C into itself and is compact (i.e. it is continuous from C into C and $Q(C)$ is relatively compact in $C(o,T_o;H)$). Then, by Schauder's Fixed Point Theorem Q has at least one fixed point f in C. If we shall succeed to carry out this plan this will complete the proof because f is a fixed point of Q iff u_f is an integral solution of (1) on $[o,T_o]$.
Now the details. We begin with the construction of the subset C. To this aim let $r > o$ be such that

$$\{u\ ;\ u \in H,\ \|u - u_o\| \leqslant r\} \subset U.$$

Since B is compact, there exists $M > o$ such that

$$\|B(t,u)\| \leqslant M$$

for every $t \in [o,T]$ and $u \in U$, $\|u - u_o\| \leqslant r$. Let us define

$$P_M := \{f\ ;\ f \in L^1(o,T;H),\ \|f(t)\| \leqslant M \text{ for a.e. } t \in [o,T]\}.$$

Clearly P_M is uniformly integrable (being bounded in $L^\infty(o,T;H)$) and therefore, by Theorem 2 there exists $T_o \in]o,T]$ such that the restriction of the family $\{u_f \; ; \; f \in P_M\}$ to $[o,T_o]$ is relatively compact in the space $C(o,T_o;H_w)$. Diminish T_o if necessary in order to have also

$$\|S(t)u_o - u_o\| + T_o M \leqslant r \tag{8}$$

for every $t \in [o,T_o]$, where M and r are as above. Put

$$C_r := \operatorname{conv}\{B(t,u) \; ; \; (t,u) \in [o,T_o] \times U, \; \|u - u_o\| \leqslant r\}$$

where conv X denotes the closed convex hull of the set X. Obviously C_r is nonempty, convex and due to Mazur's Theorem it is also compact. Finally, let us define

$$C := \{f \; ; \; f \in C(o,T_o;H), \; f(t) \in C_r \text{ for } t \in [o,T_o]\}.$$

We easily see that C is nonempty ($B(\cdot,u_o) \in C$), convex, closed and bounded in $C(o,T_o;H)$.
To prove that the operator Q defined above maps C into itself it suffices to show that

$$\|u_f(t) - u_o\| \leqslant r$$

for each $f \in C$ and $t \in [o,T_o]$. Since by (4) and (8) we have

$$\|u_f(t) - u_o\| \leqslant \|S(t)u_o - u_o\| + \int_o^t \|f(s)\| ds \leqslant \|S(t)u_o - u_o\| + T_o M \leqslant r$$

for each $f \in C$ and $t \in [o,T_o]$, this is certainly the case, and therefore $Q(C) \subset C$. In addition Q is continuous being the superposition of two continuous mappings.
At this point let us observe that we may assume with no loss of generality that H is separable. Indeed, since C_r is separable being a compact metric space, we may confine ourselves to work in the closed linear space spanned by C_r and thus we may assume that H itself is a real, separable Hilbert space.
Now to show that Q(C) is relatively compact in $C(o,T_o;H)$ let (f_n) be an arbitrary sequence in C. Since $C \subset P_M$ and $\{u_f \; ; \; f \in P_M\}$ is relatively compact in $C(o,T_o;H_w)$, the set $\{u_{f_n} \; ; \; n \in N\}$ is also relatively compact in $C(o,T_o;H_w)$. Inasmuch as H is separable, $C(o,T_o;H_w)$ is metrizable and thus the sequence (u_{f_n}) has at least one Cauchy subsequence in $C(o,T_o;H_w)$. Denote this subsequence also by (u_{f_n}) and let us observe that by virtue of (3) we have

$$\|u_{f_n}(t) - u_{f_{n+p}}(t)\|^2 \leq 2\int_0^{T_0} |(u_{f_n}(s) - u_{f_{n+p}}(s), f_n(s) - f_{n+p}(s))|ds$$

for each $n,p \in N$. Since for each $n,p \in N$ and $s \in [o,T_0]$ $f_n(s) - f_{n+p}(s)$ belongs to $C_r - C_r$ which is compact because C_r is so, and $u_{f_n} - u_{f_{n+p}} \longrightarrow o$ in $C(o,T_0;H_w)$ when $n,p \longrightarrow \infty$, an appeal to Lemma 1 shows that (u_{f_n}) is a Cauchy sequence in $C(o,T_0;H)$. Finally, as B is continuous, (Qf_n) is a Cauchy sequence in $C(o,T_0;H)$. But (f_n) was arbitrary in C and therefore Q(C) is relatively compact in the strong topology of the space $C(o,T_0;H)$ as claimed. Q.E.D.

4. An example.

Let us consider the following second order nonlinear initial boundary value problem

$$\begin{aligned} &u_{tt} - u_{xx} + \beta(u_t) \ni f(t,x,u) \text{ for a.a. } (t,x) \in]o,T[\times]o,1[\\ &u(t,o) = u(t,1) = o \text{ for a.a. } t \in]o,T[\\ &u(o,x) = u_0(x),\ u_t(o,x) = v_0(x) \text{ for a.a. } x \in]o,1[\ , \end{aligned} \tag{9}$$

where $\beta \subset R \times R$ is a maximal monotone graph with $o \in \beta(o)$, while f is a continuous function from $[o,T] \times [o,1] \times R$ into R. Using a standard device we rewrite (9) as a first order system of partial differential equations

$$\begin{aligned} &u_t - v = o && \text{for a.a. } (t,x) \in]o,T[\times]o,1[\\ &v_t - u_{xx} + \beta(v) \ni f(t,x,u) && \text{for a.a. } (t,x) \in]o,T[\times]o,1[\\ &u(t,o) = u(t,1) = o \text{ for a.a. } t \in]o,T[\\ &u(o,x) = u_0(x),\ v(o,x) = v_0(x) \text{ for a.a. } x \in]o,1[\ . \end{aligned} \tag{1o}$$

Furthermore, set

$$H := \begin{matrix} H_0^1(o,1) \\ \times \\ L^2(o,1) \end{matrix}$$

which endowed with the inner product $(\cdot,\cdot)_H$ defined by

$$\left(\begin{pmatrix} u \\ v \end{pmatrix}, \begin{pmatrix} \bar{u} \\ \bar{v} \end{pmatrix} \right)_H := \int_0^1 u'(x)\bar{u}'(x)dx + \int_0^1 v(x)\bar{v}(x)dx$$

for each $\begin{pmatrix} u \\ v \end{pmatrix}, \begin{pmatrix} \bar{u} \\ \bar{v} \end{pmatrix} \in H$ is a real Hilbert space. Now, let us define

$A : D(A) \subset H \longrightarrow 2^H$ by

$$A\begin{pmatrix} u \\ v \end{pmatrix} : = \left\{ \begin{pmatrix} -v \\ -u'' + \beta(v) \end{pmatrix} \right\}$$

for each $\begin{pmatrix} u \\ v \end{pmatrix} \in D(A)$, where

$$D(A) : = \left\{ \begin{pmatrix} u \\ v \end{pmatrix} \in H \; ; \; u \in H^2(o,1), \; v \in H_o^1(o,1), \; \beta(v) \in L^2(o,1) \right\} \; .$$

It is well-known that A is m-accretive. See for instance [1, p. 268]. Next we write (1o) as an abstract nonlinear evolution in H of the form

$$\begin{aligned} &\frac{dw}{dt}(t) + Aw(t) \ni B(t,w(t)), \; o \leqslant t \leqslant T \\ &w(o) = w_o \end{aligned} \qquad (11)$$

where $[w(t)](x) : = \begin{pmatrix} u(t,x) \\ v(t,x) \end{pmatrix}$ for a.a. $(t,x) \in]o,T[\times]o,1[$, $w_o(x) : = \begin{pmatrix} u_o(x) \\ v_o(x) \end{pmatrix}$ and $B : [o,T] \times H \longrightarrow H$ is defined by

$$[B(t,w)](x) : = \begin{pmatrix} o \\ f(t,x,u(x)) \end{pmatrix}$$

for each $\begin{pmatrix} u \\ v \end{pmatrix} \in H$, $t \in [o,T]$ and for a.a. $x \in]o,1[$.

Using Theorem 1 we can now prove.

<u>Theorem 3.</u> <u>Assume that</u> $\beta \subset R \times R$ <u>is a maximal monotone graph with</u> o <u>in</u> $\beta(o)$ <u>and for which there exist</u> $c_i > o$, $i = 1,2$ <u>such that</u>

$$\sup\{ |y| \; ; \; y \in \beta(v) \} \leqslant c_1 v^2 + c_2$$

<u>for each</u> $v \in D(\beta) = R$. <u>Assume further that</u> $f : [o,T] \times [o,1] \times R \longrightarrow R$ <u>is continuous</u>. <u>Then for each</u> $w_o \in H$ <u>there exists</u> $T_o \in]o,T]$ <u>such that the problem</u> (11) <u>has at least one integral solution defined on</u> $[o,T_o]$.

The complete proof of Theorem 3 may be found in [9]. However, for the reader's convenience we shall sketch it below without giving details. First of all let us observe that B is a compact mapping because $H_o^1(o,1)$ is compactly embedded in $C(o,1) \subset L^2(o,1)$ and f is continuous. Next, from (1o) and the quadratic growth of β we conclude that for each bounded subset D in D(A) the family $\{ S(\cdot)w \; ; \; w \in D \}$ is equicontinuous in the space $C(o,T;H_w)$ at each $t \in]o,T]$. To prove this we have to use functionals on H whose components belong to $C_o^\infty(o,1)$ which is dense in both $L^2(o,1)$ and in $H_o^1(o,1)$, to integrate by parts in (1o) and to take into account the growth condition on β in order to get a priori estimates. Then the conclusion follows from Theorem 1.

Finally, we note that none of the results in [3-7] can be applied in the proof of Theorem 3. Additional details and other examples may be found in [9].

Acknowledgement. The author would like to express his warmest thanks to the organizers, especially to Professors A. Favini and A. Venni, for their invitation to give this talk during the Conference on DIFFERENTIAL EQUATIONS IN BANACH SPACES held at Bologna, July 2-5, 1985, for their support and very kind hospitality.

REFERENCES

[1] BARBU, V. Nonlinear semigroups and differential equations in Banach spaces, Editura Academiei Bucureşti România, Noordhoff Leyden The Netherlands, 1976.

[2] EDWARDS, R. E. Functional Analysis, Holt, Rinehart and Winston, New York, 1965.

[3] GUTMAN, S. Compact perturbations of m-accretive operators in general Banach spaces, SIAM J. Math. Anal., 13(1982), p. 789-8oo.

[4] GUTMAN, S. Evolutions governed by m-accretive plus compact operators, Nonlinear Anal., 7(1983), p. 7o7-717.

[5] MARTIN, R. H., Jr. Remarks on ordinary differential equations involving dissipative plus compact operators, J. London Math. Soc., 1o(1975), p. 61-65.

[6] SCHECHTER, E. Evolution generated by semilinear dissipative plus compact operators, Trans. Amer. Math. Soc. 275(1983), p. 297-3o8.

[7] SCHECHTER, E. Perturbations of regularizing maximal monotone operators, Israel J. Math., 43(1982), p. 49-61.

[8] VRABIE, I. I. A compactness criterion in C(o,T;X) for subsets of solutions of nonlinear evolution equations governed by accretive operators, Rend. Sem. Mat. Univ. e Politec. Torino, 43(1985), p. 149-158.

[9] VRABIE, I. I. Compactness methods for nonlinear evolutions, Pitman Monographs Series, Boston-London-Melbourne, to appear.

COSINE FAMILIES OF OPERATORS AND APPLICATIONS

Michiaki Watanabe

Faculty of General Education
Niigata University
Niigata, 950-21 Japan

In this article, we shall present our recent results concerning cosine families of operators (or operator cosine functions) in Banach spaces. About twenty years ago, the fundamental generation theorem was established by the mathematicians in Italy and other countries.

However, as is known, its application to differential equations has been somehow neglected. The present author himself is concerned about the matter and also very much interested in it.

Here, we shall select three topics and discuss them mainly from the viewpoint of the applicability to differential equations in Banach spaces :

1. New proof of generation theorem ;
2. Perturbation of generator ;
3. Tool for equation $u_t = \Delta\phi(u)$.

In Section 2, applications to ordinary differential operators are given. In Section 3, concrete cosine families are used as convenient tools to do with the difference scheme for the quasi-linear diffusion equation.

Let's begin with definitions and notations. Let $(X, \|\cdot\|)$ be a Banach space, and $B(X_1, X_2)$ be the totality of bounded linear operators from a Banach space X_1 to another X_2 .

The subset $\{C(t): t \in R\}$ of $B(X,X)$ is called a (strongly continuous) *cosine family* in X if :

$$C(t+s) + C(t-s) = 2\,C(t)C(s) \quad \text{for} \quad t,s \in R\ ;$$

$$C(0) = I \ (\text{identity})\ ;$$

$$t \to C(t)x \ \text{ is continuous in } X \text{ for each } x \in X.$$

The associated *sine family* $\{S(t):\ t \in R\}$ is defined by

$$S(t)x = \int_0^t C(r)x\,dr \quad \text{for} \quad x \in X \quad \text{and} \quad t \in R\ ,$$

and the operator defined by

$$Au = \lim_{h\to 0} 2h^{-2}(C(h) - I)u,$$

$$D(A) = \{u \in X :\ \lim_{h\to 0} h^{-2}(C(h) - I)u \ \text{ exists}\}$$

is called the (infinitesimal) *generator* of $\{C(t):\ t \in R\}$.

1. NEW PROOF OF GENERATION THEOREM

First, we shall give another proof of the famous theorem on cosine family generation, established by Da Prato - Giusti[6], Sova[12] and Fattorini[7] independently. Let's recall the necessary and sufficient condition for an operator A in X to be the generator of a cosine family in X :

(A) The domain $D(A)$ of A is dense in X ; there are constants $\omega > 0$ and $M \geqq 1$ such that the set $\{z^2 : z > \omega\}$ is included in the resolvent set of A , and

$$\|D(z,n,\omega)z(z^2 - A)^{-1}x\| \leqq M\|x\|$$

for $x \in X,\ z > \omega$ and $n = 0,1,\ldots,$

where $D(z,n,\omega) = (1/n!)(z-\omega)^{n+1}(d/dz)^n$. The necessity is clear from Laplace transform of $C(t)x$, but the sufficiency is rather difficult to show.

<u>PROPOSITION 1 ([15]).</u> Under the condition (A), the

closure V of $D(A)$ in the norm :

$$|u| = \|u\| + \sup_{z>\omega, n\geq 0} \|D(z,n,\omega)A(z^2 - A)^{-1}u\|$$

becomes a Banach space such that

$$\|D(z,n,\omega)A(z^2 - A)^{-1}u\| \leq |u|,$$
$$|D(z,n,\omega)z(z^2 - A)^{-1}u| \leq M|u|,$$
$$|D(z,n,\omega)(z^2 - A)^{-1}x| \leq (M^2 + 2M + M/\omega)\|x\|$$

hold for $x \in X$, $u \in V$, $z > \omega$ and $n = 0,1,\ldots$.

PROOF. Since

$$D^n(z^2 - A)^{-1} = \sum_{k=0}^{n} (-1)^k z^{-k-1} D^{n-k} z(z^2 - A)^{-1} ,$$

where $D^k = (1/k!)(d/dz)^k$, the estimate in (A) implies

$$\|D(z,n,\omega)(z^2 - A)^{-1}x\| \leq (M/\omega)\|x\| \quad \text{for } x \in X.$$

Therefore, for each $u \in D(A)$, $|u|$ is finite. Again using this estimate, we can show that V is complete under $|\cdot|$.

The first and the second estimate are clear from the definition of $|\cdot|$. Only the last is slightly difficult to obtain. Repeated differentiation in w and z of

$$A(w^2 - A)^{-1}(z^2 - A)^{-1}$$
$$= w(w^2- A)^{-1}z(z^2- A)^{-1} - (w + z)^{-1}\{z(z^2- A)^{-1} + w(w^2- A)^{-1}\},$$

and somewhat tedious calculation gives , by (A) ,

$$\|D(w,m,\omega)A(w^2 - A)^{-1}D(z,n,\omega)(z^2 - A)^{-1}x\|$$
$$\leq \{M^2 + M \sum_{k=0}^{n} \frac{(m+k)!}{m!k!}t^k(1 - t)^{m+1} + M \sum_{j=0}^{m} \frac{(n+j)!}{n!j!}s^j(1 - s)^{n+1}\}\|x\|,$$

where $t = (z - \omega)/(w + z)$ and $s = (w - \omega)/(w + z)$. Noting that the parts $\sum_{k=0}^{n}$ and $\sum_{j=0}^{m}$ are smaller than 1 , and taking the supremum in w and m, we obtain

$$
\begin{aligned}
&|D(z,n,\omega)(z^2 - A)^{-1}x| \\
\leqq\ & \|D(z,n,\omega)(z^2 - A)^{-1}x\| + (M^2 + M + M)\|x\| \\
\leqq\ & (M/\omega + M^2 + 2M)\|x\|. \qquad \text{Q.E.D.}
\end{aligned}
$$

The proposition implies :

The domain $D(A) \times V$ of $\begin{pmatrix} 0 & I \\ A & 0 \end{pmatrix}$ is dense in $V \times X$;

$$|(z - \begin{pmatrix} 0 & I \\ A & 0 \end{pmatrix})^{-n}|_{V\times X \to V\times X} \leqq (M^2 + 3M + M/\omega)(|z| - \omega)^{-n}$$

for $|z| > \omega$ and $n = 1,2,\ldots$.

Therefore, we obtain that $\begin{pmatrix} 0 & I \\ A & 0 \end{pmatrix}$ with domain $D(A) \times V$ generates the one parameter group in $(V \times X, |\cdot| + \|\cdot\|)$:

$$\{e^{t\begin{pmatrix} 0 & I \\ A & 0 \end{pmatrix}} : t \in R\}$$

by the Hille-Yosida-Phillips theorem. Thanks to the Kisyński theorem[8] , which indicates a relation between cosine family generation and group generation, we can conclude under (A) that A is the generator of the cosine family $\{C(t): t \in R\}$ defined by

$$C(t)x = \Pi_2 e^{t\begin{pmatrix} 0 & I \\ A & 0 \end{pmatrix}}(0,x) \quad \text{for } x \in X \text{ and } t \in R,$$

where $\Pi_2 : V \times X \to X$ (projection), and that the space V coincides with the space :

$$E = \{u \in X : t \to C(t)u \text{ is of class } C^1 \text{ in } X\} ;$$

$$|u|_E = \|u\| + \max_{0 \leqq r \leqq 1} \|(d/dr)C(r)u\| .$$

<u>REMARK.</u> By Proposition 1, we can know the space V by information concerning A only, and hence we can convert the problem for $d^2u/dt^2 = Au$ into the first order system in $V \times X$ which only involves information about A :

$$(d/dt)\begin{pmatrix} u \\ v \end{pmatrix} = \begin{pmatrix} 0 & I \\ A & 0 \end{pmatrix}\begin{pmatrix} u \\ v \end{pmatrix} .$$

2. PERTURBATION OF GENERATOR

Next, we shall discuss conditions on a linear operator B in X, under which if A is a generator, so is $A + B$; and then apply them to ordinary differential operators.

<u>THEOREM 2 ([17]).</u> Let A satisfy the condition (A) with $\omega > 0$, and B satisfy the condition :

(B) $\begin{cases} D(B) \supset D(A) ; \\ \text{there is a constant } M_0 \geqq 1 \text{ such that } B(z^2 - A)^{-1} \\ \text{is strongly infinitely differentiable in } z, \text{ and} \\ \text{satisfies} \\ \qquad \|D(z,n,\omega)B(z^2 - A)^{-1}x\| \leqq M_0\|x\| \\ \text{for } x \in X,\ z > \omega \text{ and } n = 0,1,\ldots . \end{cases}$

Then, $A + B$ is a generator.

<u>PROOF.</u> Choose constants ω_1 and M_1 so large that

$$M_1 > M \quad \text{and} \quad \omega_1 \geqq \omega + M_1 M_0 (M_1 - M)^{-1}.$$

Then, we have for $z > \omega_1$ and $n = 1,2,\ldots$

$$z(z^2 - A - B)^{-1} = z(z^2 - A)^{-1}(I - B(z^2 - A)^{-1})^{-1} ;$$

$$D^n z(z^2 - A - B)^{-1} = \{\sum_{k=1}^{n} D^{n-k} z(z^2 - A - B)^{-1} \cdot D^k B(z^2 - A)^{-1}$$

$$+ D^n z(z^2 - A)^{-1}\} (I - B(z^2 - A)^{-1})^{-1}.$$

Thus, we can show by induction that $A + B$ satisfies the condition (A) with ω_1 and M_1. Q.E.D.

<u>COROLLARY 3 ([13]).</u> If B belongs to $B(V,X)$, then $A + B$ is a generator in X.

<u>PROOF.</u> Recalling the last estimate in Prop. 1, we have

$$\|D(z,n,\omega)B(z^2 - A)^{-1}x\| = \|B\cdot D(z,n,\omega)(z^2 - A)^{-1}x\|$$
$$\leqq |B|_{V\to X}\ (M^2 + 2M + M/\omega)\|x\|.$$

<u>COROLLARY (Fattorini).</u> Let B satisfy :

B is closed in X ; $D(B) \supset S(t)X$ for all $t \in R$;

$t \to BS(t)x$ is continuous in X for each $x \in X$.

Then, $A + B$ is a generator.

<u>PROOF.</u> Integration in t, and differentiation in s of $C(t + s) + C(t - s) = 2\,C(t)C(s)$ gives

$$C(t + s)u - C(t - s)u = 2\,S(t)(d/ds)C(s)u \quad \text{for } u \in E.$$

Using this equality, we obtain

$$Bu = 2^{-1}\,BS(2)u - 2\int_0^1 BS(r)\cdot(d/dr)C(r)u\,dr ,$$

which implies that B belongs to $B(V,X)$, $V = E$. Q.E.D.

<u>REMARK.</u> Cor. 3 is generalized to the time-dependent case :

(C) $d^2u/dt^2 = Au + B(t)u$, $t \in R$; $u(0) = f$ and $u'(0) = g$.

Under the condition :

$B(t)$ belongs to $B(V,X)$ for each $t \in R$;

$t \to B(t)u$ is of class C^1 in X for each $u \in V$,

we can convert the problem (C) into the first order problem :

$$\frac{d}{dt}\begin{pmatrix}u\\v\end{pmatrix} = \left\{\begin{pmatrix}0 & I\\A & 0\end{pmatrix} + \begin{pmatrix}0 & 0\\B(t) & 0\end{pmatrix}\right\}\begin{pmatrix}u\\v\end{pmatrix},\ t \in R\ ;\ \begin{pmatrix}u(0)\\v(0)\end{pmatrix} = \begin{pmatrix}f\\g\end{pmatrix}$$

and then apply a theorem on time-dependent bounded perturbation for groups, essentially due to [11, Theorem 6.2]. See [18]. The above condition makes it possible for $B(t)$ to become a *differential operator* (cf. [9]).

APPLICATION. Let's consider the ordinary differential operator of second order :

$$a(x)(d/dx)^2 + b(x)(d/dx) + c(x) \quad , p \leq x \leq q$$

with Neumann boundary condition $u'(p) = u'(q) = 0$.

Set $I = [p,q]$ for $-\infty < p < q < \infty$. Let $C(I)$ be the totality of continuous functions on I , a Banach space with norm $|u| = \max_{x\in I}|u(x)|$, and let $C^1(I)$ be the Banach space $\{u \in C(I) : u' \in C(I)\}$ with norm $|u|_1 = |u| + |u'|$.

Let's define the operators A and B in $C(I)$ by

$$Au = a(x)u'' + (a'(x)/2)u' ,$$

$$D(A) = \{u \in C(I): u'' \in C(I), u'(p) = u'(q) = 0\} ;$$

$$Bu = (b(x) - a'(x)/2)u' + c(x)u, \ D(B) = C^1(I)$$

under the condition :

$$a(x) > 0 \text{ for all } x \in I ;$$

$a(\cdot)$, $a'(\cdot)$, $b(\cdot)$ and $c(\cdot)$ are continuous on I .

PROPOSITION 4. (i) ([17]) The operators A and B in $C(I)$ satisfy the conditions (A) and (B) with an arbitrary $\omega > 0$, $M = 1$ and

$$M_0 = \max_{x\in I}|(b(x) - a'(x)/2)a(x)^{-1/2}| + \omega^{-1} \max_{x\in I}|c(x)|.$$

(ii) ([18]) The space V coincides with

$$\{u \in C^1(I) : u'(p) = u'(q) = 0\} \text{ with norm } |\cdot|_1$$

and B belongs to $B(V,X)$.

PROOF. (i) Let's denote the operators A and B with $a(x) \equiv 1$ by A_1 and B_1 respectively. Then, a direct computation gives

$$z(z^2 - A_1)^{-1}f(x)$$

$$= \sum_{k=0}^{\infty} e^{-(2k+1)z(q-p)} \{e^{z(x-p)} + e^{-z(x-p)}\}$$

$$\cdot\ 2^{-1} \int_p^q \{e^{z(q-r)} + e^{-z(q-r)}\} f(r)\, dr$$

$$-\ 2^{-1} \int_p^x \{e^{z(x-r)} - e^{-z(x-r)}\} f(r)\, dr\ ;$$

$$(d/dx)(z^2 - A_1)^{-1}f(x)$$

$$= \sum_{k=0}^{\infty} e^{-(2k+1)z(q-p)} \{e^{z(x-p)} - e^{-z(x-p)}\}$$

$$\cdot\ 2^{-1} \int_p^q \{e^{z(q-r)} + e^{-z(q-r)}\} f(r)\, dr$$

$$-\ 2^{-1} \int_p^x \{e^{z(x-r)} + e^{-z(x-r)}\} f(r)\, dr$$

for $z > 0$ and $f \in C(I)$. After repeated differentiation in z and careful estimation of integrals, we obtain

$$\left.\begin{array}{r} |D(z,n,0)z(z^2 - A_1)^{-1}f| \\ |D(z,n,0)(d/dx)(z^2 - A_1)^{-1}f| \end{array}\right\} \leqq z^{n+1}/n! \int_0^\infty e^{-zr}r^n|f|dr = |f|.$$

Recalling the proof of Prop. 1 , we obtain

$$|D(z,n,\omega)(z^2 - A_1)^{-1}f| \leqq \omega^{-1}|f| \quad \text{for an arbitrary} \quad \omega > 0\ .$$

Thus, we have for $z > \omega$ and $f \in C(I)$

$$|D(z,n,\omega)B_1(z^2 - A_1)^{-1}f| \leqq (\max_{x\in I}|b(x)| + \omega^{-1}\max_{x\in I}|c(x)|)|f|.$$

We can now deduce the desired properties of A and B by the change of variable :

$$x \to \int_p^x a(r)^{-1/2}\, dr.$$

(ii) After similar calculations, we have for $f \in C^1(I)$

$$\left.\begin{array}{l} |D(z,n,0)(d/dx)z(z^2 - A_1)^{-1}f| \\ |D(z,n,0)(d/dx)^2(z^2 - A_1)^{-1}f| \end{array}\right\} \leqq |f'| .$$

Summing up the above estimates, we have

$$|(z - \begin{pmatrix} 0 & I \\ A_1 & 0 \end{pmatrix})^{-n}|_{C^1\times C\to C^1\times C} \leqq (2 + \omega^{-1})(|z| - \omega)^{-n}$$

for $|z| > \omega$ with an arbitrary $\omega > 0$ and $n = 1,2,\ldots$. Thus, the closure in $C^1(I)$ of $D(A_1)$ $(= D(A))$, which equals the set : $\{u \in C^1(I) : u'(p) = u'(q) = 0\}$, must coincide with V by the Kisyński theorem.

Now, B is defined exactly on V and satisfies

$$|B|_{V\to C(I)} \leqq \max_{x\in I}|b(x) - a'(x)/2| + \max_{x\in I}|c(x)|.$$

REMARK. The case of Dirichlet boundary condition was dealt with by Sova[12] , but the author does not know the detailed theory. Using Theorem 2 , we can now deal with this problem systematically by a similar method to that used in the proof of (i). We expect that Theorem 2 will be applied to some *partial differential operators*.

3. TOOL FOR EQUATION $u_t = \Delta\phi(u)$

Finally, we shall deal with the quasilinear diffusion equation : $u_t = \Delta\phi(u)$ through the difference scheme :

$$h^{-1}(u(t+h,x) - u(t,x))$$

$$= \sum_{i=1}^{N} L^{-2}\{\phi(u(t,x+Le_i)) - 2\phi(u(t,x)) + \phi(u(t,x-Le_i))\},$$

where $e_i = (0,\ldots,0,\overset{i}{1},0,\ldots,0)$, and ϕ is assumed to be a nondecreasing *locally Lipschitz continuous* function on R with $\phi(0) = 0$.

Based on the results concerning semilinear equation : $\Delta u - \phi^{-1}(u) = f$ developed in [1], the associated semigroup in $L^1(R^N)$ was discussed by [2] together with its properties.

We shall present here a simple and direct method for generation and representation of the associated semigroup by

$$e^{t\Delta\phi}\, u = \lim_{h\downarrow 0} C_{h,m}^{[t/h]}\, u ,$$

where $C_{h,m}$ is defined for each fixed positive integer m by

$$\text{(D)}\quad \begin{cases} h^{-1}(C_{h,m} - I) = \sum_{i=1}^{N} 2L^{-2}(C_i(L) - I)\phi(\cdot), \\ \qquad L > 0,\ h > 0,\ L^2 = 2Nh\cdot M_m , \\ M_m = \sup_{|r|,|s|\leqq m} (\phi(r) - \phi(s))/(r - s) . \end{cases}$$

In the above, the subset $\{C_i(t): t \in R\}$ of $B(L^1(R^N),L^1(R^N))$ defined for each $i = 1,\dots,N$ by

$$C_i(t)u(x) = 2^{-1}(u(x + te_i) + u(x - te_i))$$

becomes a (strongly continuous) cosine family in $L^1(R^N)$, and plays important roles together with its generator A_i .

The following is our key lemma. Let's denote simply by L_m the set $\{u \in L^1(R^N) \cap L^\infty(R^N): \|u\|_\infty \leqq m\}$.

<u>LEMMA 5.</u> Set $A_t = \sum_{i=1}^{N} 2t^{-2}(C_i(t) - I)$. Then,

(1) $A = \sum_{i=1}^{N} A_i$ is closable, $-\bar{A}$ is *m-accretive* and

$$(I - \lambda A_t)^{-1}u \to (I - \lambda\bar{A})^{-1}u \quad \text{in } L^1(R^N) \text{ as } t \downarrow 0$$

for $\lambda > 0$ and $u \in L^1(R^N)$;

(2) $\int_{R^N} \operatorname{sgn}(u - v)\cdot A_t(\phi(u) - \phi(v))\, f(x)\, dx \leqq M_m\|u - v\|_1\|\Delta f\|_\infty$

for $u,v \in L_m$, and for any $f \in C^2(R^N) \cap L^\infty(R^N)$ satisfying $f(x) \geqq 0$ for all $x \in R^N$ and $\Delta f \in L^\infty(R^N)$.

PROOF. The proof of (1) is standard. We have only to note that $\bar{A}$ is the infinitesimal generator of the semigroup:

$$\{e^{tA_1} \cdots e^{tA_N} : t > 0\} \quad \text{in} \quad L^1(R^N),$$

where $e^{tA_i} u = (\pi t)^{-1/2} \int_0^\infty e^{-r^2/(4t)} C_i(r)u \, dr, \ i = 1,\ldots,N.$

The inequality (2) can easily be obtained from

$\operatorname{sgn}(u)\cdot A_t u \leqq A_t|u|$, a kind of *Kato's inequality*.

Q.E.D.

Now, let's consider the operator A_ϕ defined by

$$A_\phi u = \bar{A}\cdot\phi(u), \ D(A_\phi) = \{u \in L^1(R^N) \cap L^\infty(R^N) : \phi(u) \in D(\bar{A})\}.$$

THEOREM 6 ([14]). $-A_\phi$ is a densely defined *accretive* operator in $L^1(R^N)$ satisfying the *range condition* :

$$R(I - \lambda A_\phi) \supset L^1(R^N) \cap L^\infty(R^N) \quad \text{for} \quad \lambda > 0.$$

At the same time ,

$$(I - \lambda h^{-1}(C_{h,m} - I))^{-1}u \to (I - \lambda A_\phi)^{-1}u \quad \text{as} \quad h \downarrow 0$$

in $L^1(R^N)$ for $u \in L_m$ with each fixed m .

To give the proof, we need two lemmas.

LEMMA 7. For each m and h, $C_{h,m}$ maps L_m into itself and satisfies, for $u,v \in L_m$,

$$\|C_{h,m}u\|_p \leqq \|u\|_p (p = 1,\infty) \ ; \ \|C_{h,m}u - C_{h,m}v\|_1 \leqq \|u - v\|_1;$$

$$(C_{h,m}u)_y = C_{h,m}u_y \quad \text{where} \quad u_y(x) = u(x + y) \quad \text{for} \quad y \in R^N.$$

The proof is clear from (D). We have only to note that

$r \to r - 2NhL^{-2}\phi(r)$ is nondecreasing and for $r,s \in [-m,m]$

$$|r - s - 2NhL^{-2}(\phi(r) - \phi(s))| + 2NhL^{-2}|\phi(r) - \phi(s)| = |r - s|.$$

LEMMA 8. The operator $J^{\lambda}_{h,m} = (I - \lambda h^{-1}(C_{h,m} - I))^{-1}$ is well defined for $\lambda > 0$, maps L_m into itself and satisfies, for $u,v \in L_m$,

(3) $$\|J^{\lambda}_{h,m}u\|_p \leqq \|u\|_p \quad (p = 1,\infty) ;$$

(4) $$\|J^{\lambda}_{h,m}u - (J^{\lambda}_{h,m}u)_y\|_1 \leqq \|u-u_y\|_1 \quad \text{for } y \in R^N ;$$

(5) $$\int_{|x|>\rho} |J^{\lambda}_{h,m}u| \, dx \leqq \int_{|x|>\rho/2} |u| \, dx + C\, M_m\, \lambda\rho^{-1}\|u\|_1$$

for any $\rho > 1$ and a constant $C \geqq 0$ independent of λ, h and ρ.

PROOF. The inequalities (3) and (4) are simple consequences of Lemma 7.

Replacing u, v and $f(x)$ in (2) by $J^{\lambda}_{h,m}u$, 0 and $g(2|x|/\rho - 1)$ respectively, we obtain (5) with

$$C = 4(\|g''\|_\infty + (N - 1)\|g'\|_\infty),$$

where $g \in C^2(R)$ with values in $[0,1]$; 0 on $(-\infty,0]$ and 1 on $[1,\infty)$. Q.E.D.

PROOF OF THEOREM 6. Step 1. Take $u \in L^1(R^N) \cap L^\infty(R^N)$. Then, u belongs to some L_m. By the Fréchet-Kolmogorov theorem, Lemma 8 implies that the set $\{J^{\lambda}_{h,m}u : h > 0\}$ is precompact in $L^1(R^N)$.

Step 2. The equality for $\mu > 0$,

$$(I - \mu A_L)^{-1}\lambda^{-1}(J^{\lambda}_{h,m}u - u) = \mu^{-1}((I - \mu A_L)^{-1} - I)\cdot\phi(J^{\lambda}_{h,m}u)$$

together with (1) implies that the limit $w_{\lambda,m}$ of $J^{\lambda}_{h_n,m}u$ as $h_n \downarrow 0$, if exists , must satisfy : $w_{\lambda,m} - \lambda A_\phi w_{\lambda,m} = u$.

But, $-A_\phi$ is accretive by (2) with $f(x) \equiv 1$, and hence $w_{\lambda,m} = (I - \lambda A_\phi)^{-1}u$.

Step 3. The density of $D(A_\phi)$ is shown by a similar argument for the set $\{(I - \lambda A_\phi)^{-1}u : 0 < \lambda \leqq 1\}$ instead. Indeed, Lemma 8 remains true with $J^\lambda_{h,m}$ replaced by $(I - \lambda A_\phi)^{-1}$. Q.E.D.

Theorem 6 shows that A_ϕ generates a semigroup in the sense of Crandall-Liggett. At the same time, the semigroup is represented by the product formula :

$$e^{tA_\phi} u = \lim_{h \downarrow 0} C_{h,m}^{[t/h]} u \quad \text{in} \quad L^1(R^N) \quad \text{for} \quad u \in L_m$$

uniformly on every bounded subinterval of $[0,\infty)$, with the aid of nonlinear *Chernoff's formula* given by Brézis and Pazy[4, Theorem 3.2]. An idea similar to (D) has been employed in [3], however, as an approximation scheme for the equation.

We shall finish this article with another product formula.

Let A^i_ϕ , $i = 1,\ldots,N$ be operators defined by

$$A^i_\phi u = A_i \cdot \phi(u), \; D(A^i_\phi) = \{u \in L^1(R^N) \cap L^\infty(R^N) : \phi(u) \in D(A_i)\}.$$

Then, recalling the above discussion, we see that A^i_ϕ generate semigroups $\{e^{tA^i_\phi} : t > 0\}$ in $L^1(R^N)$ given by

$$e^{tA^i_\phi} u = \lim_{h \downarrow 0} (C^i_{h,m})^{[t/h]} u \quad \text{for} \quad u \in L_m \; ,$$

where $C^i_{h,m}$ are defined for each fixed m , instead of (D) , by

$$h^{-1}(C^i_{h,m} - I) = 2L^{-2}(C_i(L) - I)\phi(\cdot) \; ,$$

$$L > 0 \; , \quad h > 0 \; , \quad L^2 = 2h \cdot M_m \; .$$

THEOREM 9. Uniformly on every bounded subinterval of $[0,\infty)$

$$e^{tA_\phi} u = \lim_{h \downarrow 0} \{e^{hA^1_\phi} \cdots e^{hA^N_\phi}\}^{[t/h]} u \quad \text{for} \quad u \in L^1(R^N) \cap L^\infty(R^N).$$

OUTLINE OF PROOF. The proof is carried out by showing as in the above that $J_t^\lambda u$ converges in $L^1(R^N)$ to $(I - \lambda A_\phi)^{-1}u$ as $t \downarrow 0$ for $u \in L_m$ and $\lambda > 0$, where $J_t^\lambda = (I - \lambda t^{-1}(U(t) - I))^{-1}$ and $U(t) = e^{tA_\phi^1} \cdots e^{tA_\phi^N}$.

To apply the Fréchet-Kolmogorov theorem to the sequence $\{J_{t_n}^\lambda u\}_{n=1}^\infty$ for $t_n \downarrow 0$, we use instead of (2)

$$\int_{R^N} \operatorname{sgn}(u)\cdot t^{-1}(U(t) - I)uf(x)\,dx \leqq M_m\|u\|_1 \sum_{i=1}^N \|(\partial/\partial x_i)^2 f\|_\infty ,$$

which is derived from N applications of

$$(6) \quad \int_{R^N} \operatorname{sgn}(u)\cdot t^{-1}(e^{tA_\phi^i} - I)uf(x)\,dx \leqq M_m\|u\|_1\|(\partial/\partial x_i)^2 f\|_\infty .$$

Next, the equalities in $L^1(R^N)$

$$(7) \quad e^{tA_\phi^i}u - u = A_i\int_0^t \phi(e^{rA_\phi^i}u)\,dr \quad \text{for } t > 0$$

give

$$\begin{aligned} \lambda^{-1}(J_t^\lambda u - u) &= t^{-1}(U(t) - I)\, J_t^\lambda u \\ &= \sum_{i=1}^N A_i \cdot t^{-1}\int_0^t \phi(e^{rA_\phi^i}\, e^{tA_\phi^{i+1}} \cdots e^{tA_\phi^N} J_t^\lambda u)\,dr. \end{aligned}$$

This implies that the limit w_λ of $J_{t_n}^\lambda u$ as $t_n \downarrow 0$, if exists, satisfies $w_\lambda - \lambda A_\phi w_\lambda = u$.

The proof of (6) and (7) can be done as in [16].

REMARK. Theorem 9 suggests us that the solution of the equation $u_t = \Delta\phi(u)$ is constructed by means of the equations : $u_t = (\partial/\partial x_i)^2\phi(u)$, $i = 1,\ldots,N$ (cf. [5]).

As for Theorem 6, the author feels it reasonable that cosine families work effectively in the study of the quasilinear diffusion equation. Indeed, in the classical study of linear diffusion or wave equations, the use of trigonometric functions has been indispensable.

REFERENCES

[1] Bénilan,Ph., Brézis,H., Crandall,M.G., A semilinear elliptic equation in $L^1(R^N)$, Ann. Scoula Norm. Pisa Cl. Sci. (4)2(1975), 523-555.

[2] Bénilan,Ph., Crandall,M.G., The continuous dependence on ϕ of solutions of $u_t - \Delta\phi(u) = 0$, Indiana Univ. Math. J. 30(1981), 161-177.

[3] Berger,A.E., Brézis,H., Rogers,C.W., A numerical method for solving the problem $u_t - \Delta f(u) = 0$, RAIRO Anal. Numér. 13(1979), 297-312.

[4] Brézis,H., Pazy,A., Convergence and approximation of semigroups of nonlinear operators in Banach spaces, J. Funct. Anal. 9(1972), 63-74.

[5] Coron, J-M., Formules de Trotter pour une équation d'évolution quasilinéaire du l er ordre, J. Math. Pures Appl. 9(1982), 91-112.

[6] Da Prato,G., Giusti,E., Una caratterizzazione dei generatori di funzioni coseno astratti, Boll. Unione Mat. Italiana 22(1967), 357-362.

[7] Fattorini,H.O., Ordinary differential equations in linear topological spaces,II, J. Diff. Equa. 6(1969), 50-70.

[8] Kisyński,J., On cosine operator functions and one-parameter groups of operators, Studia Math. 44(1972), 93-105.

[9] Lutz,D., On bounded time-dependent perturbation of operator cosine functions, Aequa. Math. 23(1981), 197-203.

[10] Oharu,S., Takahashi,T., A convergence theorem of nonlinear semigroups and its application to first order quasilinear equations, J. Math. Soc. Japan 26 (1974), 124-160.

[11] Phillips,R.S., Perturbation theory for semi-groups of linear operators, Trans. Amer. Math. Soc. 74 (1953), 199-221.

[12] Sova, M., Cosine operator functions, Rozprawy Mat. 49 (1966), 1-47.

[13] Watanabe,M., A perturbation theory for abstract evolution equations of second order, Proc. Japan Acad. 58 (1982), 143-146.

[14] Watanabe,M., An approach by difference to a quasi-linear parabolic equation, Proc. Japan Acad. 59(1983), 375-378.

[15] Watanabe,M., A new proof of the generation theorem of cosine families in Banach spaces, Houston J. Math. 10(1984), 285-290.

[16] Watanabe,M., Trotter's product formula for semigroups generated by quasilinear elliptic operators, Proc. Amer. Math. Soc. 92(1984), 509-514.

[17] Serizawa,H., Watanabe,M., Perturbation for cosine families in Banach spaces, Houston J. Math. to appear.

[18] Serizawa,H., Watanabe,M., Time-dependent perturbation for cosine families in Banach spaces, Houston J. Math. to appear.

REGULARITY OF FUNCTIONS ON AN INTERVAL WITH VALUES IN THE SPACE OF FRACTIONAL POWERS OF OPERATORS

A. Yagi
Department of Mathematics
Osaka University
Toyonaka, Osaka, 560 JAPAN

1. Introduction.

Let $A(\cdot):[0,T] \to A(t)$ be a function on an interval $[0,T]$ with values linear operators $A(t)$ acting in a Hilbert space X. In several important cases it is verified that, though the domain $\mathcal{D}(A(t))$ of $A(t)$ is dependent on t, there exists $0 < \theta < 1$ such that the domain $\mathcal{D}(A(t)^\theta)$ of the fractional power $A(t)^\theta$ is independent of t. We are then concerned with a problem of verifying regularity of the function $A(\cdot)^\theta$.

Such a problem arises in the study of abstract evolution equations. For instance, consider a second order linear evolution equation

$$\frac{d^2u}{dt^2} + A(t)u = f(t) , \quad 0 \leqq t \leqq T$$

in X. As we allow the domain $\mathcal{D}(A(t))$ to vary with t, this equation itself is not easy to handle. But, if we verify independence of t of the domain $\mathcal{D}(A(t)^{1/2})$ of the square root together with strong differentiability of the function $A(\cdot)^{1/2}$, then it is possible by setting $v_0 = iA(t)^{1/2}u$ and $v_1 = du/dt$ to reduce the equation to a first order equation

$$\frac{d}{dt}\begin{pmatrix} v_0 \\ v_1 \end{pmatrix} + \frac{1}{i}\begin{pmatrix} 0 & A(t)^{1/2} \\ A(t)^{1/2} & 0 \end{pmatrix}\begin{pmatrix} v_0 \\ v_1 \end{pmatrix} + \begin{pmatrix} A(t)^{1/2}dA(t)^{-1/2}/dt & 0 \\ 0 & 0 \end{pmatrix}\begin{pmatrix} v_0 \\ v_1 \end{pmatrix}$$

$$= \begin{pmatrix} 0 \\ f(t) \end{pmatrix}, \quad 0 \leqq t \leqq T$$

in the product space $X \times X$. The domain of the coefficient operator is now independent of t (Of course treatment is extremely easy, see Arosio [1], McIntosh [4], Yagi [9,11]). Similarly, it is known that, if $A(\cdot)^{\theta}$ is Hölder continuous for a suitable exponent $0 < \theta < 1$ in a linear evolution equation of parabolic type

$$du/dt + A(t)u = f(t)\ , \quad 0 \leqq t \leqq T\ ,$$

then a unique evolution operator can be constructed (see Kato [3], Sobolevskii [7,8]).

In this note we study conditions on the function $A(\cdot)$ under which the fractional power $A(\cdot)^{\theta}$ is Hölder continuous or strongly continuously differentiable, and we state their applications to partial differential operators.

2. Hölder continuity.

Let X be a Hilbert space. By $\mathcal{T}(X;M,\omega)$ $(0 \leqq \omega < \pi,\ M \geqq 1)$ we denote the space of all linear operators of type (ω, M) acting in X. The subspace of $\mathcal{T}(X;M,\omega)$ consisting of operators the inverses of which are bounded on X is denoted by $\mathcal{T}_0(X;M,\omega)$ (cf. [14, Chap.2, Sec.3]).

Theorem 1. Let $A(\cdot)$ be a function defined on $[0,T]$ with values in $\mathcal{T}_0(X;M,\omega)$ $(0 \leqq \omega < \pi)$. Let the following conditions be satisfied:

i) For each $0 \leqq t \leqq T$, the purely imaginary power $A(t)^{iy}$ $(-\infty < y < +\infty)$ of $A(t)$ is a bounded operator on X.

ii) $A(\cdot)^{-1}$ is a strongly continuous function with values in $\mathcal{L}(X)$: $A(\cdot)^{-1} \in \mathcal{C}([0,T];\mathcal{L}_s(X))$.

iii) There exist $0 < h < 1$ and a finite number of $0 < \alpha_k < 1$ $(1 \leqq k \leqq \ell)$ such that

$$|(\{A(t)^{-1} - A(s)^{-1}\}f,g)| \leqq N|t-s|^{h} \sum_{k=1}^{\ell} \|A(s)^{\alpha_k - 1} f\| \, \|A(t)^{*-\alpha_k} g\|\ , \quad f,g \in X$$

for any $0 \leq s,t \leq T$.

Then, for any $0 < \theta < \gamma = \mathrm{Min}\{\alpha_k;\ 1 \leq k \leq \ell\}$, *the domain* $\mathcal{D}(A(t)^{\theta})$ *of the fractional power is independent of* t, *and the function* $A(\cdot)^{\theta}$ *is Hölder continuous*: $A(\cdot)^{\theta} \in \mathcal{C}^h([0,T];\mathcal{L}(\mathcal{D}(A(0)^{\theta}),X))$.

Proof. Let $0 < \theta < \gamma$. For arbitrary $f \in X$ and $g \in \mathcal{D}(A(t)^{*\theta})$, consider the scalar product

$$(\{A(t)^{-\theta} - A(s)^{-\theta}\}f, A(t)^{*\theta}g) =$$

$$\frac{\sin\theta\pi}{\pi}\int_0^{\infty} \lambda^{-\theta}(\{A(t)^{-1} - A(s)^{-1}\}A(s)(\lambda+A(s))^{-1}f, A(t)^{*\,1+\theta}(\lambda+A(t)^{*})^{-1}g)d\lambda.$$

It is immediate from iii) that

$$|(\{A(t)^{-\theta} - A(s)^{-\theta}\}f, A(t)^{*\theta}g)| \leq$$

$$\frac{N}{\pi}|t-s|^h \sum_{k=1}^{\ell} \int_0^{\infty} \lambda^{-\theta}\|A(s)^{\alpha_k}(\lambda+A(s))^{-1}f\|\,\|A(t)^{*\beta_k}(\lambda+A(t)^{*})^{-1}g\|d\lambda$$

where $\beta_k = 1 - \alpha_k + \theta$. We here use a fact (established in [10]): Condition i) implies that

$$\{\int_0^{\infty} \lambda^{1-2\alpha}\|A(t)^{\alpha}(\lambda+A(t))^{-1}f\|^2 d\lambda\}^{1/2} \leq M_{\alpha}\|f\| , \quad f \in X$$

$$\{\int_0^{\infty} \lambda^{1-2\beta}\|A(t)^{*\beta}(\lambda+A(t)^{*})^{-1}g\|^2 d\lambda\}^{1/2} \leq M^{*}_{\beta}\|g\| , \quad g \in X$$

for any $0 < \alpha,\beta < 1$. Then we obtain that

$$|(\{A(t)^{-\theta} - A(s)^{-\theta}\}f, A(t)^{*\theta}g)| \leq C_{\theta}|t-s|^h\|f\|\,\|g\| ,$$

hence the result.

3. Strong differentiability.

Theorem 2. Let $A(\cdot)$ *be a function on* $[0,T]$ *with values in*

$\mathcal{T}_0(X;M,\omega)$ $(0 \leqq \omega < \pi)$. <u>Assume the following conditions</u>:

i) <u>For each</u> $0 \leqq t \leqq T$, $A(t)^{iy}$ $(-\infty < y < +\infty)$ <u>is a bounded operator on</u> X.

ii) $A(\cdot)^{-1}$ <u>is a strongly continuously differentiable function with values in</u> $\mathcal{L}(X)$: $A(\cdot)^{-1} \in \mathcal{C}^1([0,T];\mathcal{L}_s(X))$.

iii) <u>There exist a finite number of</u> $0 < \alpha_k < 1$ $(1 \leqq k \leqq \ell)$ <u>such that</u>

$$|(\{dA(t)^{-1}/dt\}f,g)| \leqq N \sum_{k=1}^{\ell} \|A(t)^{\alpha_k - 1} f\| \, \|A(t)^{*-\alpha_k} g\| , \quad f,g \in X$$

<u>for any</u> $0 \leqq t \leqq T$.

<u>Then</u>, <u>for any</u> $0 < \theta < \gamma = \mathrm{Min}\{\alpha_k;\ 1 \leqq k \leqq \ell\}$, <u>the domain</u> $\mathcal{D}(A(t)^\theta)$ <u>is independent of</u> t <u>and the function</u> $A(\cdot)^\theta$ <u>is strongly continuously differentiable</u>: $A(\cdot)^\theta \in \mathcal{C}^1([0,T];\mathcal{L}_s(\mathcal{D}(A(0)^\theta),X))$.

Proof. By the same kind of proof as in Theorem 1 we obtain that

$$|(\{dA(t)^{-\theta}/dt\}f,A(t)^{*\theta}g)| \leqq C_\theta \|f\| \|g\| , \quad f \in X,\ g \in \mathcal{D}(A(t)^{*\theta})$$

for any $0 < \theta < \gamma$. Then the independence of $\mathcal{D}(A(t)^\theta)$ is an immediate consequence of this. To see the strong differentiability, we first note that $A(\cdot)^\theta dA(\cdot)^{-\theta}/dt J_n(\cdot) \in \mathcal{C}([0,T];\mathcal{L}_s(X))$ for any $n = 1,2,\cdots$, where $J_n(t) = (1+n^{-1}A(t))^{-1}$ is the Yosida regularization of $A(t)$. But, since $J_n(t)$ is strongly convergent to identity of X with uniformity in t as $n \to \infty$, it follows that $A(\cdot)^\theta dA(\cdot)^{-\theta}/dt \in \mathcal{C}([0,T];\mathcal{L}_s(X))$. Then, as a unique solution of the integral equation

$$A(t)^\theta A(0)^{-\theta} = I - \int_0^t A(r)^\theta \{dA(r)^{-\theta}/dr\} A(r)^\theta A(0)^{-\theta} dr,$$

$A(t)^\theta A(0)^{-\theta}$ $(0 \leqq t \leqq T)$ is strongly continuously differentiable.

4. Applications.

Our abstract results in the preceding sections apply to partial differential operators. Here we describe two examples.

Let $A(t)$ be the L^2-realization of an elliptic partial differential operator

$$A(t,x;D) = \sum_{|\alpha| \leqq 2m} a_\alpha(t,x)D^\alpha$$

on a bounded region $\Omega \subset \mathbb{R}^n$ under boundary conditions

$$B_j(t,x;D) = \sum_{|\beta| \leqq m_j} b_{j\beta}(t,x)D^\beta \quad , \quad j = 1,\cdots,m$$

on the boundary $\partial\Omega$. For each $0 \leqq t \leqq T$, we assume that $A(t) \in \mathcal{I}_0(L^2(\Omega);M,\omega)$ with some $0 \leqq \omega < \pi$. Then according to [10] (or Seeley [5,6]) it is verified that

$$\mathcal{D}(A(t)^\theta) = \{u \in H^{2m\theta}(\Omega);\ B_j(t,x;D)u = 0 \ \text{ on } \ \partial\Omega \ \text{ for } \ m_j < 2m\theta - 1/2$$

$$\text{and} \ \ B_j(t,x;D)u \in L^2_{-1/2}(\Omega) \ \text{ for } \ m_j = 2m\theta - 1/2\} .$$

In particular, $\mathcal{D}(A(t)^\theta) \equiv H_\theta(\Omega)$ is independent of t for any $0 < \theta < \gamma_B$ where

$$\gamma_B = \frac{\operatorname{Min}\{m_j \neq 0;\ 1 \leqq j \leqq m\} + 1/2}{2m} .$$

We have (for the proof see [11]):

<u>Theorem 3</u>. <u>Let</u> $a_\alpha \in \mathcal{E}^h([0,T];\mathcal{E}^\infty(\bar{\Omega}))$ [<u>resp</u>. $a_\alpha \in \mathcal{E}^1([0,T];\mathcal{E}^\infty(\bar{\Omega}))$] <u>for</u> $|\alpha| \leqq 2m$ <u>and</u> $b_{j\beta} \in \mathcal{E}^h([0,T];\mathcal{E}^\infty(\partial\Omega))$ [<u>resp</u>. $b_{j\beta} \in \mathcal{E}^1([0,T];\mathcal{E}^\infty(\partial\Omega))$] <u>for</u> $|\beta| \leqq m_j$ $(1 \leqq j \leqq m)$ <u>with some</u> $0 < h < 1$. <u>Then, for any</u> $0 < \theta < \gamma_B$, $A(\cdot)^\theta \in \mathcal{E}^h([0,T];\mathcal{L}_s(H_\theta(\Omega),L^2(\Omega)))$ [<u>resp</u>. $A(\cdot)^\theta \in \mathcal{E}^1([0,T];\mathcal{L}_s(H_\theta(\Omega),L^2(\Omega)))$].

Let next $A(t)$ be the Hamiltonian in $\mathbb{R}^3$ with a potential $V(t,\cdot) \in H^{-1/2}(\mathbb{R}^3)$:

$$\begin{cases} \mathcal{D}(A(t)) = \{u \in H^1(\mathbb{R}^3);\ -\Delta u + V(t,x)u \in H^0(\mathbb{R}^3)\} \\ A(t)u = -\Delta u + V(t,x)u + \beta u\ . \end{cases}$$

We can then prove (see [11]):

Theorem 4. For any $0 < \theta < 3/4$, $\mathcal{D}(A(t)^\theta) \equiv H^{2\theta}(\mathbb{R}^3)$ is independent of t. Moreover, let $V \in \mathcal{C}^h([0,T];H^{-1/2}(\mathbb{R}^3))$ [resp. $V \in \mathcal{C}^1([0,T];H^{-/2}(\mathbb{R}^3))$] for some $0 < h < 1$, then, for any $0 < \theta < 3/4$, $A(\cdot)^\theta \in \mathcal{C}^h([0,T];\mathcal{L}_s(H^{2\theta}(\mathbb{R}^3),H^0(\mathbb{R}^3)))$ [resp. $A(\cdot)^\theta \in \mathcal{C}^1([0,T];\mathcal{L}_s(H^{2\theta}(\mathbb{R}^3),H^0(\mathbb{R}^3)))$] .

References

[1] Arosio A., Abstract linear hyperbolic equations with variable domain. Ann. Mat. Pura Appl. 135, 173-218 (1983)

[2] Furuya K., Analyticity of solutions of quasilinear evolution equations. Osaka J. Math. 18, 669-698 (1981)

[3] Kato T., Abstract evolution equations of parabolic type in Banach and Hilbert spaces. Nagoya Math. J. 5, 93-125 (1961)

[4] McIntosh A., Square roots of elliptic operators. J. Fun. Anal. 61, 307-327 (1985)

[5] Seeley R., Norms and domains of the complex powers A_B^z. Amer. J. Math. 93, 299-309 (1971)

[6] Seeley R., Interpolation in L^p with boundary conditions. Studia Math. 44, 47-60 (1972)

[7] Sobolevskii P.E., First order differential equations in Hilbert space with a variable positive-definite self-adjoint operator whose fractional power has a constant domain of definition. Doklady Akad. Nauk 123, 984-987 (1958)

[8] Sobolevskii P.E., Parabolic equations in Banach space with an unbounded variable operator, a fractional power of which has a constant domain of definition. Soviet Math. Dokl. 2, 545-548 (1961)

[9] Yagi A., Differentiability of families of the fractional powers of self-adjoint operators associated with sesquilinear forms. Osaka J. Math. 20, 265-284 (1983)

[10] Yagi A., Coïncidence entre des espaces d'interpolation et des domaines de puissances fractionnaires d'opérateurs. C. R. Acad. Sc. Paris 299, 173-176 (1984)

[11] Yagi A., Applications of the purely imaginary powers of operators in Hilbert space. To appear.

[12] Muramatu T., Theory of Interpolation Spaces and Linear Operators. Kinokuniya, Tokyo, 1985 (in Japanese)

[13] Simon B., Quantum Mechanics for Hamiltonians Defined as Quadra-

tic Forms. Princeton Univ. Press, Princeton, 1971
[14] Tanabe H., Evolution Equations. Iwanami, Tokyo, 1975 (in Japanese). English translation, Equations of Evolution. Pitman, London, 1979
[15] Triebel H., Interpolation Theory, Function Spaces, Differential Operators. Noth Holland, Amsterdam, 1978